## Dependence Logic

Dependence is a common phenomenon, wherever one looks: ecological systems, astronomy, human history, stock markets – but what is the logic of dependence? This book is the first to carry out a systematic logical study of this important concept, giving on the way a precise mathematical treatment of Hintikka's independence friendly logic. Dependence logic adds the concept of dependence to first order logic. Here the syntax and semantics of dependence logic are studied, dependence logic is given an alternative game theoretic semantics, and sharp results about its complexity are proven. This is a textbook suitable for a special course in logic in mathematics, philosophy, and computer science departments, and contains over 200 exercises, many of which have a full solution at the end of the book. It is also accessible to readers with a basic knowledge of logic, who are interested in new phenomena in logic.

LONDON MATHEMATICAL SOCIETY STUDENT TEXTS

Managing editor: Professor J. W. Bruce,
Department of Mathematics, University of Hull, UK

LONDON MATHEMATICAL SOCIETY STUDENT TEXTS 70

# Dependence Logic

## A New Approach to Independence Friendly Logic

JOUKO VÄÄNÄNEN

*University of Amsterdam*

*and*

*University of Helsinki*

**CAMBRIDGE**
UNIVERSITY PRESS

# CAMBRIDGE
## UNIVERSITY PRESS

Shaftesbury Road, Cambridge CB2 8EA, United Kingdom

One Liberty Plaza, 20th Floor, New York, NY 10006, USA

477 Williamstown Road, Port Melbourne, VIC 3207, Australia

314–321, 3rd Floor, Plot 3, Splendor Forum, Jasola District Centre, New Delhi – 110025, India

103 Penang Road, #05–06/07, Visioncrest Commercial, Singapore 238467

Cambridge University Press is part of Cambridge University Press & Assessment,
a department of the University of Cambridge.

We share the University's mission to contribute to society through the pursuit of
education, learning and research at the highest international levels of excellence.

www.cambridge.org
Information on this title: www.cambridge.org/9780521700153

First published 2007

*A catalogue record for this publication is available from the British Library*

ISBN   978-0-521-87659-9   Hardback
ISBN   978-0-521-70015-3   Paperback

# Contents

# Preface

This book is based on lectures I gave at the Department of Mathematics and Statistics, University of Helsinki, during the academic year 2005–2006. I am indebted to the students who followed the course, in particular to Åsa Hirvonen, Meeri Kesälä, Ville Nurmi, Eero Raaste, and Ryan Siders. Thanks also go to Ville Nurmi for suggesting numerous corrections to the text, compiling the solutions to the exercises in the course, and for allowing me to include the solutions in this book. I am very grateful to Wilfrid Hodges for many useful discussions on dependence. I thank the Newton Institute (Cambridge, UK) for inviting me for the five weeks, during which time the final manuscript was prepared. The preparation of the manuscript was partially supported by grant 40734 of the Academy of Finland. I wish to thank Peter Thompson of Cambridge University Press for all the arrangements concerning publishing, and I am deeply grateful to Juliette Kennedy for her generous help in all stages of writing this book.

# Preface

# 1

# Introduction

Dependence is a common phenomenon, wherever one looks: ecological systems, astronomy, human history, stock markets. With global warming, the dependence of life on earth on the actions of mankind has become a burning issue. But what is the logic of dependence? In this book we set out to make a systematic logical study of this important concept.

Dependence manifests itself in the presence of multitude. A single event cannot manifest dependence, as it may have occurred as a matter of chance. Suppose one day it blows from the west and it rains. There need not be any connection between the wind and the rain, just as if one day it rains and it is Friday the 13th. But over a whole year we may observe that we can tell whether rain is expected by looking at the direction of the wind. Then we would be entitled to say that in the observed location and in the light of the given data, whether it rains depends on the direction of the wind. One would get a more accurate statement about dependence by also observing other factors, such as air pressure.

Dependence logic adds the concept of dependence to first order logic. In ordinary first order logic the meaning of the identity

$$x = y \qquad (1.1)$$

is that the values of $x$ and $y$ are the same. This is a trivial form of dependence. The meaning of

$$fx = y \qquad (1.2)$$

is that the interpretation of the function symbol $f$ maps the value of $x$ to the value of $y$. This is an important form of dependence, one where we actually know the mapping which creates the dependence. Note that the dependence

1

may be more subtle, as in

$$f x z = y.$$

Here $y$ certainly depends on $x$ but also on $z$. In this case we say that $y$ depends on both $x$ and on $z$, but is determined by the two together.

We introduce the new atomic formulas

$$=(x, y), \tag{1.3}$$

the meaning of which is that the values of $x$ and $y$ depend on each other in the particular way that values of $x$ completely determine the values of $y$. Note the difference between Eqs. (1.1), (1.2) and (1.3). The first says that $x$ determines $y$ in the very strong sense of $y$ being identical with $x$. The second says that $x$ determines $y$ via the mapping $f$. Finally, the third says there is *some* way in which $x$ determines $y$, but we have no idea what that is.

The dependence in Eq. (1.3) is quite common in daily life. We have data that show that weather depends on various factors such as air pressure and air temperature, and we have a good picture of the mathematical equations that these data have to satisfy, but we do not know how to solve these equations, and therefore we do not know how to compute the weather when the critical parameters are given. We could say that the weather obeys dependence of the kind given in Eq. (1.3) rather than of the kind in Eq. (1.2). Historical events typically involve dependencies of the type in Eq. (1.3), as we do not have a perfect theory of history which would explain why events happen the way they do. Human genes undoubtedly determine much of the development of an individual, but we do not know how; we can just see the results.

In order to study the logic of dependence we need a framework involving multitude, such as multiple records of historical events, day to day observations of weather and stock transactions. This seems to lead us to study statistics or database theory. These are, however, wrong leads. If we observe that a lamp is lit up four times in a row when we turn a switch, but also that once the lamp does not light up even if we turned the switch (Fig. 1.1), we have to conclude that the light is not completely determined by the switch, as it is by the combined effect of the switch and the plug. From the point of view of dependence, statistical data or a database are relevant only to the extent that they record *change*.

In first order logic the order in which quantifiers are written determines the mutual dependence relations between the variables. For example, in

$$\forall x_0 \exists x_1 \forall x_2 \exists x_3 \phi$$

the variable $x_1$ depends on $x_0$, and the variable $x_3$ depends on both $x_0$ and $x_2$. In dependence logic we write down explicitly the dependence relations

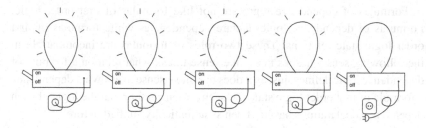

Fig. 1.1. Does the switch determine whether the lamp is lit?

between variables and by so doing make it possible to express dependencies not otherwise expressible in first order logic.

The first step in this direction was taken by Henkin with his partially ordered quantifiers, e.g.

$$\begin{pmatrix} \forall x_0 & \exists x_1 \\ \forall x_2 & \exists x_3 \end{pmatrix} \phi,$$

where $x_1$ depends only on $x_0$ and $x_3$ depends only on $x_2$. The remarkable observation about the extension $L(H)$ of first order logic by this quantifier, made by Ehrenfeucht, was that $L(H)$ is not axiomatizable.

The second step was taken by Hintikka and Sandu, who introduced the slash-notation

$$\forall x_0 \exists x_1 \forall x_2 \exists x_3 / \forall x_0 \phi,$$

where $\exists x_3 / \forall x_0$ means that $x_3$ is "independent" of $x_0$ in the sense that a choice for the value of $x_3$ should not depend on the value of $x_0$. The observation of Hintikka and Sandu was that we can add slashed quantifiers $\exists x_3 / \forall x_0$ coherently to first order logic if we give up some of the classical properties of negation, most notably the Law of Excluded Middle. They called their logic *independence friendly logic*.

We take the further step of writing down explicitly the mutual dependence relationships between variables. Thus we write

$$\forall x_0 \exists x_1 \forall x_2 \exists x_3 (=(x_2, x_3) \wedge \phi) \tag{1.4}$$

to indicate that $x_3$ depends on $x_2$ only. The new atomic formula $= (x_2, x_3)$ has the explicit meaning that $x_3$ depends on $x_2$ and on nothing else. This results in a logic which we call *dependence logic*. It is equivalent in expressive power to the logic of Hintikka and Sandu in the sense that there are truth-preserving translations from one to the other. In having the ability to express dependence on the atomic level it is more general.

Formulas of dependence logic are not like formulas of first order logic. Formulas of dependence logic declare dependencies while formulas of first order logic state relations. These two roles of formulas are incompatible in the following sense. It does not make sense to ask what relation a formula of dependence logic defines, just as it does not make sense to ask what dependence a formula of first order logic states. It seems to the author that the logic of such dependence declarations has not been systematically studied before.

At the end of this book we introduce a stronger logic called *team logic*, reminiscent of the extended independence friendly logic of Hintikka. Team logic is, unlike dependence logic and independence friendly logic, closed under the usual Boolean operations and it satisfies the Law of Excluded Middle.

## Historical remarks

The possibility of extending first order logic by partially ordered quantifiers was presented by Henkin [14], where also Ehrenfeucht's result, referred to above, can be found. Independence friendly logic was introduced by Hintikka and Sandu [16] (see also ref. [17]) and advocated strongly by Hintikka in ref. [19]. Hodges [21, 22] gave a compositional semantics for independence friendly logic and we very much follow his approach. Further properties of this semantics are proved in refs. [4], [23] and [41]. Cameron and Hodges [5] showed that there are limitations to the extent to which the semantics can be simplified from the one given in ref. [21]. Connections between independence friendly logic, set theory and second order logic are discussed in ref. [40].

# 2

# Preliminaries

## 2.1 Relations

An $n$-tuple is a sequence $(a_1, \ldots, a_n)$ with $n$ components $a_1, \ldots, a_n$ in this order. A special case is the empty sequence $\emptyset$, which corresponds to the case $n = 0$. A *relation* on a set $M$ is a set $R$ of $n$-tuples of elements of $M$ for some fixed $n$, where $n$ is the *arity* of $R$. The simplest examples are the usual *identity* relations on a set $M$:

$$\{(x, x) : x \in M\},$$
$$\{(x, x, y) : x, y \in M\},$$
$$\{(x, y, x) : x, y \in M\},$$
$$\{(x, y, y) : x, y \in M\},$$
$$\{(x, x, x) : x \in M\}.$$

Two special relations are the empty relation $\emptyset$, which is the same in any arity, and the unique 0-ary relation $\{\emptyset\}$. We think of a function $f : M \to M$ as a relation $\{(x, f(x)) : x \in M\}$ on $M$.

## 2.2 Vocabularies and structures

A vocabulary is a set $L$ of constant, relation and function symbols. We use $c$ to denote constant symbols, $R$ to denote relation symbols, and $f$ to denote function symbols in a vocabulary, possibly with subindexes. Each symbol $s$ in $L$ has an *arity* $\#_L(s)$, which is a natural number. The arity of constant symbols is zero. The arity of a relation symbol may be zero. We use $x_0, x_1, \ldots$ to denote variables.

An $L$-structure $\mathcal{M}$ is a non-empty set $M$, the *domain* of $\mathcal{M}$, endowed with an element $c^{\mathcal{M}}$ of $M$ for each $c \in L$, an $\#_L(R)$-ary relation $R^{\mathcal{M}}$ on $M$ for

$R \in L$, and an $\#_L(f)$-ary function $f^{\mathcal{M}}$ on $M$ for $f \in L$. The $L$–structures $\mathcal{M}$ and $\mathcal{M}'$ are *isomorphic* if there is a bijection $\pi : M \to M'$ such that $\pi(c^{\mathcal{M}}) = c^{\mathcal{M}'}$ and for all $a_1, \ldots, a_{\#_L(R)} \in M$ we have $(a_1, \ldots, a_{\#_L(R)}) \in R^{\mathcal{M}}$ if and only if $(\pi(a_1), \ldots, \pi(a_{\#_L(R)})) \in R^{\mathcal{M}'}$, and $f^{\mathcal{M}'}(\pi(a_1), \ldots, \pi(a_{\#_L(f)})) = \pi(f^{\mathcal{M}}(a_1, \ldots, a_{\#_L(f)}))$. In this case we say that $\pi$ is an *isomorphism* from $\mathcal{M}$ to $\mathcal{M}'$, denoted $\pi : \mathcal{M} \cong \mathcal{M}'$.

If $\mathcal{M}$ is an $L$-structure and $\mathcal{M}'$ is an $L'$-structure such that $L' \subseteq L$, $c^{\mathcal{M}} = c^{\mathcal{M}'}$ for $c \in L'$, $R^{\mathcal{M}} = R^{\mathcal{M}'}$ for $c \in L'$, and $f^{\mathcal{M}} = f^{\mathcal{M}'}$ for $f \in L'$, then $\mathcal{M}'$ is said to be a *reduct* of $\mathcal{M}$ (to the vocabulary $L'$), denoted $\mathcal{M}' = \mathcal{M}{\upharpoonright}L'$, and $\mathcal{M}$ is said to be an *expansion* of $\mathcal{M}'$ (to the vocabulary $L$). If $\mathcal{M}$ is an $L$-structure and $a \in M$, then the expansion $\mathcal{M}'$ of $\mathcal{M}$, denoted $(\mathcal{M}, a)$, to a vocabulary $L \cup \{c\}$, where $c \notin L$, is defined by $c^{\mathcal{M}'} = a$; $(\mathcal{M}, a_1, \ldots, a_n)$ is defined similarly.

## 2.3 Terms and formulas

Constant symbols of $L$ and variables are $L$-*terms*; if $t_1, \ldots, t_n$ are $L$-terms, then $f t_1 \ldots t_n$ is an $L$-term for each $f$ in $L$ of arity $n$. The set $\mathrm{Var}(t)$ of *variables* of a term $t$ is simply the set of variables that occur in $t$. If $\mathrm{Var}(t) = \emptyset$, then $t$ is called a *constant term*. For example, $fc$ is a constant term. Every constant term $t$ has a definite *value* $t^{\mathcal{M}}$ in any $L$-structure $\mathcal{M}$, defined inductively as follows: if $t$ is a constant symbol, $t^{\mathcal{M}}$ is defined already. Otherwise, $(f t_1 \ldots t_n)^{\mathcal{M}} = f^{\mathcal{M}}(t_1^{\mathcal{M}}, \ldots, t_n^{\mathcal{M}})$.

Any function $s$ from a finite set $\mathrm{dom}(s)$ of variables into the domain $M$ of an $L$-structure $\mathcal{M}$ is called an *assignment* of $\mathcal{M}$. Set theoretically, $s = \{(a, s(a)) : a \in \mathrm{dom}(s)\}$. The restriction $s{\upharpoonright}A$ of $s$ to a set $A$ is the function $\{(a, s(a)) : a \in \mathrm{dom}(s) \cap A\}$. An assignment $s$ assigns a *value* $t^{\mathcal{M}}\langle s \rangle$ in $M$ to any $L$-term $t$ such that $\mathrm{Var}(t) \subseteq \mathrm{dom}(s)$ as follows: $c^{\mathcal{M}}\langle s \rangle = c^{\mathcal{M}}$, $x_n^{\mathcal{M}}\langle s \rangle = s(x_n)$, and $(f t_1 \ldots t_n)^{\mathcal{M}}\langle s \rangle = f^{\mathcal{M}}(t_1^{\mathcal{M}}\langle s \rangle, \ldots, t_n^{\mathcal{M}}\langle s \rangle)$.

The *veritas* symbol $\top$ is an $L$-formula. Strings $t_i := t_j$ and $R t_1 \ldots t_n$ are *atomic $L$-formulas* whenever $t_1, \ldots, t_n$ are $L$-terms and $R$ is a relation symbol in $L$ with arity $n$. We sometimes write $(t_i = t_j)$ for clarity. Atomic $L$-formulas are $L$-formulas. If $\phi$ and $\psi$ are $L$-formulas, then $(\phi \vee \psi)$ and $\neg\phi$ are $L$-formulas. If $\phi$ is an $L$-formula and $n \in \mathbb{N}$, then $\exists x_n \phi$ is an $L$-formula. We use $(\phi \wedge \psi)$ to denote $\neg(\neg\phi \vee \neg\psi)$, $(\phi \to \psi)$ to denote $(\neg\phi \vee \psi)$, $(\phi \leftrightarrow \psi)$ to denote $((\phi \to \psi) \wedge (\psi \to \phi))$, and $\forall x_n \phi$ to denote $\neg\exists x_n \neg\phi$. Formulas defined in this way are called *first order*. An $L$-formula is *quantifier free* if it has no quantifiers.

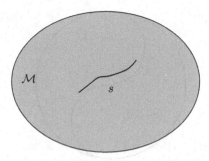

Fig. 2.1. A model and an assignment.

A formula, possibly containing occurrences of the shorthands $\wedge$ and $\forall$, is in *negation normal form* if it has negations in front of atomic formulas only.

The set $\mathrm{Fr}(\phi)$ of *free variables* of a formula $\phi$ is defined as follows:

$$\mathrm{Fr}(t_1 = t_2) = \mathrm{Var}(t_1) \cup \mathrm{Var}(t_2),$$
$$\mathrm{Fr}(Rt_1 \dots t_n) = \mathrm{Var}(t_1) \cup \dots \cup \mathrm{Var}(t_n),$$
$$\mathrm{Fr}(\phi \vee \psi) = \mathrm{Fr}(\phi) \cup \mathrm{Fr}(\psi),$$
$$\mathrm{Fr}(\neg\phi) = \mathrm{Fr}(\phi),$$
$$\mathrm{Fr}(\exists x_n \phi) = \mathrm{Fr}(\phi) \setminus \{x_n\}.$$

If $\mathrm{Fr}(\phi) = \emptyset$, we call $\phi$ an *L-sentence*.

## 2.4 Truth and satisfaction

Truth in first order logic can be defined in different equivalent ways. The most common approach is the following, based on the more general concept of *satisfaction* of *L*-formulas. There is an alternative game theoretic definition of truth, most relevant for this book, and we will introduce it in Chapter 5. In the definition below the concept of an assignment $s$ satisfying an *L*-formula $\phi$ in an *L*-structure, denoted $\mathcal{M} \models_s \phi$, is defined by giving a sufficient condition for $\mathcal{M} \models_s \phi$ in terms of subformulas of $\phi$.

For quantifiers we introduce the concept of a *modified* assignment. If $s$ is an assignment and $n \in \mathbb{N}$, then $s(a/x_n)$ is the assignment which agrees with $s$ everywhere except that it maps $x_n$ to $a$. In other words, $\mathrm{dom}(s(a/x_n)) = \mathrm{dom}(s) \cup \{x_n\}$, $s(a/x_n)(x_i) = s(x_i)$ when $x_i \in \mathrm{dom}(s) \setminus \{x_n\}$, and $s(a/x_n)(x_n) = a$.

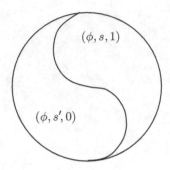

Fig. 2.2. Truth and falsity.

We define $\mathcal{T}$ as the smallest set such that:

(P1)  if $t_1^{\mathcal{M}}\langle s \rangle = t_2^{\mathcal{M}}\langle s \rangle$, then $(t_1 = t_2, s, 1) \in \mathcal{T}$;

(P2)  if $t_1^{\mathcal{M}}\langle s \rangle \neq t_2^{\mathcal{M}}\langle s \rangle$, then $(t_1 = t_2, s, 0) \in \mathcal{T}$;

(P3)  if $(t_1^{\mathcal{M}}\langle s \rangle, \ldots, t_n^{\mathcal{M}}\langle s \rangle) \in R^{\mathcal{M}}$, then $(Rt_1 \ldots t_n, s, 1) \in \mathcal{T}$;

(P4)  if $(t_1^{\mathcal{M}}\langle s \rangle, \ldots, t_n^{\mathcal{M}}\langle s \rangle) \notin R^{\mathcal{M}}$, then $(Rt_1 \ldots t_n, s, 0) \in \mathcal{T}$;

(P5)  if $(\phi, s, 1) \in \mathcal{T}$ or $(\psi, s, 1) \in \mathcal{T}$, then $(\phi \vee \psi, s, 1) \in \mathcal{T}$;

(P6)  if $(\phi, s, 0) \in \mathcal{T}$ and $(\psi, s, 0) \in \mathcal{T}$, then $(\phi \vee \psi, s, 0) \in \mathcal{T}$;

(P7)  if $(\phi, s, 1) \in \mathcal{T}$, then $(\neg\phi, s, 0) \in \mathcal{T}$;

(P8)  if $(\phi, s, 0) \in \mathcal{T}$, then $(\neg\phi, s, 1) \in \mathcal{T}$;

(P9)  if $(\phi, s(a/x_n), 1) \in \mathcal{T}$ for some $a$ in $M$, then $(\exists x_n\phi, s, 1) \in \mathcal{T}$;

(P10)  if $(\phi, s(a/x_n), 0) \in \mathcal{T}$ for all $a$ in $M$, then $(\exists x_n\phi, s, 0) \in \mathcal{T}$.

Finally we define $\mathcal{M} \models_s \phi$ if $(\phi, s, 1) \in \mathcal{T}$. A formula $\psi$ is said to be a *logical consequence* of another formula $\phi$, in symbols $\phi \Rightarrow \psi$, if for all $\mathcal{M}$ and $s$ such that $\mathcal{M} \models_s \phi$ we have $\mathcal{M} \models_s \psi$. A formula $\psi$ is said to be *logically equivalent* to another formula $\phi$, in symbols $\phi \equiv \psi$, if $\phi \Rightarrow \psi$ and $\psi \Rightarrow \phi$.

**Exercise 2.1** *Prove for all first order $\phi$: $(\phi, s, 1) \in \mathcal{T}$ or $(\phi, s, 0) \in \mathcal{T}$.*

**Exercise 2.2** *Prove that for no first order $\phi$ and for no $s$ we have $(\phi, s, 1) \in \mathcal{T}$ and $(\phi, s, 0) \in \mathcal{T}$.*

**Exercise 2.3** *Prove for all first order $\phi$: $(\neg\phi, s, 1) \in \mathcal{T}$ if and only if $(\phi, s, 1) \notin \mathcal{T}$.*

We define two operations $\phi \mapsto \phi^{\mathrm{p}}$ and $\phi \mapsto \phi^{\mathrm{d}}$ by simultaneous induction, using the shorthands $\phi \wedge \psi$ and $\forall x_n \phi$, as follows:

$$\phi^{\mathrm{d}} = \neg\phi \text{ if } \phi \text{ atomic},$$
$$\phi^{\mathrm{p}} = \phi \text{ if } \phi \text{ atomic},$$
$$(\neg\phi)^{\mathrm{d}} = \phi^{\mathrm{p}},$$
$$(\neg\phi)^{\mathrm{p}} = \phi^{\mathrm{d}},$$
$$(\phi \vee \psi)^{\mathrm{d}} = \phi^{\mathrm{d}} \wedge \psi^{\mathrm{d}},$$
$$(\phi \vee \psi)^{\mathrm{p}} = \phi^{\mathrm{p}} \vee \psi^{\mathrm{p}},$$
$$(\exists x_n \phi)^{\mathrm{d}} = \forall x_n \phi^{\mathrm{d}},$$
$$(\exists x_n \phi)^{\mathrm{p}} = \exists x_n \phi^{\mathrm{p}}.$$

We call $\phi^{\mathrm{d}}$ the *dual* of $\phi$. The basic result concerning duality in first order logic is that $\phi \equiv \phi^{\mathrm{p}}$ and $\neg\phi \equiv \phi^{\mathrm{d}}$. Thus the dual operation is a mechanical way for translating a formula $\phi$ to one which is logically equivalent to the negation of $\phi$, without actually adding negation anywhere except in front of atomic formulas. Note that the dual of a formula in negation normal form is again in negation normal form. This is important because negation does not, *a priori*, preserve the negation normal form, unlike the other logical operations $\wedge, \vee, \exists, \forall$.

**Exercise 2.4** *Show that $\phi^{\mathrm{p}}$ and $\phi^{\mathrm{d}}$ are always in negation normal form.*

**Exercise 2.5** *Prove $(\phi^{\mathrm{d}})^{\mathrm{d}} = \phi^{\mathrm{p}}$ and $(\phi^{\mathrm{p}})^{\mathrm{p}} = \phi^{\mathrm{p}}$.*

**Exercise 2.6** *Compute $(\phi^{\mathrm{p}})^{\mathrm{d}}$ and $(\phi^{\mathrm{d}})^{\mathrm{p}}$.*

**Exercise 2.7** *Prove $\phi \equiv \phi^{\mathrm{p}}$ and $\neg\phi \equiv \phi^{\mathrm{d}}$ for any $\phi$ in first order logic.*

Both $\phi \mapsto \phi^{\mathrm{d}}$ and $\phi \mapsto \phi^{\mathrm{p}}$ preserve logical equivalence. Thus if we define the formula $\phi^*$, for any first order formula $\phi$ written in negation normal form, to be the result of replacing each logical operation in $\phi$ by its dual (i.e. $\wedge$ by $\vee$ and vice versa, $\forall$ by $\exists$ and vice versa), then any logical equivalence $\phi \equiv \psi$ gives rise to another logical equivalence $\phi^* \equiv \psi^*$. This is the *Principle of Duality*.

**Exercise 2.8** *Prove that $\phi \equiv \psi$ implies $\phi^* \equiv \psi^*$.*

In terms of game theoretic semantics, which we discuss in Chapter 5, the dual of a sentence, in being logically equivalent to its negation, corresponds to permuting the roles of the players.

# 3

# Dependence logic

Dependence logic introduces the concept of *dependence* into first order logic by adding a new kind of atomic formula. We call these new atomic formulas *atomic dependence formulas*. The definition of the semantics for dependence logic is reminiscent of the definition of the semantics for first order logic, presented in Chapter 2. But instead of defining satisfaction for assignments, we follow ref. [21] and jump one level up considering *sets* of assignments. This leads us to formulate the semantics of dependence logic in terms of the concept of the *type* of a set of assignments.

The reason for the transition to a higher level is, roughly speaking, that one cannot manifest dependence, or independence for that matter, in a single assignment. To see a pattern of dependence, one needs a whole set of assignments.

This is because dependence notions can be best investigated in a context involving repeated actions by agents presumably governed by some possibly hidden rules. In such a context dependence is manifested by recurrence, and independence by lack of it.

Our framework consists of three components:

teams, agents, and features.

Teams are sets of agents. Agents are objects with features. Features are like variables which can have any value in a given fixed set.

If we have $n$ features and $m$ possible values for each feature, we have altogether $m^n$ different agents. Teams are simply subsets of this space of all possible agents.

Although our treatment of dependence logic is entirely mathematical, our intuition of dependence phenomena comes from real life examples, thinking of different ways dependence manifests itself in the real world. Statisticians certainly have much to say about this, but when we go deeper into the logic of dependence we see that the crucial concept is determination, not mere dependence. Another way in which dependence differs from statistics is that we study total dependence, not statistically significant dependence. It would seem reasonable to define probabilistic dependence logic, but we will not go into that here.

## 3.1 Examples and a mathematical model for teams

In practical examples, a feature is anything that can be in the domain of a function: color, length, weight, prize, profession, salary, gender, etc. To be specific, we use variables, $x_0, x_1, \ldots$ to denote features. If features are variables, then agents are *assignments*. When we define dependence logic, we use the variable $x_n$ to refer to the value $s(x_n)$ of the feature $x_n$ in an agent $s$.

(i) A team of seven agents with features $\{x_0, x_1, x_2\}$ as domain and $\mathbb{Q}$ as the set of possible values of the features could look like Table 3.1. One can think of this as a set of seven possible assignments to the variables $x_0, x_1, x_2$ in the model $(\mathbb{Q}, <)$. Some of the assignments satisfy $x_0 < x_1$ and they all satisfy $x_2 < x_1$.

(ii) Consider teams of soccer players. In this case the players are the agents. The number assigned to each player as well as the colors of their shirts and shorts are the features, denoted by variables $x_0, x_1, x_2$, respectively. Teams are sets of players in the usual sense of the word "team." Table 3.2 depicts a team. If we counted only the color of the players' shirts and shorts as features, we would obtain the generated team of three agents depicted in Table 3.3.

(iii) Databases are good examples of teams. By a database we mean in this context a table of data arranged in columns and rows. The columns are the features, the rows are the agents, and the set consisting of the rows is the

Table 3.1.

|       | $x_0$ | $x_1$ | $x_2$ |
|-------|-------|-------|-------|
| $s_0$ | 1.5   | 4     | 0.51  |
| $s_1$ | 2.1   | 4     | 0.55  |
| $s_2$ | 2.1   | 4     | 0.53  |
| $s_3$ | 5.1   | 4     | 0.54  |
| $s_4$ | 8.9   | 4     | 0.53  |
| $s_5$ | 21    | 4     | 0.54  |
| $s_6$ | 100   | 4     | 0.54  |

Table 3.2. *Soccer players as a team*

|  | (Player) $x_0$ | (Shirt) $x_1$ | (Shorts) $x_2$ |
|---|---|---|---|
| $s_0$ | 1 | yellow | white |
| $s_1$ | 2 | yellow | white |
| $s_2$ | 3 | yellow | white |
| $s_3$ | 4 | yellow | white |
| $s_4$ | 5 | red | white |
| $s_5$ | 6 | red | black |
| $s_6$ | 7 | red | black |

Table 3.3. *A generated team*

|  | (Shirt) $x_1$ | (Shorts) $x_2$ |
|---|---|---|
| $s_0$ | yellow | white |
| $s_1$ | red | white |
| $s_2$ | red | black |

Table 3.4. *A database*

| | Fields | | | |
|---|---|---|---|---|
| Record | $x_1$ | $x_2$ | ... | $x_n$ |
| 1 | 52 | 24 | ... | 1 |
| 2 | 68 | 362 | ... | 0 |
| 3 | 11 | 7311 | ... | 1 |
| ... | ... | ... | ... | ... |
| $k$ | 101 | 43 | ... | 1 |

team. In database theory the columns are often called fields or attributes, and the rows are called records or tuples. Table 3.4 is an example of a database. If the row number (1 to $k$ in Table 3.4) is counted as a feature, then all rows are different agents. Otherwise rows with identical values in all the features are identified, resulting an a team called the *generated team*, i.e. the team generated by the particular database. Table 3.5 depicts a database and the generated team.

(iv) The game history team. Imagine a game where players make moves, following a strategy they have chosen with a certain goal in mind. We are thinking of games in the sense of von Neumann and Morgenstern's *Theory of Games and Economic Behavior* [43]. Examples of such games are board and card games, business games, games related to social behavior, etc. We think of the moves of the game as features. If during a game 350 moves are made by the players, then we have 350 features. Plays, i.e. sequences of moves of the game that comprise an entire play of the game, are the agents. Any collection of plays is a team. A team may arise, for example, as follows: two players play a certain game 25 times, thus producing 25 sequences of moves. A team of 25 agents is created.

It may be desirable to know answers to the following kinds of questions.
(a) What is the strategy that a player is following, or is he or she following any strategy at all?
(b) Is a player using information about his or her (or other players') moves that he or she is committed not to use? This issue is closely related to game theoretic semantics of dependence logic treated in Chapter 5.

The following game illustrates how a player can use information that may be not admitted. There are two players **I** and **II**. (For ease of reference, we always refer to player **I** as "he" and player **II** as "she".) Player **I** starts by choosing an integer $n$. Then **II** chooses an integer $m$. After this **II** makes another move and chooses, this time without seeing $n$, an integer $l$. So player **II** is committed to choose $l$ without seeing $n$, even if she saw $n$ when she picked $m$. One may ask, how can she forget a number she has seen once, but if the number has many digits this is quite plausible. Player **II** wins if $l > n$. In other words, **II** has the impossible looking task of choosing an integer $l$ which is bigger than an integer $n$ that she is not allowed to know. Her trick, which we call the signalling-strategy, is to store information about $n$ into $m$ and then choose $l$ only on the basis of what $m$ is. Table 3.5 shows an example of a game history team in this

Table 3.5. *Game history and the generated team*

| Play | I | II | II | | $x_0$ | $x_1$ | $x_2$ |
|------|-----|-----|-----|-------|-------|-------|-------|
| 1 | 1 | 1 | 2 | | | | |
| 2 | 40 | 40 | 41 | | | | |
| 3 | 2 | 2 | 3 | | | | |
| 4 | 0 | 0 | 1 | $s_0$ | 0 | 0 | 1 |
| 5 | 1 | 1 | 2 | $s_1$ | 1 | 1 | 2 |
| 6 | 2 | 2 | 3 | $s_2$ | 2 | 2 | 3 |
| 7 | 40 | 40 | 41 | $s_3$ | 40 | 40 | 41 |
| 8 | 100 | 100 | 101 | $s_4$ | 100 | 100 | 101 |

Table 3.6. *Suspicious game history and the generated team*

| Play | I | II | II |
|------|-----|---|-----|
| 1 | 1 | 0 | 2 |
| 2 | 40 | 0 | 41 |
| 3 | 2 | 0 | 3 |
| 4 | 0 | 0 | 1 |
| 5 | 1 | 0 | 2 |
| 6 | 2 | 0 | 3 |
| 7 | 40 | 0 | 41 |
| 8 | 100 | 0 | 101 |

| | $x_0$ | $x_1$ | $x_2$ |
|------|-----|---|-----|
| $s_0$ | 0 | 0 | 1 |
| $s_1$ | 1 | 0 | 2 |
| $s_2$ | 2 | 0 | 3 |
| $s_3$ | 40 | 0 | 41 |
| $s_4$ | 100 | 0 | 101 |

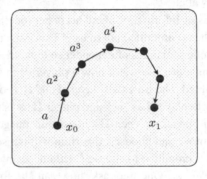

Fig. 3.1. Dependence of $x_1$ on $x_0$.

game. We can see that player **II** has been using the signalling-strategy. If we instead observed the behavior of Table 3.6, we could doubt whether **II** is obeying the rules, as her second move seems clearly to depend on the move of **I** which she is not supposed to see.

(v) Every formula $\phi(x_1, \ldots, x_n)$ of any logic and structure $\mathcal{M}$ together give rise to the team of all assignments that satisfy $\phi(x_1, \ldots, x_n)$ in $\mathcal{M}$. In this case the variables are the features and the assignments are the agents. This (possibly quite large) team manifests the dependence structure $\phi(x_1, \ldots, x_n)$ that expresses in $\mathcal{M}$. If $\phi$ is the first order formula $x_0 = x_1$, then $\phi$ expresses the very strong dependence of $x_1$ on $x_0$, namely of $x_1$ being equal to $x_0$. The team of assignments satisfying $x_0 = x_1$ in a structure is the set of assignments $s$ which give to $x_0$ the same value as to $x_1$. If $\phi$ is the infinitary formula

$$(x_0 = x_1) \vee (x_0 \cdot x_0 = x_1) \vee (x_0 \cdot x_0 \cdot x_0 = x_1) \vee \ldots,$$

then $\phi$ expresses the dependence of $x_1$ on $x_0$ of being in the set $\{x_0, x_0 \cdot x_0, x_0 \cdot x_0 \cdot x_0, \ldots\}$. See Fig. 3.1.

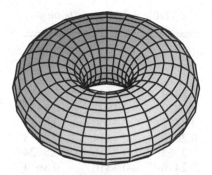

Fig. 3.2. Picture of a torus team.

(vi) Every first order sentence $\phi$ and structure $\mathcal{M}$ together give rise to teams consisting of assignments that arise in the semantic game (see Section 5.1) of $\phi$ and $\mathcal{M}$. The semantic game is a game for two players, **I** and **II**, in which **I** tries to show that $\phi$ is not true in $\mathcal{M}$ and **II** tries to show that $\phi$ is indeed true in $\mathcal{M}$. The game proceeds according to the structure of $\phi$. At conjunctions player **I** chooses a conjunct. At universal quantifiers player **I** chooses a value for the universally bound variable. At disjunctions player **II** chooses a disjunct. At existential quantifiers player **II** picks up a value for the existentially bound variable. At negations the players exchange roles. Thus the players build up move by move an assignment $s$. When an atomic formula is met, player **II** wins if the formula is true in $\mathcal{M}$ under the assignment $s$, otherwise player **I** wins. See Section 5.1 for details. If $M \models \phi$ and the winning strategy of **II** is $\tau$ in this semantic game, a particularly interesting team consists of all plays of the semantic game in which **II** uses $\tau$. This team is interesting because the strategy $\tau$ can be read off from the team. We can view the study of teams of plays in this game as a generalization of the study of who wins the semantic game. The semantic game of dependence logic is treated in Chapter 5.

(vii) Space team. Let us consider the three-dimensional Euclidean space $\mathbb{R}^3$. Let $S$ be a surface in $\mathbb{R}^3$, e.g. the torus

$$S = \{(\cos v \cos u, \cos v \sin u, \sin v) : u, v \in [0, 2\pi]\}.$$

The set $S$ is a team in which the three coordinates $x$, $y$, and $z$ are the features, and the points on the surface are the agents (see Fig. 3.2 and Table 3.7).

We now give a mathematical model for teams.

Table 3.7. *A torus team*

| x | y | z |
|---|---|---|
| ... | ... | ... |
| 0.2919 | 0.4546 | 0.8415 |
| 0.2829 | 0.4504 | 0.8468 |
| 0.2739 | 0.4460 | 0.8521 |
| 0.2650 | 0.4414 | 0.8573 |
| 0.2563 | 0.4366 | 0.8624 |
| 0.2476 | 0.4316 | 0.8674 |
| 0.2390 | 0.4265 | 0.8724 |
| 0.2305 | 0.4212 | 0.8772 |
| 0.2222 | 0.4157 | 0.8820 |
| 0.2139 | 0.4101 | 0.8866 |
| 0.2057 | 0.4042 | 0.8912 |
| 0.1977 | 0.3983 | 0.8957 |
| 0.1898 | 0.3922 | 0.9001 |
| 0.1820 | 0.3859 | 0.9044 |
| 0.1744 | 0.3794 | 0.9086 |
| 0.1669 | 0.3729 | 0.9128 |
| 0.1595 | 0.3661 | 0.9168 |
| 0.1522 | 0.3592 | 0.9208 |
| ... | ... | ... |

**Definition 3.1** *An* agent *is any function s from a finite set* dom(s) *of variables, also called* features, *to a fixed set M. The set* dom(s) *is called the* domain *of s, and the set M is called the* codomain *of s. A* team *is any set X of agents with the same domain, called the* domain of X *and denoted by* dom(X), *and the same codomain, likewise called the* codomain *of X. A team with codomain M is called a team of M. If V is a finite set of variables, we use* Team(M, V) *to denote the set of all teams of M with domain V.*

Since we have defined teams as *sets*, not *multisets*, of assignments, one assignment can occur only once in a team. Allowing multisets would not, however, change anything essential in this study.

## 3.2  Formulas as types of teams

We define a logic which has an atomic formula for expressing dependence. We call this logic the *dependence logic* and denote it by $\mathcal{D}$. In Section 3.6 we will recover independence friendly logic as a fragment of dependence logic.

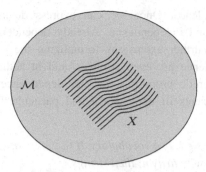

Fig. 3.3. A model and a team.

In first order logic the meaning of a formula is derived from the concept of an assignment satisfying the formula. In dependence logic the meaning of a formula is based on the concept of a team being of the *(dependence) type* of the formula.

Recall that teams are sets of agents (assignments) and that agents are functions from a finite set of variables, called the domain of the agent into an arbitrary set called the codomain of the agent (Definition 3.1). In a team the domain of all agents is assumed to be the same set of variables, just as the codomain of all agents is assumed to be the same set ( Fig. 3.3).

Our atomic dependence formulas have the following form:

$$=(t_1, \ldots, t_n).$$

The intuitive meaning of this is as follows:

the value of the term $t_n$ depends only on the values of the terms $t_1, \ldots, t_{n-1}$.

As singular cases, we have

$$=(),$$

which we take to be universally true, and

$$=(t),$$

which declares that the value of the term $t$ depends on nothing, i.e. is constant. Note that $=(x_1)$ is quite non-trivial and indispensable if we want to say that all agents are similar as far as feature $x_1$ is concerned. Such similarity is manifested by the team of Table 3.6, where all agents have value 0 in their feature $x_1$.

Actually, our atomic formulas express determination rather than dependence. The reason for this is that determination is a more basic concept than dependence. Once we can express determination, we can define dependence and independence. Already dependence logic has considerable strength. Further extensions formalizing the concepts of dependence and independence are even stronger, and in addition lack many of the nice model theoretic properties that our dependence logic enjoys. We will revisit the concepts of dependence and particularly independence in Section 8.2

**Definition 3.2** *Suppose $L$ is a vocabulary. If $t_1, \ldots, t_n$ are $L$-terms and $R$ is a relation symbol in $L$ with arity $n$, then strings*

$$t_i = t_j,$$
$$=(t_1, \ldots, t_n),$$
$$Rt_1 \ldots t_n$$

*are $L$-formulas of dependence logic $\mathcal{D}$. They are called* atomic *formulas. If $\phi$ and $\psi$ are $L$-formulas of $\mathcal{D}$, then*

$$(\phi \vee \psi)$$

*and*

$$\neg \phi$$

*are $L$-formulas of $\mathcal{D}$. If $\phi$ is an $L$-formula of $\mathcal{D}$ and $n \in \mathbb{N}$, then*

$$\exists x_n \phi$$

*is an $L$-formula of $\mathcal{D}$.*

As is apparent from Definition 3.2, the syntax of dependence logic $\mathcal{D}$ is very similar to that of first order logic, the only difference being the inclusion of the new atomic formulas $=(t_1, \ldots, t_n)$. We use

$$(\phi \wedge \psi)$$

to denote $\neg(\neg\phi \vee \neg\psi)$,

$$(\phi \rightarrow \psi)$$

to denote $(\neg\phi \vee \psi)$,

$$(\phi \leftrightarrow \psi)$$

to denote $((\phi \rightarrow \psi) \wedge (\psi \rightarrow \phi))$, and

$$\forall x_n \phi$$

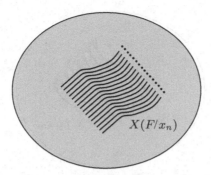

Fig. 3.4.  A supplement team.

to denote $\neg \exists x_n \neg \phi$. A formula of dependence logic which does not contain any atomic formulas of the form $=(t_1, \ldots, t_n)$ is called *first order*. The *veritas* symbol $\top$ is definable as $=()$. We call this also first order as we took $\top$ as a special symbol in Chapter 2.

The set $\mathrm{Fr}(\phi)$ of *free variables* of a formula $\phi$ is defined otherwise as for first order logic, except that we have the new case

$$\mathrm{Fr}(=(t_1, \ldots, t_n)) = \mathrm{Var}(t_1) \cup \cdots \cup \mathrm{Var}(t_n).$$

If $\mathrm{Fr}(\phi) = \emptyset$, we call $\phi$ an *L-sentence* of dependence logic.

We define now two important operations on teams, the supplement and the duplication operations. The supplement operation adds a new feature to the agents in a team, or alternatively changes the value of an existing feature.

Suppose a strategy officer of a company comes to a director with a plan for a team to design a new product. The director asks: "What about the language skills of the team members?" The strategy officer answers: "No problem, I can supplement the language skills to the team description, and then you will see that the team is really of the type we need." This is the idea of supplementing a team (see Fig. 3.4).

**Definition 3.3** *If $M$ is a set, $X$ is a team with $M$ as its codomain and $F : X \rightarrow M$, we let $X(F/x_n)$ denote the supplement team $\{s(F(s)/x_n) : s \in X\}$.*

A duplicate team is obtained by duplicating the agents of a team until all possibilities occur as far as a particular feature is concerned.

Suppose a strategy officer of a company comes to a director with a plan for a team for a company wide committee. The strategy officer says, "I decided we need a programmer, an analyst, and a sales person. I chose such people from

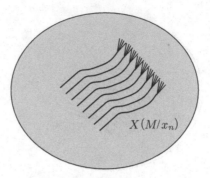

Fig. 3.5. A duplicate team.

each of our five departments." This is the idea of duplicating a team. The team to be duplicated consisted of three agents with just one feature with values in the set programmer, analyst, sales person. The duplicated team has 15 agents corresponding to a programmer, an analyst, and a sales person from each of the five departments. As this example indicates, in real life examples the features do not always have values in the same set, as in our mathematical model. We could rectify this by considering many-sorted structures, but this would lead to unnecessarily complicated notation.

**Definition 3.4** *If $M$ is a set and $X$ is a team of $M$ we use $X(M/x_n)$ to denote the* duplicate *team $\{s(a/x_n) : a \in M, s \in X\}$. See Fig. 3.5.*

We are ready to define the semantics of dependence logic.

**Definition 3.5** *Suppose $L$ is a vocabulary and $\mathcal{M}$ is an $L$-structure. We define the set $\mathcal{T}$, or more exactly $\mathcal{T}_{\mathcal{M}}$, called the* fundamental predicate *of $\mathcal{M}$, of triples $(\phi, X, d)$, where $\phi$ is an $L$-formula of dependence logic, $\mathrm{Fr}(\phi) \subseteq \mathrm{dom}(X)$ and $d \in \{0, 1\}$, as the smallest set such that*

(D1) *if $t_1^{\mathcal{M}}\langle s \rangle = t_2^{\mathcal{M}}\langle s \rangle$ for all $s \in X$, then $(t_1 = t_2, X, 1) \in \mathcal{T}$;*
(D2) *if $t_1^{\mathcal{M}}\langle s \rangle \neq t_2^{\mathcal{M}}\langle s \rangle$ for all $s \in X$, then $(t_1 = t_2, X, 0) \in \mathcal{T}$;*
(D3) *if $t_n^{\mathcal{M}}\langle s \rangle = t_n^{\mathcal{M}}\langle s' \rangle$ for all $s, s' \in X$ such that*

$$t_1^{\mathcal{M}}\langle s \rangle = t_1^{\mathcal{M}}\langle s' \rangle, \ldots, t_{n-1}^{\mathcal{M}}\langle s \rangle = t_{n-1}^{\mathcal{M}}\langle s' \rangle,$$

*then $(=(t_1, \ldots, t_n), X, 1)) \in \mathcal{T}$;*
(D4) *$(=(t_1, \ldots, t_n), \emptyset, 0) \in \mathcal{T}$;*
(D5) *if $(t_1^{\mathcal{M}}\langle s \rangle, \ldots, t_n^{\mathcal{M}}\langle s \rangle) \in R^{\mathcal{M}}$ for all $s \in X$, then $(Rt_1 \ldots t_n, X, 1) \in \mathcal{T}$;*
(D6) *if $(t_1^{\mathcal{M}}\langle s \rangle, \ldots, t_n^{\mathcal{M}}\langle s \rangle) \notin R^{\mathcal{M}}$ for all $s \in X$, then $(Rt_1 \ldots t_n, X, 0) \in \mathcal{T}$;*

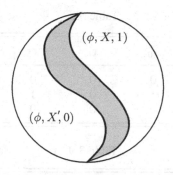

Fig. 3.6. Truth and falsity.

(D7) *if* $(\phi, X, 1) \in \mathcal{T}$, $(\psi, Y, 1) \in \mathcal{T}$, *and* $\mathrm{dom}(X) = \mathrm{dom}(Y)$, *then* $(\phi \vee \psi, X \cup Y, 1) \in \mathcal{T}$;

(D8) *if* $(\phi, X, 0) \in \mathcal{T}$ *and* $(\psi, X, 0) \in \mathcal{T}$, *then* $(\phi \vee \psi, X, 0) \in \mathcal{T}$;

(D9) *if* $(\phi, X, 1) \in \mathcal{T}$, *then* $(\neg\phi, X, 0) \in \mathcal{T}$;

(D10) *if* $(\phi, X, 0) \in \mathcal{T}$, *then* $(\neg\phi, X, 1) \in \mathcal{T}$;

(D11) *if* $(\phi, X(F/x_n), 1) \in \mathcal{T}$ *for some* $F : X \to M$, *then* $(\exists x_n \phi, X, 1) \in \mathcal{T}$;

(D12) *if* $(\phi, X(M/x_n), 0) \in \mathcal{T}$, *then* $(\exists x_n \phi, X, 0) \in \mathcal{T}$.

*Finally, we define*

$$X \text{ is of type } \phi \text{ in } \mathcal{M}, \text{ denoted } \mathcal{M} \models_X \phi$$

*if* $(\phi, X, 1) \in \mathcal{T}$. *Furthermore,*

$$\phi \text{ is true in } \mathcal{M}, \text{ denoted } \mathcal{M} \models \phi,$$

*if* $\mathcal{M} \models_{\{\emptyset\}} \phi$, *and*

$$\phi \text{ is valid, denoted } \models \phi,$$

*if* $\mathcal{M} \models \phi$ *for all* $\mathcal{M}$.

Note that,

$$\mathcal{M} \models_X \neg\phi \text{ if } (\phi, X, 0) \in \mathcal{T};$$
$$\mathcal{M} \models \neg\phi \quad \text{if } (\phi, \{\emptyset\}, 0) \in \mathcal{T}. \text{ Then we say that } \phi \text{ is } false \text{ in } \mathcal{M}.$$

We will see in a moment that it is not true in general that $(\phi, X, 1) \in \mathcal{T}$ or $(\phi, X, 0) \in \mathcal{T}$. Likewise it is not true in general that $\mathcal{M} \models \phi$ or $\mathcal{M} \models \neg\phi$, nor that $\mathcal{M} \models \phi \vee \neg\phi$. In other words, no sentence can be both true and false in a model, but some sentences can be neither true nor false in a model. This gives our logic a nonclassical flavor. See Fig. 3.6.

Table 3.8.

|       | $x_0$ | $x_1$ | $x_2$ | $x_3$ |
|-------|-------|-------|-------|-------|
| $s_0$ | 0     | 0     | 0     | 1     |
| $s_1$ | 1     | 0     | 1     | 0     |
| $s_2$ | 0     | 0     | 0     | 1     |

Table 3.9.

|       | $x_0$  | $x_1$  | $x_2$ | $x_3$ |
|-------|--------|--------|-------|-------|
| $s_0$ | red    | yellow | white | white |
| $s_1$ | white  | red    | white | white |
| $s_2$ | black  | red    | black | white |

**Example 3.6** *Let $\mathcal{M}$ be a structure with $M = \{0, 1\}$. Consider the team given in Table 3.8. This team is of type $=(x_1)$, since $s_i(x_1) = 0$ for all $i$. This team is of type $x_0 = x_2$, as $s_i(x_0) = s_i(x_2)$ for all $i$. This team is of type $\neg x_0 = x_3$, as $s_i(x_0) \neq s_i(x_3)$ for all $i$. This team is of type $=(x_0, x_1)$, as $s_i(x_0) = s_j(x_0)$ implies $s_i(x_3) = s_j(x_3)$. This team is not of type $=(x_1, x_2)$, as $s_0(x_1) = s_1(x_1)$, but $s_0(x_2) \neq s_1(x_2)$. Finally, this team is of type $=(x_0) \vee =(x_0)$, as it can be represented as the union $\{s_0, s_2\} \cup \{s_1\}$, where $\{s_0, s_2\}$ and $\{s_1\}$ are both of type $=(x_0)$.*

**Example 3.7** *Let $\mathcal{M}$ be a structure with $M = \{yellow, white, red, black\}$. Consider the team in Table 3.9. This team is of type $=(x_3)$, since $s_i(x_3) = white$ for all $i$. This team is also of type $=(x_0, x_1)$, as $s_i(x_0) = s_j(x_0)$ implies $i = j$ and hence $s_i(x_1) = s_j(x_1)$. This team is not of type $=(x_1, x_2)$, as $s_1(x_1) = s_2(x_1)$, but $s_1(x_2) \neq s_2(x_2)$.*

Definition 3.5 refers to a set $\mathcal{T}$ of triples. Such a smallest set $\mathcal{T}$ always exists. For example, the set of all possible triples $(\phi, X, d)$ satisfies (D1)–(D12), and the intersection of all sets $\mathcal{T}$ with (D1)–(D12) has again the properties (D1)–(D12).

We may also construct the minimal $\mathcal{T}$ step by step starting from $\mathcal{T}_0 = \emptyset$ and letting $\mathcal{T}_{n+1}$ be an extension of $\mathcal{T}_n$ satisfying the consequents if $\mathcal{T}_n$ satisfies the antecedents of (D1)–(D12). Then the minimal $\mathcal{T}$ is $\cup_n \mathcal{T}_n$.

Proposition 8 shows that the implications of Definition 3.5 can all be reversed, making $\mathcal{T}$ a fixed point of the inductive definition of truth.

**Proposition 3.8** *The set $\mathcal{T}$ satisfies:*

(E1) $(t_1 = t_2, X, 1) \in \mathcal{T}$ *if and only if for all $s \in X$ we have $t_1^{\mathcal{M}}\langle s \rangle = t_2^{\mathcal{M}}\langle s \rangle$;*

(E2) $(t_1 = t_2, X, 0) \in \mathcal{T}$ *if and only if for all $s \in X$ we have $t_1^{\mathcal{M}}\langle s \rangle \neq t_2^{\mathcal{M}}\langle s \rangle$;*

(E3) $(=(t_1, \ldots, t_n), X, 1) \in \mathcal{T}$ *if and only if for all $s, s' \in X$ such that $t_1^{\mathcal{M}}\langle s \rangle = t_1^{\mathcal{M}}\langle s' \rangle, \ldots, t_{n-1}^{\mathcal{M}}\langle s \rangle = t_{n-1}^{\mathcal{M}}\langle s' \rangle$, we have $t_n^{\mathcal{M}}\langle s \rangle = t_n^{\mathcal{M}}\langle s' \rangle$;*

(E4) $(=(t_1, \ldots, t_n), X, 0) \in \mathcal{T}$ *if and only if $X = \emptyset$;*

(E5) $(Rt_1 \ldots t_n, X, 1) \in \mathcal{T}$ *if and only if for all $s \in X$ we have $(t_1^{\mathcal{M}}\langle s \rangle, \ldots, t_n^{\mathcal{M}}\langle s \rangle) \in R^{\mathcal{M}}$;*

(E6) $(Rt_1 \ldots t_n, X, 0) \in \mathcal{T}$ *if and only if for all $s \in X$ we have $(t_1^{\mathcal{M}}\langle s \rangle, \ldots, t_n^{\mathcal{M}}\langle s \rangle) \notin R^{\mathcal{M}}$;*

(E7) $(\phi \vee \psi, X, 1) \in \mathcal{T}$ *if and only if $X = Y \cup Z$ such that $\mathrm{dom}(Y) = \mathrm{dom}(Z)$,*
$(\phi, Y, 1) \in \mathcal{T}$ *and $(\psi, Z, 1) \in \mathcal{T}$;*

(E8) $(\phi \vee \psi, X, 0) \in \mathcal{T}$ *if and only if $(\phi, X, 0) \in \mathcal{T}$ and $(\psi, X, 0) \in \mathcal{T}$;*

(E9) $(\neg\phi, X, 0) \in \mathcal{T}$ *if and only if $(\phi, X, 1) \in \mathcal{T}$;*

(E10) $(\neg\phi, X, 1) \in \mathcal{T}$ *if and only if $(\phi, X, 0) \in \mathcal{T}$;*

(E11) $(\exists x_n \phi, X, 1) \in \mathcal{T}$ *if and only if $(\phi, X(F/x_n), 1) \in \mathcal{T}$ for some $F : X \to \mathcal{M}$;*

(E12) $(\exists x_n \phi, X, 0) \in \mathcal{T}$ *if and only if $(\phi, X(M/x_n), 0) \in \mathcal{T}$.*

*Proof* Suppose $(\theta, X, d)$ is a triple for which one of the claims (E1)–(E12) fails. Note that we can read uniquely from the triple whose claim it is that fails. Let $\mathcal{T}' = \mathcal{T} \setminus \{(\theta, X, d)\}$. We show that $\mathcal{T}'$ satisfies (D1)–(D12), which contradicts the minimality of $\mathcal{T}$. We leave as an exercise the fact that $\mathcal{T}'$ satisfies (D1)–(D6). We can see that $\mathcal{T}'$ satisfies (D7): if $(\phi, X, 1) \in \mathcal{T}'$ and $(\psi, Y, 1) \in \mathcal{T}'$, then $((\phi \vee \psi), X \cup Y, 1) \in \mathcal{T}'$, unless $((\phi \vee \psi), X, 1) = (\theta, X, 1)$. Since $\theta$ is a disjunction, (E7) in such a case fails. Thus $(\phi, X, 1) \notin \mathcal{T}$ or $(\psi, Y, 1) \notin \mathcal{T}$. But this contradicts the assumption that $(\phi, X, 1) \in \mathcal{T}'$ and $(\psi, Y, 1) \in \mathcal{T}'$. Also $\mathcal{T}'$ satisfies (D8): if $(\phi, X, 0) \in \mathcal{T}'$ and $(\psi, X, 0) \in \mathcal{T}'$, then $((\phi \vee \psi), X, 0) \in \mathcal{T}'$, unless $((\phi \vee \psi), X, 0) = (\theta, X, d)$. Since $\theta$ is a disjunction, (E8) in such a case fails. Thus $(\phi, X, 0) \notin \mathcal{T}$ or $(\psi, X, 0) \notin \mathcal{T}$. But this contradicts the assumption that $(\phi, X, 0) \in \mathcal{T}'$ and $(\psi, X, 0) \in \mathcal{T}'$. We leave the other cases as an exercise. $\square$

Note that

- $(\phi \wedge \psi, X, 1) \in \mathcal{T}$ if and only if $(\phi, X, 1) \in \mathcal{T}$ and $(\psi, X, 1) \in \mathcal{T}$;
- $(\phi \wedge \psi, X, 0) \in \mathcal{T}$ if and only if $X = Y \cup Z$ such that $\mathrm{dom}(Y) = \mathrm{dom}(Z)$, $(\phi, Y, 0) \in \mathcal{T}$, and $(\psi, Z, 0) \in \mathcal{T}$;
- $(\forall x_n \phi, X, 1) \in \mathcal{T}$, if and only if $(\phi, X(M/x_n), 1) \in \mathcal{T}$;
- $(\forall x_n \phi, X, 0) \in \mathcal{T}$ if and only if $(\phi, X(F/x_n), 0) \in \mathcal{T}$ for some $F : X \to M$.

*Dependence logic*

It may seem strange to define (D4) as $(=(t_1, \ldots, t_n), \emptyset, 0) \in \mathcal{T}$. Why not allow $(=(t_1, \ldots, t_n), X, 0) \in \mathcal{T}$ for non-empty $X$? The reason is that if we negate "for all $s, s' \in X$ such that $t_1^{\mathcal{M}}\langle s \rangle = t_1^{\mathcal{M}}\langle s' \rangle, \ldots, t_{n-1}^{\mathcal{M}}\langle s \rangle = t_{n-1}^{\mathcal{M}}\langle s' \rangle$, we have $t_n^{\mathcal{M}}\langle s \rangle = t_n^{\mathcal{M}}\langle s' \rangle$," maintaining the analogy with (D2) and (D6), we get "for all $s, s' \in X$ we have $t_1^{\mathcal{M}}\langle s \rangle = t_1^{\mathcal{M}}\langle s' \rangle, \ldots, t_{n-1}^{\mathcal{M}}\langle s \rangle = t_{n-1}^{\mathcal{M}}\langle s' \rangle$ and $t_n^{\mathcal{M}}\langle s \rangle \neq t_n^{\mathcal{M}}\langle s' \rangle$," which is only possible if $X = \emptyset$.

Some immediate observations can be made using Definition 3.5. We first note that the empty team $\emptyset$ is of the type of any formula, as $(\phi, \emptyset, 1) \in \mathcal{T}$ holds for all $\phi$. In fact, we have the following lemma.

**Lemma 3.9** *For all $\phi$ and $\mathcal{M}$ we have $(\phi, \emptyset, 1) \in \mathcal{T}$ and $(\phi, \emptyset, 0) \in \mathcal{T}$.*

*Proof* Inspection of Definition 3.5 reveals that all the necessary implications hold vacuously when $X = \emptyset$. □

In other words, the empty team is for all $\phi$ of type $\phi$ and of type $\neg \phi$. Since the type of a team is defined by reference to all agents in the team, the empty team ends up having all types, just as it is usually agreed that the intersection of an empty collection of subsets of a set $M$ is the set $M$ itself. A consequence of this is that there are no formulas $\phi$ and $\psi$ of dependence logic such that $\mathcal{M} \models_X \phi$ implies $\mathcal{M} \not\models_X \psi$, for all $\mathcal{M}$ and all $X$. Namely, letting $X = \emptyset$ would yield a contradiction.

The following test is very important and will be used repeatedly in the sequel. Closure downwards is a fundamental property of types in dependence logic.

**Proposition 3.10 (Closure Test)** *Suppose $Y \subseteq X$. Then $\mathcal{M} \models_X \phi$ implies $\mathcal{M} \models_Y \phi$.*

*Proof* Every condition from (E1) to (E12) in Proposition 3.8 is closed under taking a subset of $X$. So if $(\phi, X, 1) \in \mathcal{T}$ and $Y \subseteq X$, then $(\phi, Y, 1) \in \mathcal{T}$. □

The intuition behind the Closure Test is as follows. To witness the failure of dependence we need a counterexample, two or more assignments that manifest the failure. The smaller the team, the fewer the number of counterexamples. In a one-agent team, a counterexample to dependence is no longer possible. On the other hand, the bigger the team, the more likely it is that some lack of dependence becomes exposed. In the maximal team of all possible assignments no dependence is possible, unless the universe has just one element.

**Corollary 3.11** *There is no formula $\phi$ of dependence logic such that for all $X \neq \emptyset$ and all $\mathcal{M}$ we have*

$$\mathcal{M} \models_X \phi \iff \mathcal{M} \not\models_X =(x_0, x_1).$$

*Proof* Suppose for a contradiction $\mathcal{M}$ has at least two elements $a, b$. Let $X$ consist of $s = \{(x_0, a), (x_1, a)\}$ and $s' = \{(x_0, a), (x_1, b)\}$. Now $\mathcal{M} \not\models_X =(x_0, x_1)$, so $\mathcal{M} \models_X \phi$. By the Closure Test, $\mathcal{M} \models_{\{s\}} \phi$, whence $\mathcal{M} \not\models_{\{s\}} =(x_0, x_1)$. But this is clearly false. □

We can replace "all $\mathcal{M}$" by "some $\mathcal{M}$ with more than one element in the universe" in Corollary 3.11. Note that in particular we do not have for all $X \neq \emptyset$: $\mathcal{M} \models_X \neg =(x_0, x_1) \iff \mathcal{M} \not\models_X =(x_0, x_1)$.

**Example 3.12** *Every team $X$, the domain of which contains $x_i$ and $x_j$, is of type $x_i = x_j \vee \neg x_i = x_j$, as we can write $X = Y \cup Z$, where $Y = \{s \in X : s(x_i) = s(x_j)\}$ and $Z = \{s \in X : s(x_i) \neq s(x_j)\}$. Note that then $Y$ is of type $x_i = x_j$, and $Z$ is of type $x_i \neq x_j$.*

The following example formalizes the idea that we can always choose $x_1$ such that it only depends on $x_0$, if there are no other requirements. The most obvious way to establish the dependence is to choose $x_1$ identical to $x_0$.

**Example 3.13** $\models \forall x_0 \exists x_1(=(x_0, x_1))$. *To prove this, fix $\mathcal{M}$. Let $X$ be the set of $s : \{x_0, x_1\} \to M$ with $s(x_1) = s(x_0)$. We use Definition 3.5. By (D3), $(=(x_0, x_1), X, 1) \in \mathcal{T}$, for if $s, s' \in X$ and $s(x_0) = s'(x_0)$, then $s(x_1) = s'(x_1)$. Let $Y$ be the set of $s : \{x_0\} \to M$. If we let $F(s) = s(x_0)$ then by (D11) $(\exists x_1(=(x_0, x_1)), Y, 1) \in \mathcal{T}$. Thus $(\forall x_0 \exists x_1(=(x_0, x_1)), \{\emptyset\}, 1) \in \mathcal{T}$.*

In Example 3.14 there is another variable in the picture, but it does not matter.

**Example 3.14** $\models \forall x_0 \forall x_2 \exists x_1(=(x_0, x_1))$. *To prove this, fix $\mathcal{M}$. Let $X$ be the set of assignments $s : \{x_0, x_1, x_2\} \to M$ with $s(x_1) = s(x_0)$. We use Definition 3.5. By (D3), $(=(x_0, x_1), X, 1) \in \mathcal{T}$. Let $Y$ be the set of all $s : \{x_0, x_1\} \to M$. If we let $F(s) = s(x_0)$, by (D11) $(\exists x_1(=(x_0, x_1)), Y, 1) \in \mathcal{T}$. Thus the triple $(\forall x_0 \forall x_2 \exists x_1(=(x_0, x_1)), \{\emptyset\}, 1)$ is in $\mathcal{T}$.*

The following example confirms the intuition that functions convey dependence. If a model has a function $f^{\mathcal{M}}$, then we can pick for every $a \in M$ an element $b \in M$ which depends only on $a$ via the function $f^{\mathcal{M}}$.

**Example 3.15** *Let $f$ be a function symbol of the vocabulary. Then always $\models \forall x_0 \exists x_1(=(x_0, x_1) \wedge x_1 = f x_0)$. To prove this, fix $\mathcal{M}$. Let $X$ be the set of $s : \{x_0, x_1\} \to M$ with $s(x_1) = f^{\mathcal{M}}(s(x_0))$. We use Definition 3.5. By (D1), $((x_1 = f x_0), X, 1) \in \mathcal{T}$. By (D3), $(=(x_0, x_1), X, 1) \in \mathcal{T}$. Thus by (D7), $((=(x_0, x_1) \wedge x_1 = f x_0), X, 1) \in \mathcal{T}$. Let $Y$ be the set of $s : \{x_0\} \to M$. If we let $F(s) = f^{\mathcal{M}}(s(x_0))$ we get from (D11) $(\exists x_1(=(x_0, x_1) \wedge x_1 = f x_0), Y, 1) \in \mathcal{T}$. Thus $(\forall x_0 \exists x_1(=(x_0, x_1) \wedge x_1 = f x_0), \{\emptyset\}, 1) \in \mathcal{T}$.*

The following example is another way of saying that the identity $x_0 = x_1$ implies the dependence of $x_1$ on $x_0$. This is the most trivial kind of dependence.

**Example 3.16** $\models \forall x_0 \forall x_1 (x_0 = x_1 \rightarrow =(x_0, x_1))$. *To prove this, fix* $\mathcal{M}$. *Let* $X$ *be the set of all assignments* $s : \{x_0, x_1\} \rightarrow M$. *Let* $Y = \{s \in X : s(x_0) = s(x_1)\}$ *and* $Z = X \setminus Y$. *Now* $(x_0 = x_1, Z, 0) \in \mathcal{T}$, *i.e.* $(\neg(x_0 = x_1), Z, 1) \in \mathcal{T}$ *and* $(=(x_0, x_1), Y, 1) \in \mathcal{T}$. *Thus,* $((x_0 = x_1 \rightarrow =(x_0, x_1)), X, 1) \in \mathcal{T}$.

It is important to take note of a difference between universal quantification in first order logic and universal quantification in dependence logic. It is perfectly possible to have a formula $\phi(x_0)$ of dependence logic of the empty vocabulary with just $x_0$ free such that for a new constant symbol $c$ we have

$$\models \phi(c)$$

and still

$$\not\models \forall x_0 \phi(x_0),$$

as the following example shows. For this example, remember that $=(x_1)$ is the type "$x_1$ is constant" of a team in which all agents have the same value for their feature $x_1$. Recall also the definition of the expansion $(\mathcal{M}, a)$ in Section 2.2.

**Example 3.17** *Suppose* $\mathcal{M}$ *is a model of the empty*[1] *vocabulary with at least two elements in its domain. Let* $\phi$ *be the sentence* $\exists x_1 (=(x_1) \wedge c = x_1)$ *of dependence logic. Then*

$$(\mathcal{M}, a) \models \exists x_1 (=(x_1) \wedge c = x_1) \tag{3.1}$$

*for all expansions of* $(\mathcal{M}, a)$ *of* $\mathcal{M}$ *to the vocabulary* $\{c\}$. *To prove Eq. (3.1) suppose we are given an element* $a \in M$. *We can define* $F_a(\emptyset) = a$ *and then* $(\mathcal{M}, a) \models_{\{\{(x_1, a)\}\}} (=(x_1) \wedge c = x_1)$, *where we have used* $\{\emptyset\}(F_a/x_1) = \{\{(x_1, a)\}\}$. *However,*

$$\mathcal{M} \not\models \forall x_0 \exists x_1 (=(x_1) \wedge x_0 = x_1). \tag{3.2}$$

*To prove Eq. (3.2) suppose the contrary, that is*

$$\mathcal{M} \models_{\{\emptyset\}} \forall x_0 \exists x_1 (=(x_1) \wedge x_0 = x_1).$$

*Then*

$$\mathcal{M} \models_{\{\{(x_0, a)\} : a \in M\}} \exists x_1 (=(x_1) \wedge x_0 = x_1),$$

---

[1] The empty vocabulary has no constant, relation or function symbols. Structures for the empty vocabulary consist of merely a non-empty set as the universe.

*where we have written* $\{\emptyset\}(M/x_0)$ *out as* $\{\{(x_0, a)\} : a \in M\}$. *Let* $F : \{\{(x_0, a)\} : a \in M\} \to M$ *such that*

$$\mathcal{M} \models_{\{\{(x_0,a),(x_1,G(a))\}:a\in M\}} (=(x_1) \wedge x_0 = x_1), \qquad (3.3)$$

*where* $G(a) = F(\{(x_0, a)\})$ *and* $\{\{(x_0, a)\} : a \in M\}(F/x_1)$ *has been written as* $\{\{(x_0, a), (x_1, G(a))\} : a \in M\}$. *In particular,*

$$\mathcal{M} \models_{\{\{(x_0,a),(x_1,G(a))\}:a\in M\}} =(x_1),$$

*which means that* $F$ *has to have a constant value. Since* $M$ *has at least two elements, the consequence*

$$\mathcal{M} \models_{\{\{(x_0,a),(x_1,G(a))\}:a\in M\}} x_0 = x_1$$

*of Eq. (3.3) contradicts (D1).*

**Example 3.18** *A team* $X$ *is of type* $\exists x_0 (=(x_2, x_0) \wedge x_0 = x_1)$ *if and only if every* $s \in X$ *can be modified to* $s(a_s/x_0)$ *such that for all* $s \in X$

(i) $a_s = s(x_1)$,
(ii) $a_s$ *depends only on* $s(x_2)$ *in* $X$,

*if and only if* $X$ *is of type* $=(x_2, x_1)$.

**Example 3.19** *A team* $X$ *is of type* $\exists x_0 (= (x_2, x_0) \wedge \neg(x_0 = x_1))$ *if every* $s \in X$ *can be modified to* $s(a_s/x_0)$ *in such a way that*

(i) $a_s \neq s(x_1)$,
(ii) $a_s$ *is dependent only on* $s(x_2)$ *in* $X$.

*This means that we have to be able to determine what* $s(x_1)$ *is, in order to avoid it, only on the basis of what* $s(x_2)$ *is. (See also Exercise 3.15.)*

**Example 3.20** *A team* $X$ *with domain* $\{x_1\}$ *is of type* $\exists x_0 (=(x_0) \wedge Rx_0x_1)$ *if every* $s \in X$, *which is really just* $s(x_1)$, *can be modified to* $s(a_s/x_0)$ *such that* $s(a_s/x_0)$ *satisfies* $Rx_0x_1$ *and* $a_s$ *is the same for all* $s \in X$. *That is, there is one* $a$ *such that* $\{s(a/x_0) : s \in X\}$ *is of type* $Rx_0x_1$.

**Exercise 3.1** *Suppose* $L = \{R\}$, $\#_L(R) = 2$. *Show that every team* $X$, *the domain of which contains* $x_i$ *and* $x_j$, *is of type* $Rx_ix_j \vee \neg Rx_ix_j$.

**Exercise 3.2** *Describe teams* $X \in \text{Team}(M, \{x_0, x_1\})$ *of type*

(i) $=(x_0, x_1)$,
(ii) $=(x_1, x_0)$,
(iii) $=(x_0, x_0)$.

**Exercise 3.3** *Let $L$ be the vocabulary $\{c, f\}$. Depending on the model $\mathcal{M}$, which teams $X \in \text{Team}(M, \{x_0\})$ are of the following types?*

  (i) $=(c, c)$,
 (ii) $=(x_0, c)$,
(iii) $=(c, x_0)$,
(iv) $=(c, fx_0)$.

**Exercise 3.4** *Let $\mathcal{M} = (\mathbb{N}, +, \cdot, 0, 1)$. Which teams $X \in \text{Team}(M, \{x_0, x_1\})$ are of the following types?*

  (i) $=(x_0, x_0 + x_1)$,
 (ii) $=(x_0 \cdot x_0, x_1 \cdot x_1)$.

**Exercise 3.5** *Let $L$ be the vocabulary $\{f, g\}$. Which teams $X \in \text{Team}(M, \{x_0, x_1\})$ are of the following types?*

  (i) $=(fx_0, x_0)$,
 (ii) $=(fx_1, x_0)$,
(iii) $=(fx_0, gx_1)$.

**Exercise 3.6** *Let $L$ be the vocabulary $\{f, g\}$. Describe teams $X \in \text{Team}(M, \{x_0, x_1, x_2\})$ of type*

  (i) $=(x_0, x_1, x_2)$,
 (ii) $=(x_0, x_0, x_2)$.

**Exercise 3.7** *Let $\mathcal{M} = (\mathbb{N}, +, \cdot, 0, 1)$ and*

$$X_n = \{\{(0, a), (1, b)\} : 1 < a \le n, 1 < b \le n, a \le b\}.$$

*Show that $X_5$ is of type $=(x_0 + x_1, x_0 \cdot x_1, x_0)$. This is also true for $X_n$ for any $n$, but is slightly harder to prove.*

**Exercise 3.8** $\models \forall x_1 \ldots \forall x_n(=(x_1, \ldots, x_n, x_i))$, *if* $1 \le i \le n$.

**Exercise 3.9** $\models \forall x_0 \forall x_1(x_1 = c \to =(x_0, x_1))$.

**Exercise 3.10** $\models \forall x_0 \forall x_1(x_1 = fx_0 \to =(x_0, x_1))$.

**Exercise 3.11** *For which of the following formulas $\phi$ is it true that for all $X \ne \emptyset$: $\mathcal{M} \models_X \neg\phi \iff \mathcal{M} \not\models_X \phi$:*

  (i) $(=(x_0, x_1) \wedge \neg x_0 = x_1)$,
 (ii) $(=(x_0, x_1) \to x_0 = x_1)$,
(iii) $(=(x_0, x_1) \vee \neg x_0 = x_1)$?

**Exercise 3.12** *Show that the set $\mathcal{T}$ of all triples $(\phi, X, d)$, where $\phi$ is an L-formula of dependence logic, $\mathrm{Fr}(\phi) \subseteq \mathrm{dom}(X)$ and $d \in \{0, 1\}$, satisfies (D1)– (D12) of Definition 3.5.*

**Exercise 3.13** *Show that the family of all sets $\mathcal{T}$ satisfying (D1)–(D12) of Definition 3.5 is closed under arbitrary intersections.*

**Exercise 3.14** *Finish the proof of Proposition 3.8.*

**Exercise 3.15** *The following appealing claim is wrong. Why? Suppose $\mathcal{M}$ is an L-structure with $\geq 2$ elements. Then a team $X$ is of type $\exists x_0 (=(x_2, x_0) \wedge x_0 = x_1)$ if and only if it is of type $\exists x_0 (=(x_2, x_0) \wedge \neg (x_0 = x_1))$.*

**Exercise 3.16** *The following appealing claim is wrong. Why? Suppose $\mathcal{M}$ is an L-structure with $\geq 2$ elements. Then a team $X$ is of type $\exists x_0 (=(x_0) \wedge x_0 = x_1)$ if and only if it is of type $\exists x_0 (=(x_0) \wedge \neg (x_0 = x_1))$.*

**Exercise 3.17** *The following appealing claim is wrong. Why? Suppose $L = \{f\}$, $\#(f) = 1$, and $\mathcal{M}$ is an L-structure. Then a team with domain $\{x_0\}$ is of type $\exists x_1 (=(fx_0, x_1) \wedge fx_1 = fx_0)$ if and only if the mapping $s(x_0) \mapsto f^{\mathcal{M}}(s(x_0))$ is one-to-one for $s \in X$.*

**Exercise 3.18** *The following appealing claim is wrong. Why? Suppose $L = \{R\}$, $\#(R) = 2$, and $\mathcal{M}$ is an L-structure. Then a team $X$ with domain $\{x_0\}$, which we identify with a subset of $M$, is of type $\exists x_1 (=(x_0, x_1) \wedge R x_0 x_1)$ if and only if $R^{\mathcal{M}}$ is a function with domain $X$.*

**Exercise 3.19** *([21]) This exercise shows that the Closure Test is the best we can do. Let $L$ be the vocabulary of one n-ary predicate symbol $R$. Let $\mathcal{M}$ be a finite set and $m \in \mathbb{N}$. Suppose $S$ is a set of teams of $M$ with domain $\{x_1, \ldots, x_m\}$ such that $S$ is closed under subsets. Find an interpretation $R^{\mathcal{M}} \subseteq M^n$ and a formula $\phi$ of $\mathcal{D}$ such that a team $X$ with domain $\{x_1, \ldots, x_k\}$ is of type $\phi$ in $\mathcal{M}$ if and only if $X \in S$.*

**Exercise 3.20** *Use the method of ref. [5], mutatis mutandis, to show that there is no compositional semantics for dependence logic in which the meanings of formulas are sets of assignments (rather than sets of teams) and which agrees with Definition 3.5 for sentences.*

## 3.3 Logical equivalence and duality

The concept of logical consequence and the derived concept of logical equivalence are both defined below in a semantic form. In first order logic there is also

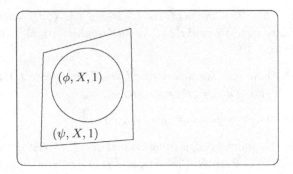

Fig. 3.7. Logical consequence $\phi \Rightarrow \psi$.

a proof theoretic (or syntactic) concept of logical consequence and it coincides with the semantic concept. This fact is referred to as the Gödel Completeness Theorem. In dependence logic we have only semantic notions. There are obvious candidates for syntactic concepts but they are not well understood yet. For example, it is known that the Gödel Completeness Theorem fails badly (see Section 4.6).

**Definition 3.21** *$\psi$ is a logical consequence of $\phi$,*

$$\phi \Rightarrow \psi,$$

*if for all $\mathcal{M}$ and all $X$ with $\mathrm{dom}(X) \supseteq \mathrm{Fr}(\phi) \cup \mathrm{Fr}(\psi)$ and $\mathcal{M} \models_X \phi$ we have $\mathcal{M} \models_X \psi$. Further, $\psi$ is a strong logical consequence of $\phi$,*

$$\phi \Rightarrow^* \psi,$$

*if for all $\mathcal{M}$ and for all $X$ with $\mathrm{dom}(X) \supseteq \mathrm{Fr}(\phi) \cup \mathrm{Fr}(\psi)$ and $\mathcal{M} \models_X \phi$ we have $\mathcal{M} \models_X \psi$, and all $X$ with $\mathrm{dom}(X) \supseteq \mathrm{Fr}(\phi) \cup \mathrm{Fr}(\psi)$ and $\mathcal{M} \models_X \neg\psi$ we have $\mathcal{M} \models_X \neg\phi$. Also, $\psi$ is* logically equivalent *with $\phi$,*

$$\phi \equiv \psi,$$

*if $\phi \Rightarrow \psi$ and $\psi \Rightarrow \phi$. Lastly, $\psi$ is* strongly logically equivalent *with $\phi$,*

$$\phi \equiv^* \psi,$$

*if $\phi \Rightarrow^* \psi$ and $\psi \Rightarrow^* \phi$. See Figs. 3.7 and 3.8.*

Note that $\phi \Rightarrow^* \psi$ if and only if $\phi \Rightarrow \psi$ and $\neg\psi \Rightarrow \neg\phi$. Thus the fundamental concept is $\phi \Rightarrow \psi$ and $\phi \Rightarrow^* \psi$ reduces to it. Note also that $\phi$ and $\psi$ are logically equivalent if and only if for all $X$ with $\mathrm{dom}(X) \supseteq \mathrm{Fr}(\phi) \cup \mathrm{Fr}(\psi)$

$$(\phi, X, 1) \in \mathcal{T} \text{ if and only if } (\psi, X, 1) \in \mathcal{T},$$

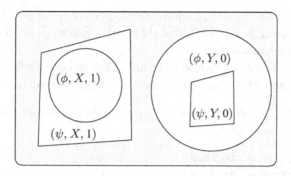

Fig. 3.8. Strong logical consequence $\phi \Rightarrow^* \psi$.

and $\phi$ and $\psi$ are strongly logically equivalent if and only if for all $X$ with $\text{dom}(X) \supseteq \text{Fr}(\phi) \cup \text{Fr}(\psi)$ and all $d$,

$$(\phi, X, d) \in \mathcal{T} \text{ if and only if } (\psi, X, d) \in \mathcal{T}.$$

We have some familiar looking strong logical equivalences in propositional logic, reminiscent of axioms of semigroups with identity. In Lemma 3.22 we group the equivalences according to duality.

**Lemma 3.22** *The following strong logical equivalences hold in dependence logic:*

(i) $\neg\neg\phi \equiv^* \phi$;
(ii) (a) $(\phi \wedge \top) \equiv^* \phi$,
    (b) $(\phi \vee \top) \equiv^* \top$;
(iii) (a) $(\phi \wedge \psi) \equiv^* (\psi \wedge \phi)$,
    (b) $(\phi \vee \psi) \equiv^* (\psi \vee \phi)$,
(iv) (a) $(\phi \wedge \psi) \wedge \theta \equiv^* \phi \wedge (\psi \wedge \theta)$,
    (b) $(\phi \vee \psi) \vee \theta \equiv^* \phi \vee (\psi \vee \theta)$;
(v) (a) $\neg(\phi \vee \psi) \equiv^* (\neg\phi \wedge \neg\psi)$,
    (b) $\neg(\phi \wedge \psi) \equiv^* (\neg\phi \vee \neg\psi)$.

*Proof* We prove only Claim (iii) (b) and leave the rest to the reader. By (E8), $(\phi \vee \psi, X, 0) \in \mathcal{T}$ if and only if $((\phi, X, 0) \in \mathcal{T}$ and $(\psi, X, 0) \in \mathcal{T})$ if and only if $(\psi \vee \phi, X, 0) \in \mathcal{T}$. Suppose then $(\phi \vee \psi, X, 1) \in \mathcal{T}$. By (E7) there are $Y$ and $Z$ such that $X = Y \cup Z$, $(\phi, Y, 1) \in \mathcal{T}$, and $(\psi, Z, 1) \in \mathcal{T}$. By (D7), $(\psi \vee \phi, X, 1) \in \mathcal{T}$. Conversely, if $(\psi \vee \phi, X, 1) \in \mathcal{T}$, then there are $Y$ and $Z$ such that $X = Y \cup Z$, $(\psi, Y, 1) \in \mathcal{T}$, and $(\phi, Z, 1) \in \mathcal{T}$. By (D7), $(\phi \vee \psi, X, 1) \in \mathcal{T}$. $\square$

Many familiar propositional equivalences fail on the level of strong equivalence, in particular the Law of Excluded Middle, weakening laws, and distributivity laws. See Exercise 3.21.

We also have some familiar looking strong logical equivalences for quantifiers. In Lemma 3.23 we again group the equivalences according to duality.

**Lemma 3.23** *The following strong logical equivalences and consequences hold in dependence logic:*

(i) (a) $\exists x_m \exists x_n \phi \equiv^* \exists x_n \exists x_m \phi$,
   (b) $\forall x_m \forall x_n \phi \equiv^* \forall x_n \forall x_m \phi$;
(ii) (a) $\exists x_n (\phi \vee \psi) \equiv^* (\exists x_n \phi \vee \exists x_n \phi)$,
   (b) $\forall x_n (\phi \wedge \psi) \equiv^* (\forall x_n \phi \wedge \forall x_n \phi)$;
(iii) (a) $\neg \exists x_n \phi \equiv^* \forall x_n \neg \phi$,
   (b) $\neg \forall x_n \phi \equiv^* \exists x_n \neg \phi$;
(iv) (a) $\phi \Rightarrow^* \exists x_n \phi$,
   (b) $\forall x_n \phi \Rightarrow^* \phi$.

A useful method for proving logical equivalences is the method of substitution. In first order logic this is based on the strong compositionality[2] of the semantics. The same is true for dependence logic.

**Definition 3.24** *Suppose $\theta$ is a formula in the vocabulary $L \cup \{P\}$, where $P$ is an n-ary predicate symbol. Let $\mathrm{Sb}(\theta, P, \phi(x_1, \ldots, x_n))$ be obtained from $\theta$ by replacing $Pt_1 \ldots t_n$ everywhere by $\phi(t_1, \ldots, t_n)$.*

**Lemma 3.25 (Preservation of equivalence under substitution)** *Suppose $\phi_0(x_1, \ldots, x_n)$ and $\phi_1(x_1, \ldots, x_n)$ are L-formulas of dependence logic such that $\phi_0(x_1, \ldots, x_n) \equiv^* \phi_1(x_1, \ldots, x_n)$. Suppose $\theta$ is a formula in the vocabulary $L \cup \{P\}$, where $P$ is an n-ary predicate symbol. Then $\mathrm{Sb}(\theta, P, \phi_0(x_1, \ldots, x_n)) \equiv^* \mathrm{Sb}(\theta, P, \phi_1(x_1, \ldots, x_n))$.*

*Proof* The proof is straightforward. We use induction on $\theta$. Let us use $\mathrm{Sb}_d(\theta)$ as a shorthand for $\mathrm{Sb}(\theta, P, \phi_d)$.

   *Atomic case.* Suppose $\theta$ is $Rt_1 \ldots t_n$. Now $\mathrm{Sb}_d(\theta) = \phi_d$. The claim follows from $\phi_0 \equiv^* \phi_1$.
   *Disjunction.* Note that $\mathrm{Sb}_d(\phi \vee \psi) = \mathrm{Sb}_d(\phi) \vee \mathrm{Sb}_d(\psi)$. Now $(\mathrm{Sb}_d(\phi \vee \psi), X, 1) \in \mathcal{T}$ if and only if $(\mathrm{Sb}_d(\phi) \vee \mathrm{Sb}_d(\psi), X, 1) \in \mathcal{T}$ if and only

---

[2] In compositional semantics, roughly speaking, the meaning of a compound formula is completely determined by the way the formula is built from parts and by the meanings of the parts.

if $(X = Y \cup Z$ such that $(\mathrm{Sb}_d(\phi), Y, 1) \in \mathcal{T}$ and $(\mathrm{Sb}_d(\psi), Z, 1) \in \mathcal{T})$. By the induction hypothesis, this is equivalent to $(X = Y \cup Z$ such that $(\mathrm{Sb}_{1-d}(\phi), Y, 1) \in \mathcal{T}$ and $(\mathrm{Sb}_{1-d}(\psi), Z, 1) \in \mathcal{T})$, i.e. $(\mathrm{Sb}_{1-d}(\phi) \vee \mathrm{Sb}_{1-d}(\psi), X, 1) \in \mathcal{T}$, which is finally equivalent to $(\mathrm{Sb}_{1-d}(\phi \vee \psi), X, 1) \in \mathcal{T}$. On the other hand, $(\mathrm{Sb}_d(\phi \vee \psi), X, 0) \in \mathcal{T}$ if and only if $(\mathrm{Sb}_d(\phi) \vee \mathrm{Sb}_d(\psi), X, 0) \in \mathcal{T}$ if and only if $((\mathrm{Sb}_d(\phi), X, 0) \in \mathcal{T}$ and $(\mathrm{Sb}_d(\psi), X, 0) \in \mathcal{T})$. By the induction hypothesis, this is equivalent to $(\mathrm{Sb}_{1-d}(\phi), X, 0) \in \mathcal{T}$ and $(\mathrm{Sb}_{1-d}(\psi), X, 0) \in \mathcal{T}$, i.e. $(\mathrm{Sb}_{1-d}(\phi) \vee \mathrm{Sb}_{1-d}(\psi), X, 0) \in \mathcal{T}$, which is finally equivalent to $(\mathrm{Sb}_{1-d}(\phi \vee \psi), X, 0) \in \mathcal{T}$.

*Negation.* $\mathrm{Sb}_e(\neg\phi) = \neg\,\mathrm{Sb}_e(\phi)$. Now $(\mathrm{Sb}_e(\neg\phi), X, d) \in \mathcal{T}$ if and only if $(\neg\,\mathrm{Sb}_e(\phi), X, d) \in \mathcal{T}$, which is equivalent to $(\mathrm{Sb}_e(\phi), X, 1 - d) \in \mathcal{T}$. By the induction hypothesis, this is equivalent to $(\mathrm{Sb}_{1-e}(\phi), X, 1 - d) \in \mathcal{T}$, i.e. $(\neg\,\mathrm{Sb}_{1-e}(\phi), X, d) \in \mathcal{T}$, and finally this is equivalent to $(\mathrm{Sb}_{1-e}(\neg\phi), X, d) \in \mathcal{T}$.

*Existential quantification.* Note that $\mathrm{Sb}_d(\exists x_n \phi) = \exists x_n \,\mathrm{Sb}_d(\phi)$. We may infer, as above, that $(\mathrm{Sb}_d(\exists x_n \phi), X, 1) \in \mathcal{T}$ if and only if $(\exists x_n \,\mathrm{Sb}_d(\phi), X, 1) \in \mathcal{T}$ if and only if there is $F : X \to M$ such that $(\mathrm{Sb}_d(\phi), X(F/x_n), 1) \in \mathcal{T}$. By the induction hypothesis, this is equivalent to the following: there is $F : X \to M$ such that $(\mathrm{Sb}_{1-d}(\phi), X(F/x_n), 1) \in \mathcal{T}$, i.e. to $(\exists x_n \,\mathrm{Sb}_{1-d}(\phi), X, 1) \in \mathcal{T}$, which is finally equivalent to $(\mathrm{Sb}_{1-d}(\exists x_n \phi), X, 1) \in \mathcal{T}$. On the other hand, $(\mathrm{Sb}_d(\exists x_n \phi), X, 0) \in \mathcal{T}$ if and only if $(\exists x_n \,\mathrm{Sb}_d(\phi), X, 0) \in \mathcal{T}$, if and only if $(\mathrm{Sb}_d(\phi), X(M/x_n), 0) \in \mathcal{T}$. By the induction hypothesis, this is equivalent to $(\mathrm{Sb}_{1-d}(\phi), X(M/x_n), 0) \in \mathcal{T}$, i.e. $(\exists x_n \,\mathrm{Sb}_{1-d}(\phi), X, 0) \in \mathcal{T}$, which is finally equivalent to $(\mathrm{Sb}_{1-d}(\exists x_n \phi), X, 0) \in \mathcal{T}$. $\qquad\square$

As we did for first order logic in Chapter 2, we define two operations $\phi \mapsto \phi^{\mathrm{p}}$ and $\phi \mapsto \phi^{\mathrm{d}}$ by simultaneous induction, using the shorthands $\phi \wedge \psi$ and $\forall x_n \phi$:

$$\phi^{\mathrm{d}} = \neg\phi \text{ if } \phi \text{ atomic,}$$
$$\phi^{\mathrm{p}} = \phi \text{ if } \phi \text{ atomic,}$$
$$(\neg\phi)^{\mathrm{d}} = \phi^{\mathrm{p}},$$
$$(\neg\phi)^{\mathrm{p}} = \phi^{\mathrm{d}},$$
$$(\phi \wedge \psi)^{\mathrm{d}} = \phi^{\mathrm{d}} \vee \psi^{\mathrm{d}},$$
$$(\phi \wedge \psi)^{\mathrm{p}} = \phi^{\mathrm{p}} \wedge \psi^{\mathrm{p}},$$
$$(\exists x_n \phi)^{\mathrm{d}} = \forall x_n \phi^{\mathrm{d}},$$
$$(\exists x_n \phi)^{\mathrm{p}} = \exists x_n \phi^{\mathrm{p}}.$$

We again call $\phi^d$ the *dual* of $\phi$, and $\phi^p$ the *negation normal form* of $\phi$. The basic results about duality in dependence logic, as in first order logic, are as given in Lemma 3.26.

**Lemma 3.26** $\phi \equiv^* \phi^p$ *and* $\neg\phi \equiv^* \phi^d$.

Thus the $\phi^p$ operation is a mechanical method for translating a formula to a strongly logically equivalent formula in negation normal form. The dual operation is a mechanical method for translating a formula $\phi$ to one which is strongly logically equivalent to the negation of $\phi$.

We will see later (Section 7.3) that there is no hope of explaining $\phi \Rightarrow \psi$ in terms of a few simple rules. There are examples of $\phi$ and $\psi$ such that to decide whether $\phi \Rightarrow \psi$ or not, one has to decide whether there are measurable cardinals in the set theoretic universe. Likewise, there are examples of $\phi$ and $\psi$ such that to decide whether $\phi \Rightarrow \psi$, one has to decide whether the Continuum Hypothesis holds.

We examine next some elementary logical properties of formulas of dependence logic. Lemma 3.27 shows that the truth of a formula depends only on the interpretations of the variables occurring free in the formula. To this end, we define $X{\restriction}V$ to be $\{s{\restriction}V : s \in X\}$.

**Lemma 3.27** *Suppose* $V \supseteq \mathrm{Fr}(\phi)$. *Then* $\mathcal{M} \models_X \phi$ *if and only if* $\mathcal{M} \models_{X{\restriction}V} \phi$.

*Proof* Key to this result is the fact that $t^{\mathcal{M}}\langle s \rangle = t^{\mathcal{M}}\langle s{\restriction}V \rangle$ whenever $\mathrm{Fr}(t) \subseteq V$. We use induction on $\phi$ to prove $(\phi, X, d) \in \mathcal{T} \iff (\phi, X{\restriction}V, d) \in \mathcal{T}$ whenever $\mathrm{Fr}(\phi) \subseteq V$. If $\phi$ is atomic, the claim is obvious, even in the case of $=(t_1, \ldots, t_n)$.

> *Disjunction.* Suppose $(\phi \vee \psi, X, 1) \in \mathcal{T}$. Then $X = Y \cup Z$ such that $(\phi, Y, 1) \in \mathcal{T}$ and $(\psi, Z, 1) \in \mathcal{T}$. By the induction hypothesis, $(\phi, Y{\restriction}V, 1) \in \mathcal{T}$ and $(\psi, Z{\restriction}V, 1) \in \mathcal{T}$. Of course, $X \restriction V = (Y \restriction V) \cup (Z \restriction V)$. Thus $(\phi \vee \psi, X{\restriction}V, 1) \in \mathcal{T}$. Conversely, suppose $(\phi \vee \psi, X{\restriction}V, 1) \in \mathcal{T}$. Then $X \restriction V = Y \cup Z$ such that $(\phi, Y, 1) \in \mathcal{T}$ and $(\phi, Z, 1) \in \mathcal{T}$. Choose $Y'$ and $Z'$ such that $Y'{\restriction}V = Y$, $Z'{\restriction}V = Z$ and $X = Y' \cup Z'$. Now we have $(\phi, Y', 1) \in \mathcal{T}$ and $(\psi, Z', 1) \in \mathcal{T}$ by the induction hypothesis, Thus $(\phi \vee \psi, X, 1) \in \mathcal{T}$. Suppose then $(\phi \vee \psi, X, 0) \in \mathcal{T}$. Then $(\phi, X, 0) \in \mathcal{T}$ and $(\psi, X, 0) \in \mathcal{T}$. By the induction hypothesis $(\phi, X{\restriction}V, 0) \in \mathcal{T}$ and $(\psi, X{\restriction}V, 0) \in \mathcal{T}$. Thus $(\phi \vee \psi, X{\restriction}V, 0) \in \mathcal{T}$. Conversely, suppose $(\phi \vee \psi, X{\restriction}V, 0) \in \mathcal{T}$. Then $(\phi, X{\restriction}V, 0) \in \mathcal{T}$ and $(\psi, X{\restriction}V, 0) \in \mathcal{T}$. Now $(\phi, X, 0) \in \mathcal{T}$ and $(\psi, X, 0) \in \mathcal{T}$ by the induction hypothesis. Thus $(\phi \vee \psi, X, 0) \in \mathcal{T}$.

*Negation.* Suppose $(\neg\phi, X, d) \in T$. Then $(\phi, X, 1 - d) \in T$. By the induction hypothesis, $(\phi, X{\restriction}V, 1 - d) \in T$. Thus $(\neg\phi, X{\restriction}V, d) \in T$. Conversely, suppose $(\neg\phi, X{\restriction}V, d) \in T$. Then $(\phi, X{\restriction}V, 1 - d) \in T$. Now we have $(\phi, X, 1 - d) \in T$ by the induction hypothesis. Thus $(\neg\phi, X, d) \in T$.

*Existential quantification.* Suppose $(\exists x_n, X, 1) \in T$. Then there is $F : X \to M$ such that $(\phi, X(F/x_n), 1) \in T$. By the induction hypothesis, $(\phi, X(F/x_n){\restriction}W, 1) \in T$, where $W = V \cup \{n\}$. Note that $X(F/x_n){\restriction}W = (X{\restriction}V)(F/x_n)$. Thus $(\exists x_n\phi, X{\restriction}V, 1) \in T$. Conversely, suppose $(\exists x_n\phi, X{\restriction}V, 1) \in T$. Then there is $F : X{\restriction}V \to M$ such that $(\phi, (X{\restriction}V)(F/x_n), 1) \in T$. Again, $X(F/x_n){\restriction}W = (X{\restriction}V)(F/x_n)$, and thus, by the induction hypothesis, $(\phi, X(F/x_n), 1) \in T$, i.e. $(\exists x_n\phi, X, 1) \in T$. □

In Lemma 3.28 we have the restriction, familiar from substitution rules of first order logic, that in substitution no free occurrence of a variable should become a bound.

**Lemma 3.28 (Change of free variables)** *Let the free variables of $\phi$ be $x_1, \ldots, x_n$. Let $i_1, \ldots, i_n$ be distinct. Let $\phi'$ be obtained from $\phi$ by replacing $x_j$ everywhere by $x_{i_j}$, where $j = 1, \ldots, n$. If $X$ is an assignment set with $\mathrm{dom}(X) = \{1, \ldots, n\}$, let $X'$ consist of the assignments $x_{i_j} \mapsto s(x_j)$, where $s \in X$. Then*

$$\mathcal{M} \models_X \phi \iff \mathcal{M} \models_{X'} \phi'.$$

Finally, we note the important fact that types are preserved by isomorphisms (recall the definition of $\mathcal{M} \cong \mathcal{M}'$ in Section 2.2).

**Lemma 3.29 (Isomorphism preserves truth)** *Suppose $\mathcal{M} \cong \mathcal{M}'$. If $\phi \in \mathcal{D}$, then $\mathcal{M} \models \phi \iff \mathcal{M}' \models \phi$.*

*Proof* Let $T$ be the fundamental predicate (see Definition 3.5) for $\mathcal{M}$ and let $T'$ be the fundamental predicate for $\mathcal{M}'$. Let $\pi : \mathcal{M} \cong \mathcal{M}'$. For any assignment $s$ for $M$, let $\pi s$ be the assignment $\pi s(x_n) = \pi(s(x_n))$. For any team $X$ for $M$, let $\pi X$ be the set of $\pi s$, where $s \in X$. We prove by induction on $\phi$ that for all teams $X$ for $M$ with the free variables of $\phi$ in its domain, and all $d \in \{0, 1\}$:

$$(\phi, X, d) \in T \iff (\phi, \pi X, d) \in T'.$$

The claim is trivial for first order atomic formulas, and is clearly preserved by negation.

Case (A) $\phi$ is $=(t_1, \ldots, t_n)$. Suppose $(\phi, X, 1) \in \mathcal{T}$, i.e. for all $s, s' \in X$ such that $t_1^{\mathcal{M}}\langle s \rangle = t_1^{\mathcal{M}}\langle s' \rangle, \ldots, t_{n-1}^{\mathcal{M}}\langle s \rangle = t_{n-1}^{\mathcal{M}}\langle s' \rangle$, we have $t_n^{\mathcal{M}}\langle s \rangle = t_n^{\mathcal{M}}\langle s' \rangle$. Let $\pi s, \pi s' \in \pi X$ such that $t_1^{\mathcal{M}'}\langle \pi s \rangle = t_1^{\mathcal{M}'}\langle \pi s' \rangle, \ldots, t_{n-1}^{\mathcal{M}'}\langle \pi s \rangle = t_{n-1}^{\mathcal{M}'}\langle \pi s' \rangle$. As $t_i^{\mathcal{M}'}\langle \pi s \rangle = \pi t_i^{\mathcal{M}}\langle s \rangle$ for all $i$, we have $t_1^{\mathcal{M}}\langle s \rangle = t_1^{\mathcal{M}}\langle s' \rangle, \ldots, t_{n-1}^{\mathcal{M}}\langle s \rangle = t_{n-1}^{\mathcal{M}}\langle s' \rangle$, hence we also have $t_n^{\mathcal{M}}\langle s \rangle = t_n^{\mathcal{M}}\langle s' \rangle$, and finally $t_n^{\mathcal{M}'}\langle \pi s \rangle = t_n^{\mathcal{M}'}\langle \pi s' \rangle$, as desired. The converse is similar. Suppose then $(\phi, X, 0) \in \mathcal{T}$. This is equivalent to $X = \emptyset$, which is equivalent to $\pi X = \emptyset$, i.e. $(\phi, \pi X, 0) \in \mathcal{T}'$.

Case (B) $\phi$ is $\psi \vee \theta$. Suppose $(\phi, X, 0) \in \mathcal{T}$. Then $(\psi, X, 0) \in \mathcal{T}$ and $(\theta, X, 0) \in \mathcal{T}$, whence $(\psi, \pi X, 0) \in \mathcal{T}'$ and $(\theta, \pi X, 0) \in \mathcal{T}'$, and finally $(\phi, \pi X, 0) \in \mathcal{T}'$. The converse is similar. Suppose then $(\phi, X, 1) \in \mathcal{T}$. Then $X = X_0 \cup X_1$ such that $(\psi, X_0, 1) \in \mathcal{T}$ and $(\theta, X_1, 1) \in \mathcal{T}$. By the induction hypothesis, $(\psi, \pi X_0, 1) \in \mathcal{T}'$ and $(\theta, \pi X_1, 1) \in \mathcal{T}'$. Of course, $\pi X = \pi X_0 \cup \pi X_1$. Thus $(\phi, \pi X, 1) \in \mathcal{T}'$. Conversely, suppose $(\phi, \pi X, 1) \in \mathcal{T}'$. Then $\pi X = X_0 \cup X_1$ such that $(\psi, \pi X_0, 1) \in \mathcal{T}'$ and $(\theta, \pi X_1, 1) \in \mathcal{T}'$. Let $Y_0 = \{s \in X : \pi s \in X_0\}$ and $Y_1 = \{s \in X : \pi s \in X_1\}$. Thus $\pi Y_0 = X_0$ and $\pi Y_1 = X_1$. By the induction hypothesis, $\mathcal{M} \models_{Y_0} \phi$ and $\mathcal{M} \models_{Y_1} \psi$. Since $X = Y_0 \cup Y_1$, we have $(\phi, X, 1) \in \mathcal{T}$.

Case (C) $\phi$ is $\exists x_n \psi$. This is left as an exercise.    $\square$

**Exercise 3.21** *Prove the following non-equivalences:*

(i) (a) $\phi \vee \neg\phi \not\equiv^* \top$,
   (b) $\phi \wedge \neg\phi \not\equiv^* \neg\top$, *but* $\phi \wedge \neg\phi \equiv \neg\top$;
(ii) $(\phi \wedge \phi) \not\equiv^* \phi$, *but* $(\phi \wedge \phi) \equiv \phi$;
(iii) $(\phi \vee \phi) \not\equiv^* \phi$;
(iv) $(\phi \vee \psi) \wedge \theta \not\equiv^* (\phi \wedge \theta) \vee (\psi \wedge \theta)$;
(v) $(\phi \wedge \psi) \vee \theta \not\equiv^* (\phi \vee \theta) \wedge (\psi \vee \theta)$.

*Note that each of these non-equivalences is actually an equivalence in first order logic.*

**Exercise 3.22** *Prove the following results familiar also from first order logic:*

(i) (a) $\exists x_n(\phi \wedge \psi) \not\equiv (\exists x_n \phi \wedge \exists x_n \psi)$, *but* $\exists x_n(\phi \wedge \psi) \Rightarrow^* (\exists x_n \phi \wedge \exists x_n \psi)$,
   (b) $\forall x_n(\phi \vee \psi) \not\equiv (\forall x_n \phi \vee \forall x_n \psi)$, *but* $(\forall x_n \phi \vee \forall x_n \psi) \Rightarrow^* \forall x_n(\phi \vee \psi)$;
(ii) $\exists x_n \forall x_m \phi \not\equiv^* \forall x_m \exists.x_n \phi$. *but* $\exists x_n \forall x_m \phi \Rightarrow^* \forall x_m \exists x_n \phi$.

**Exercise 3.23** *Prove Lemma 3.22.*

**Exercise 3.24** *Prove Lemma 3.23.*

**Exercise 3.25** *Prove Lemma 3.26.*

**Exercise 3.26** *Prove Lemma 3.28.*

**Exercise 3.27** *Prove Case C in Lemma 3.29.*

## 3.4 First order formulas

Some formulas of dependence logic can be immediately recognized as first order merely by their appearance. They simply do not have any occurrences of the dependence formulas $=(t_1, \ldots, t_n)$ as subformulas. We then appropriately call them *first order*. Other formulas may be apparently non-first order, but turn out to be logically equivalent to a first order formula. Our goal in this section is to show that for apparently first order formulas our dependence logic truth definition (Definition 3.5 with $X \neq \emptyset$) coincides with the standard first order truth definition (Section 2.4). We also give a simple criterion called the *Flatness Test* that can be used to test whether a formula of dependence logic is logically equivalent to a first order formula.

We begin by proving that a team is of a first order type $\phi$ if every assignment $s$ in $X$ satisfies $\phi$. Note the *a priori* difference between an assignment $s$ satisfying a first order formula $\phi$ and the team $\{s\}$ being of type $\phi$. We will show that these conditions are equivalent, but this indeed needs a proof.

**Proposition 3.30** *If an L-formula $\phi$ of dependence logic is first order, then:*

*(i) if $\mathcal{M} \models_s \phi$ for all $s \in X$, then $(\phi, X, 1) \in \mathcal{T}$;*
*(ii) if $\mathcal{M} \models_s \neg\phi$ for all $s \in X$, then $(\phi, X, 0) \in \mathcal{T}$.*

*Proof* We use induction as follows.

(1) If $t_1^{\mathcal{M}}\langle s \rangle = t_2^{\mathcal{M}}\langle s \rangle$ for all $s \in X$, then $(t_1 = t_2, X, 1) \in \mathcal{T}$ by (D1).
(2) If $t_1^{\mathcal{M}}\langle s \rangle \neq t_2^{\mathcal{M}}\langle s \rangle$ for all $s \in X$, then $(t_1 = t_2, X, 0) \in \mathcal{T}$ by (D2).
(3) $(=(), X, 1) \in \mathcal{T}$ by (D3).
(4) $(=(), \emptyset, 0) \in \mathcal{T}$ by (D4).
(5) If $(t_1^{\mathcal{M}}\langle s \rangle, \ldots, t_n^{\mathcal{M}}\langle s \rangle) \in R^{\mathcal{M}}$ for all $s \in X$, then $(Rt_1 \ldots t_n, X, 1) \in \mathcal{T}$ by (D5).
(6) If $(t_1^{\mathcal{M}}\langle s \rangle, \ldots, t_n^{\mathcal{M}}\langle s \rangle) \notin R^{\mathcal{M}}$ for all $s \in X$, then $(Rt_1 \ldots t_n, X, 0) \in \mathcal{T}$ by (D6).
(7) If $\mathcal{M} \models_s \neg(\phi \vee \psi)$ for all $s \in X$, then $\mathcal{M} \models_s \neg\phi$ for all $s \in X$ and $\mathcal{M} \models_s \neg\psi$ for all $s \in X$, whence $(\phi, X, 0) \in \mathcal{T}$ and $(\psi, X, 0) \in \mathcal{T}$, and finally $((\phi \vee \psi), X, 0) \in \mathcal{T}$ by (D7).

(8) If $\mathcal{M} \models_s \phi \vee \psi$ for all $s \in X$, then $X = Y \cup Z$ such that $\mathcal{M} \models \phi$ for all $s \in Y$ and $\mathcal{M} \models \psi$ for all $s \in Z$. Thus $(\psi, Y, 1) \in \mathcal{T}$ and $(\psi, Z, 1) \in \mathcal{T}$, whence $((\phi \vee \psi), Y \cup Z, 1) \in \mathcal{T}$ by (D8).

(9) If $\mathcal{M} \models_s \neg \phi$ for all $s \in X$, then $(\phi, X, 0) \in \mathcal{T}$, whence $(\neg \phi, X, 1) \in \mathcal{T}$ by (D9).

(10) If $\mathcal{M} \models_s \neg \neg \phi$ for all $s \in x$, then $(\phi, X, 1) \in \mathcal{T}$, whence $(\neg \phi, X, 0) \in \mathcal{T}$ by (D10).

(11) If $\mathcal{M} \models_s \exists x_n \phi$ for all $s \in X$, then for all $s \in X$ there is $a_s \in M$ such that $\mathcal{M} \models_{s(a_s/x_n)} \phi$. Now $(\phi, \{s\}(F/x_n), 1) \in \mathcal{T}$ for $F : X \to M$ such that $F(s) = a_s$. Thus $(\exists x_n \phi, X, 1) \in \mathcal{T}$.

(12) If $\mathcal{M} \models_s \neg \exists x_n \phi$ for all $s \in X$, then for all $a \in M$ we have for all $s \in X$ $\mathcal{M} \models_{s(a/x_n)} \neg \phi$. Now $(\phi, X(M/x_n), 0) \in \mathcal{T}$. Thus $(\exists x_n \phi, X, 0) \in \mathcal{T}$. $\qquad \square$

Now for the other direction.

**Proposition 3.31** *If an L-formula $\phi$ of dependence logic is first order, then:*

(i) *if $(\phi, X, 1) \in \mathcal{T}$, then $\mathcal{M} \models_s \phi$ for all $s \in X$;*
(ii) *if $(\phi, X, 0) \in \mathcal{T}$, then $\mathcal{M} \models_s \neg \phi$ for all $s \in X$.*

*Proof* We use induction as follows.

(1) If $(t_1 = t_2, X, 1) \in \mathcal{T}$, then $t_1^{\mathcal{M}}\langle s \rangle = t_2^{\mathcal{M}}\langle s \rangle$ for all $s \in X$ by (E1).

(2) If $(t_1 = t_2, X, 0) \in \mathcal{T}$, then $t_1^{\mathcal{M}}\langle s \rangle \neq t_2^{\mathcal{M}}\langle s \rangle$ for all $s \in X$ by (E2).

(3) $(=(), X, 1) \in \mathcal{T}$ and likewise $\mathcal{M} \models_s \top$ for all $s \in X$.

(4) $(=(), \emptyset, 0) \in \mathcal{T}$ and likewise $\mathcal{M} \models_s \neg \top$ for all (i.e. none) $s \in \emptyset$.

(5) If $(Rt_1 \ldots t_n, X, 1) \in \mathcal{T}$, then $(t_1^{\mathcal{M}}\langle s \rangle, \ldots, t_n^{\mathcal{M}}\langle s \rangle) \in R^{\mathcal{M}}$ for all $s \in X$ by (E5).

(6) If $(Rt_1 \ldots t_n, X, 0) \in \mathcal{T}$, then $(t_1^{\mathcal{M}}\langle s \rangle, \ldots, t_n^{\mathcal{M}}\langle s \rangle) \notin R^{\mathcal{M}}$ for all $s \in X$ by (E6).

(7) If $(\phi \vee \psi, X, 0) \in \mathcal{T}$, then $(\phi, X, 0) \in \mathcal{T}$ and $(\psi, X, 0) \in \mathcal{T}$ by E7, whence $\mathcal{M} \models_s \neg \phi$ for all $s \in X$ and $\mathcal{M} \models_s \neg \psi$ for all $s \in X$, whence finally $\mathcal{M} \models_s \neg(\phi \vee \psi)$ for all $s \in X$.

(8) If $(\phi \vee \psi, X, 1) \in \mathcal{T}$, then $X = Y \cup Z$ such that $(\phi, Y, 1) \in \mathcal{T}$ and $(\psi, Z, 1) \in \mathcal{T}$ by (E8), whence $\mathcal{M} \models_s \phi$ for all $s \in Y$ and $\mathcal{M} \models_s \psi$ for all $s \in Z$, and therefore $\mathcal{M} \models_s \phi \wedge \psi$ for all $s \in X$.

We leave the other cases as an exercise. $\qquad \square$

We are now ready to combine Propositions 3.30 and 3.31 in order to prove that the semantics we gave in Definition 3.5 coincide in the case of first order formulas with the more traditional semantics given in Section 2.4.

**Corollary 3.32** *Let $\phi$ be a first order L-formula of dependence logic. Then:*

(i) $\mathcal{M} \models_{\{s\}} \phi$ *if and only if* $\mathcal{M} \models_s \phi$;
(ii) $\mathcal{M} \models_X \phi$ *if and only if* $\mathcal{M} \models_s \phi$ *for all* $s \in X$.

*Proof* If $\mathcal{M} \models_{\{s\}} \phi$, then $\mathcal{M} \models_s \phi$ by Proposition 3.31. If $\mathcal{M} \models_s \phi$, then $\mathcal{M} \models_{\{s\}} \phi$ by Proposition 3.30. □

We shall now introduce a test, comparable to the Closure Test introduced above. The Closure Test was used to test which types of teams are definable in dependence logic. With our new test we can check whether a type is first order, at least up to logical equivalence.

**Definition 3.33 (Flatness[3] Test)** *We say that $\phi$ passes the* Flatness Test *if, for all $\mathcal{M}$ and $X$,*

$$\mathcal{M} \models_X \phi \iff (\mathcal{M} \models_{\{s\}} \phi \text{ for all } s \in X).$$

**Proposition 3.34** *Passing the Flatness Test is preserved by logical equivalence.*

*Proof* Suppose $\phi \equiv \psi$ and $\phi$ passes the Flatness Test. Suppose $\mathcal{M} \models_{\{s\}} \psi$ for all $s \in X$. By logical equivalence, $\mathcal{M} \models_{\{s\}} \phi$ for all $s \in X$. But $\phi$ passes the Flatness Test. So $\mathcal{M} \models_X \phi$, and therefore, by our assumption, $\mathcal{M} \models_X \psi$. □

**Proposition 3.35** *Any L-formula $\phi$ of dependence logic that is logically equivalent to a first order formula satisfies the Flatness Test.*

*Proof* Suppose $\phi \equiv \psi$, where $\psi$ is first order. Since $\psi$ satisfies the Flatness Test, $\phi$ does also, by Proposition 3.34. □

**Example 3.36** $=(x_0, x_1)$ *does not pass the Flatness Test, as the team $X = \{\{(0, 0), (1, 1)\}, \{(0, 1), (1, 1)\}\}$ in a model $\mathcal{M}$ with at least two elements 0 and 1 shows. Namely, $\mathcal{M} \not\models_X =(x_0, x_1)$, but $\mathcal{M} \models_{\{s\}} =(x_0, x_1)$ for $s \in X$. We conclude that $=(x_0, x_1)$ is not logically equivalent to a first order formula.*

**Example 3.37** $\exists x_2(=(x_0, x_2) \land x_2 = x_1)$ *does not pass the Flatness Test, as the team $X = \{s, s'\}$, $s = \{(0, 0), (1, 1)\}$, $s' = \{(0, 0), (1, 0)\}$ in a model $\mathcal{M}$ with at least two elements 0 and 1 shows. Namely, if $F : X \to M$ witnesses $\mathcal{M} \models_{X(F/x_2)} =(x_0, x_2) \land x_2 = x_1$, then $s(x_0) = s'(x_0)$, but*

$$1 = s(x_1) = F(s) = F(s') = s'(x_1) = 0,$$

---

[3] Hodges defines a flattening operation in ref. [21], hence the word 'Flatness'.

*a contradiction. We conclude that* $\exists x_2(=(x_0, x_2) \wedge x_2 = x_1)$ *is not logically equivalent to a first order formula.*

**Example 3.38** *Let* $L = \{+, \cdot, 0, 1, <\}$ *and* $\mathcal{M} = (\mathbb{N}, +, \cdot, 0, 1, <)$*, the standard model of arithmetic. The formula* $\exists x_0(=(x_0) \wedge (x_1 < x_0))$ *fails to meet the Flatness Test. To see this, we first note that if* $X = \{s\}$*, then X is of the type of the formula, as we can choose* $a_s$ *to be equal to* $s(x_1) + 1$*. On the other hand, let* $X = \{s_n : n \in \mathbb{N}\}$*, where* $s_n(x_1) = n$*. It is impossible to choose a such that* $a > s_n(x_1)$ *for all* $n \in \mathbb{N}$*.*

**Exercise 3.28** *Finish the proof of Proposition 3.31.*

**Exercise 3.29** *Find in each case a logically equivalent first order formula:*

   (i) $\exists x_0(=(x_1, x_0) \wedge P x_0)$,
  (ii) $\exists x_0(=(x_1, x_0) \wedge P x_1)$,
 (iii) $\exists x_0((=(x_2, x_0) \wedge P x_0) \rightarrow P x_1)$,
 (iv) $\exists x_0(=(x_1, x_0) \wedge R x_0 x_1)$,
  (v) $\exists x_0(=(x_0) \wedge (R x_1 x_2 \vee R x_0 x_0))$.

**Exercise 3.30** *Find in each case a logically equivalent first order formula:*

   (i) $\exists x_0(=(x_1, x_0) \wedge (f x_1 = x_1))$,
  (ii) $\exists x_0(=(x_2, x_0) \wedge (P f x_0 \wedge \neg P x_1))$.

**Exercise 3.31** *Which of the following formulas are logically equivalent to a first order formula?*

   (i) $=() \vee \neg =()$;
  (ii) $=(x_0)$;
 (iii) $=(x_0, x_0)$.

**Exercise 3.32** *Which of the following formulas are logically equivalent to a first order formula?*

   (i) $=(x_0, x_1, x_2) \wedge x_0 = x_1$;
  (ii) $(=(x_0, x_2) \wedge x_0 = x_1) \rightarrow =(x_1, x_2)$;
 (iii) $=(x_0, x_1, x_2) \vee \neg =(x_0, x_1, x_2)$.

**Exercise 3.33** *Which of the following formulas are logically equivalent to a first order formula?*

   (i) $\forall x_0 \exists x_2(=(x_0, x_2) \wedge x_2 = x_1)$;
  (ii) $\forall x_1 \exists x_2(=(x_0, x_2) \wedge x_2 = x_1)$;
 (iii) $\forall x_0 \forall x_1 \exists x_2(=(x_0, x_2) \wedge x_2 = x_1)$.

**Exercise 3.34** *Which of the following formulas are logically equivalent to a first order formula?*

  (i) $\forall x_0(x_0 = x_1 \rightarrow \; =(x_0, x_1))$;

  (ii) $\forall x_0(fx_0 = x_1 \rightarrow \; =(x_0, x_1))$;

  (iii) $\forall x_0(x_0 = fx_1 \rightarrow \; =(x_0, x_1))$;

  (iv) $\forall x_1 \forall x_0(fx_0 = fx_1 \rightarrow \; =(x_0, x_1))$.

**Exercise 3.35** *Let $L = \emptyset$ and let $\mathcal{M}$ be an $L$-structure with $M = \{0, 1\}$. Show that the following types of a team $X$ with domain $\{x_0, x_1, x_2\}$ are non-first order:*

  (i) $\exists x_0(=(x_2, x_0) \wedge \neg(x_0 = x_1))$,

  (ii) $\exists x_0(=(x_2, x_0) \wedge (x_0 = x_1 \vee x_0 = x_2))$,

  (iii) $\exists x_0(=(x_2, x_0) \wedge (x_0 = x_1 \wedge \neg x_0 = x_2))$.

**Exercise 3.36** *Let $L = \{<\}$ and let $\mathcal{M}$ be an $L$-structure $(\mathbb{N}, <)$. Show that the following types of a team $X$ with domain $\{x_0, x_1, x_2\}$ are non-first order:*

  (i) $\exists x_0(=(x_2, x_0) \wedge \neg(x_0 = x_1))$,

  (ii) $\exists x_0(x_0 < x_1 \wedge x_0 < x_2 \wedge \; =(x_2, x_0))$,

  (iii) $\exists x_0(x_0 < x_2 \wedge \neg(x_0 = x_1) \wedge \; =(x_2, x_0))$.

**Exercise 3.37** *Let $L = \{R\}$, $\#(R) = 1$. Find an $L$-structure $\mathcal{M}$ which demonstrates that the following properties of a team $X$ with domain $\{x_0, x_1, x_2\}$ are non-first order:*

  (i) $\exists x_0(Rx_0 \wedge \; =(x_1, x_0) \wedge \neg x_0 = x_2)$,

  (ii) $\exists x_0(=(x_2, x_0) \wedge (Rx_0 \leftrightarrow Rx_1))$,

  (iii) $\exists x_0(=(x_2, x_0) \wedge ((Rx_1 \wedge \neg Rx_0) \vee (\neg Rx_1 \wedge Rx_0)))$.

**Exercise 3.38** *Let $L = \{R\}$, $\#(R) = 2$. Find an $L$-structure $\mathcal{M}$ which demonstrates that the following properties of a team $X$ with domain $\{x_0, x_1, x_2\}$ are non-first order:*

  (i) $\exists x_0(=(x_2, x_0) \wedge Rx_0x_1)$,

  (ii) $\exists x_0(=(x_2, x_0) \wedge (Rx_1x_0 \wedge Rx_2x_0))$,

  (iii) $\exists x_0(=(x_2, x_0) \wedge ((Rx_1x_0) \leftrightarrow (Rx_2x_0)))$.

**Exercise 3.39** *Let $L = \{f\}$, $\#(f) = 1$. Find an $L$-structure $\mathcal{M}$ which demonstrates that the following properties of a team $X$ with domain $\{x_1\}$ are non-first order:*

  (i) $\exists x_0(=(x_0) \wedge (fx_1 = x_0))$,

  (ii) $\exists x_0 \exists x_0(=(x_0) \wedge (fx_0 = x_1))$,

  (iii) $\exists x_0 \exists x_0(=(x_0) \wedge (fx_0 = fx_1))$.

**Exercise 3.40** *A formula $\phi$ of dependence logic is* coherent *if the following holds: Any team $X$ is of type $\phi$ if and only if for every $s, s' \in X$ the pair team $\{s, s'\}$ is of type $\phi$. Note that the formula $(=(x_1, \ldots, x_n) \wedge \phi)$ is coherent if $\phi$ is. Show that for every first order $\phi$ with $\mathrm{Fr}(\phi) = \{x_1\}$, the type $\exists x_0 (=(x_1, x_0) \wedge \phi)$ is coherent.*

**Exercise 3.41** *Give an example of a formula $\phi$ of dependence logic which is not coherent (see Exercise 3.40 for the definition of coherence).*

## 3.5 The flattening technique

We now introduce a technique which may seem frivolous at first sight but proves very useful in the end. This is the process of flattening, by which we mean getting rid of the dependence formulas $=(t_1, \ldots, t_n)$. Naturally we lose something, but this method reveals whether a formula has genuine occurrences of dependence or just ersatz ones.

**Definition 3.39** *The flattening $\phi^f$ of a formula $\phi$ of dependence logic is defined by induction as follows:*

$$(t_1 = t_2)^f = (t_1 = t_2),$$
$$(Rt_1 \ldots t_n)^f = Rt_1 \ldots t_n,$$
$$(=(t_1, \ldots, t_n))^f = \top,$$
$$(\neg\phi)^f = \neg\phi^f,$$
$$(\phi \vee \psi)^f = \phi^f \vee \psi^f,$$
$$(\exists x_n \phi)^f = \exists x_n \phi^f.$$

Note that the result of flattening is always first order. The main feature of flattening is that it preserves truth.

**Proposition 3.40** *If $\phi$ is an L-formula of dependence logic, then $\phi \Rightarrow \phi^f$.*

*Proof* Inspection of Definition 3.5 reveals immediately that in each case where $(\phi, X, d) \in \mathcal{T}$, we also have $(\phi^f, X, d) \in \mathcal{T}$.                   $\square$

We can use Proposition 3.40 to prove various useful little results which are often comforting in enforcing our intuition. We first point out that although a team may be of the type of both a formula and its negation, this can only happen if the team is empty and thereby is of the type of any formula.

**Corollary 3.41** $\mathcal{M} \models_X (\phi \wedge \neg\phi)$ *if and only if $X = \emptyset$.*

*Proof* We already know that $\mathcal{M} \models_\emptyset (\phi \wedge \neg\phi)$. On the other hand, if $\mathcal{M} \models_X (\phi \wedge \neg\phi)$ and $s \in X$, then $\mathcal{M} \models_s (\phi^f \wedge \neg\phi^f)$, a contradiction. $\qquad\square$

**Corollary 3.42 (Modus Ponens)** *Suppose* $\mathcal{M} \models_X \phi \rightarrow \psi$ *and* $\mathcal{M} \models_X \phi$. *Then* $\mathcal{M} \models_X \psi$.

*Proof* $\mathcal{M} \models_X \neg\phi \vee \psi$ implies $X = Y \cup Z$ such that $\mathcal{M} \models_Y \neg\phi$ and $\mathcal{M} \models_Z \psi$. Now $\mathcal{M} \models_Y \phi$ and $\mathcal{M} \models_Y \neg\phi$, whence $Y = \emptyset$. Thus $X = Z$ and $\mathcal{M} \models_X \psi$ follows. $\qquad\square$

In general, we may conclude from Proposition 3.40 that a non-empty team cannot have the type of a formula which is contradictory in first order logic when flattened. When all the subtle properties of dependence logic are laid bare in front of us, we tend to seek solace in anything solid, anything that we know for certain from our experience in first order logic. Flattening is one solace. By simply ignoring the dependence statements $=(t_1, \ldots, t_n)$, we can recover in a sense the first order content of the formula. When we master this technique, we begin to understand the effect of the presence of dependence statements in a formula.

**Example 3.43** *No non-empty team can have the type of any of the following formulas, whatever formulas of dependence logic the formulas $\phi$ and $\psi$ are:*

$$\phi(c) \wedge \forall x_0 \neg\phi(x_0),$$
$$\forall x_0 \neg\phi \wedge \forall x_0 \neg\psi \wedge \exists x_0(\phi \vee \psi),$$
$$\neg(((\phi \rightarrow \psi) \rightarrow \phi) \rightarrow \phi).$$

*The flattenings of these formulas are, respectively, given by*

$$\phi^f(c) \wedge \forall x_0 \neg\phi^f(x_0),$$
$$\forall x_0 \neg\phi^f \wedge \forall x_0 \neg\psi^f \wedge \exists x_0(\phi^f \vee \psi^f),$$
$$\neg(((\phi^f \rightarrow \psi^f) \rightarrow \phi^f) \rightarrow \phi^f),$$

*none of which can be satisfied by any assignment in first order logic. In the last case one can use truth tables to verify this.*

As the previous example shows, the Truth Table Method, so useful in propositional calculus, also has a role in dependence logic.

**Exercise 3.42** *Let $\phi$ be the formula $\exists x_0 \forall x_1 \neg(=(x_2, x_1) \wedge (x_0 = x_1))$ of $\mathcal{D}$. Show that the flattening of $\phi$ is not a strong logical consequence of $\phi$.*

**Exercise 3.43** *Show that if $\mathcal{M} \models_X (\phi \rightarrow \psi)$ and $\mathcal{M} \models_X \neg\psi$, then $\mathcal{M} \models_X \neg\phi$.*

**Exercise 3.44** *Show that no non-empty team can have the type of any of the following formulas:*

(i) $\neg =(x_0, x_1)$,
(ii) $\neg(=(x_0, x_1) \rightarrow =(x_2, x_1))$,
(iii) $\neg =(fx_0, x_0) \vee \neg =(x_0, fx_0)$,
(iv) $\forall x_0 \exists x_1 \forall x_2 \exists x_3 \neg(\phi \rightarrow =(x_0, x_1))$.

**Exercise 3.45** *Explain the difference between teams of type $=(x_0, x_2) \wedge =(x_1, x_2)$ and teams of type $=(x_0, x_1, x_2)$.*

**Exercise 3.46** *Show that if $\models \phi \rightarrow \psi$, then $\phi \Rightarrow^* \psi$.*

**Exercise 3.47** *Show that if $\phi$ has only $x_0$ and $x_1$ free, then*

$$\forall x_0 \exists x_1 \phi \Rightarrow \forall x_0 \exists x_1(=(x_0, x_1) \wedge \phi).$$

**Exercise 3.48** *Show that the formulas $\forall x_0 \exists x_1 \forall x_2 \exists x_3(=(x_0, x_1) \wedge =(x_2, x_3) \wedge \phi)$ and $\forall x_2 \exists x_3 \forall x_0 \exists x_1(=(x_0, x_1) \wedge =(x_2, x_3) \wedge \phi)$, where $\phi$ is first order, are logically equivalent.*

**Exercise 3.49** *Prove the following:*

(i) $\exists x_n(\phi \wedge \psi) \equiv^* \phi \wedge \exists x_n \psi$ *if $x_n$ not free in $\phi$;*
(ii) $\forall x_n(\phi \vee \psi) \equiv^* \phi \vee \forall x_n \psi$ *if $x_n$ not free in $\phi$.*

**Exercise 3.50** *Prove that $\models \exists x_1(=(x_1) \wedge x_1 = c)$ but $\not\models \forall x_0 \exists x_1(=(x_1) \wedge x_1 = x_0)$.*

**Exercise 3.51 (Prenex normal form)** *A formula of dependence logic is in prenex normal form if all quantifiers are in the beginning of the formula. Use Lemma 3.23 and Exercise 3.49 to prove that every formula of dependence logic is strongly equivalent to a formula which has the same free variables and is in the prenex normal form.*

## 3.6 Dependence/independence friendly logic

We review the relation of our dependence logic $\mathcal{D}$ to the independence friendly logics of refs. [19]–[21].

The *backslashed quantifier*,

$$\exists x_n \backslash \{x_{i_0}, \dots, x_{i_{m-1}}\}\phi, \tag{3.4}$$

introduced in ref. [20], with the intuitive meaning

"there exists $x_n$, depending only on $x_{i_0} \ldots x_{i_{m-1}}$, such that $\phi$," (3.5)

can be defined in dependence logic by the following formula:

$$\exists x_n (=(x_{i_0}, \ldots, x_{i_{m-1}}, x_n) \wedge \phi).$$ (3.6)

Conversely, we can define $=(x_{i_0}, \ldots, x_{i_{m-1}}, x_m)$ in terms of Eq. (3.4) by means of the following formula:

$$\exists x_n \backslash \{x_{i_0}, \ldots, x_{i_{m-1}}\}(x_n = x_m).$$ (3.7)

Similarly, we can define $=(t_1, \ldots, t_n)$ in terms of Eq. (3.4), when $t_1, \ldots, t_n$ are terms.

*Dependence friendly logic*, denoted DF, is the fragment of dependence logic obtained by leaving out the atomic dependence formulas $=(t_1, \ldots, t_n)$ and adding all the backslashed quantifiers (Eq. (3.4)). Dependence logic and DF have the same expressive power, not just on the level of sentences, but even on the level of formulas in the following sense.

**Proposition 3.44**

(i) *For every $\phi$ in $\mathcal{D}$, there is $\phi^*$ in DF so that, for all models $\mathcal{M}$ and all teams X,*

$$\mathcal{M} \models_X \phi \iff \mathcal{M} \models_X \phi^*.$$

(ii) *For every $\psi$ in DF, there is $\psi^{**}$ in $\mathcal{D}$ so that, for all models $\mathcal{M}$ and all teams X,*

$$\mathcal{M} \models_X \psi \iff \mathcal{M} \models_X \psi^{**}.$$

We can base the study of dependence either on the atomic formulas $t_1 = t_n$, $Rt_1 \ldots t_n$, $=(t_1, \ldots, t_n)$, together with the logical operations $\neg, \vee, \exists x_n$, as we have done in this book, or on the atomic formulas $t_1 = t_n$, $Rt_1 \ldots t_n$, together with the logical operations $\neg, \vee, \exists x_n \backslash \{x_{i_0}, \ldots, x_{i_{m-1}}\}$. The results of this book remain true if $\mathcal{D}$ is replaced by DF.

The *slashed quantifier*,

$$\exists x_n / \{x_{i_0}, \ldots, x_{i_{m-1}}\}\phi,$$ (3.8)

used in ref. [21] has the following intuitive meaning:

"there exists $x_n$, independently of $x_{i_0} \ldots x_{i_{m-1}}$, such that $\phi$," (3.9)

which we take to mean

"there exists $x_n$, depending only on variables other than $x_{i_0} \ldots x_{i_{m-1}}$, such that $\phi$," (3.10)

If the other variables, referred to in Eq. (3.10) are $x_{j_0} \ldots x_{j_{l-1}}$, then Eq. (3.9) is intuitively equivalent to

$$\exists x_n \backslash \{x_{j_0}, \ldots, x_{j_{l-1}}\} \phi. \tag{3.11}$$

*Independence friendly logic*, denoted IF, is the fragment of dependence logic obtained by leaving out the atomic dependence formulas $=(t_1, \ldots, t_n)$ and adding all the slashed quantifiers from Eq. (3.8) with Eq. (3.9) (or rather Eq. (3.11)) as their meaning. Sentences of dependence logic and IF have the same expressive power in the following sense:

(i) For every sentence $\phi$ in $\mathcal{D}$, there is a sentence $\phi^*$ in IF so that, for all models $\mathcal{M}$,

$$\mathcal{M} \models \phi \iff \mathcal{M} \models \phi^*.$$

(ii) For every sentence $\psi$ in IF, there is a sentence $\psi^{**}$ in $\mathcal{D}$ so that, for all models $\mathcal{M}$,

$$\mathcal{M} \models \psi \iff \mathcal{M} \models \psi^{**}.$$

We observed that we can base the study of dependence on $\mathcal{D}$ or DF and everything will go through more or less in the same way. However, IF differs more from $\mathcal{D}$ than DF, even if the expressive power is in the above sense the same as that of $\mathcal{D}$, and even if there is the intuitive equivalence of Eqs. (3.9) and (3.11).

Dealing with Eq. (3.10) rather than Eq. (3.5) involves the complication that one has to decide whether "other variable" refers to other variables actually appearing in a formula $\phi$, or to other variables in the domain of the team $X$ under consideration. In the latter case, variables not occurring in the formula $\phi$ may still determine whether the team $X$ is of type $\phi$.

Consider, for example, the following formula:

$$\exists x_0/\{x_1\}(x_0 = x_1). \tag{3.12}$$

The teams in Table 3.10 are of the type given in Eq. (3.12) as we can let $x_0$ depend on $x_2$. The variable $x_2$, which does not occur in Eq. (3.12), *signals*

Table 3.10.

| $x_0$ | $x_1$ | $x_2$ | $x_0$ | $x_1$ | $x_2$ |
|---|---|---|---|---|---|
| 1 | 1 | 1 | 1 | 1 | 5 |
| 1 | 3 | 3 | 1 | 3 | 2 |
| 1 | 8 | 8 | 1 | 8 | 1 |

Table 3.11.

| $x_0$ | $x_1$ | $x_2$ |
|-------|-------|-------|
| 1 | 1 | 5 |
| 1 | 3 | 5 |
| 1 | 8 | 5 |

what $x_1$ is. However, the team in Table 3.11 is not of the type given in Eq. (3.12), even though all three teams agree on all variables that occur in Eq. (3.12). The corresponding formula,

$$\exists x_0 \backslash \{x_2\}(x_0 = x_1), \tag{3.13}$$

of DF avoids this as all variables that are actually used are mentioned in the formula. In this respect, DF is easier to work with than IF.

**Exercise 3.52** *Give a logically equivalent formula in $\mathcal{D}$ for the DF-formula $\exists x_2 \backslash x_1 R x_1 x_2$.*

**Exercise 3.53** *Give for both of the following $\mathcal{D}$-formulas:*

(i) $\exists x_2 \exists x_3 (=(x_0, x_2) \wedge =(x_1, x_3) \wedge R x_0 x_1 x_2 x_3)$,
(ii) $=(x_0) \vee =(x_1)$,

*a logically equivalent formula in DF.*

**Exercise 3.54** *Give for each of the following $\mathcal{D}$-sentences $\phi$:*

(i) $\forall x_0 \exists x_1 (\neg =(x_0, x_1) \wedge \neg x_1 = x_0)$,
(ii) $\forall x_0 \forall x_1 \exists x_2 (=(x_1, x_2) \wedge \neg x_2 = x_1)$.

*a sentence $\phi^*$ in IF so that $\phi$ and $\phi^*$ have the same models.*

**Exercise 3.55** *Give for both of the following IF-sentences:*

(i) $\forall x_0 \exists x_1 / \{x_0\}(x_0 = x_1)$,
(ii) $\forall x_0 \exists x_1 / \{x_0\}(x_1 \leq x_0)$.

*a first order sentence with the same models.*

**Exercise 3.56** *Give a definition of $=(t_1, \ldots, t_n)$ in DF.*

**Exercise 3.57** *Prove Proposition 3.44.*

# 4

# Examples

We now study some more complicated examples involving many quantifiers. In all these examples we use quantifiers to express the existence of some functions. There is a certain easy trick for accomplishing this which hopefully becomes apparent to the reader. The main idea is that some variables stand for arguments and some stand for values of functions that the sentence stipulates to exist.

## 4.1 Even cardinality

On a finite set $\{a_1, \ldots, a_n\}$ of even size, one can define a one-to-one function $f$ which is its own inverse and has no fixed points, as in the following picture:

Conversely, any finite set with such a function has even cardinality. In the following sentence we think of $f(x_0)$ as $x_1$ and of $f(x_2)$ as $x_3$. So $x_1$ depends only on $x_0$, and $x_3$ depends only on $x_2$, which is guaranteed by $=(x_2, x_3)$. To make sure $f$ has no fixed points we stipulate $\neg(x_0 = x_1)$. The condition $(x_1 = x_2 \to x_3 = x_0)$ says in effect $(f(x_0) = x_2 \to f(x_2) = x_0)$, i.e. $f(f(x_0)) = x_0$. Let

$$\Phi_{\text{even}} : \forall x_0 \exists x_1 \forall x_2 \exists x_3 (=(x_2, x_3) \land \neg(x_0 = x_1)$$
$$\land (x_0 = x_2 \to x_1 = x_3)$$
$$\land (x_1 = x_2 \to x_3 = x_0)).$$

We claim that the sentence $\Phi_{\text{even}}$ of dependence logic is true in a finite structure

Table 4.1. *The teams X and X(F/x$_3$), when M = {a$_1$, ..., a$_4$}.*

|        | $x_0$ | $x_1$ | $x_2$ |              | $x_0$ | $x_1$ | $x_2$ | $x_3$ |
|--------|-------|-------|-------|--------------|-------|-------|-------|-------|
| $s_1$  | $a_1$ | $a_2$ | $a_1$ | $s_1(F/x_3)$  | $a_1$ | $a_2$ | $a_1$ | $a_2$ |
| $s_2$  | $a_1$ | $a_2$ | $a_2$ | $s_2(F/x_3)$  | $a_1$ | $a_2$ | $a_2$ | $a_1$ |
| $s_3$  | $a_1$ | $a_2$ | $a_3$ | $s_3(F/x_3)$  | $a_1$ | $a_2$ | $a_3$ | $a_4$ |
| $s_4$  | $a_1$ | $a_2$ | $a_4$ | $s_4(F/x_3)$  | $a_1$ | $a_2$ | $a_4$ | $a_3$ |
| $s_5$  | $a_2$ | $a_1$ | $a_1$ | $s_5(F/x_3)$  | $a_2$ | $a_1$ | $a_1$ | $a_2$ |
| $s_6$  | $a_2$ | $a_1$ | $a_2$ | $s_6(F/x_3)$  | $a_2$ | $a_1$ | $a_2$ | $a_1$ |
| $s_7$  | $a_2$ | $a_1$ | $a_3$ | $s_7(F/x_3)$  | $a_2$ | $a_1$ | $a_3$ | $a_4$ |
| $s_8$  | $a_2$ | $a_1$ | $a_4$ | $s_8(F/x_3)$  | $a_2$ | $a_1$ | $a_4$ | $a_3$ |
| $s_9$  | $a_3$ | $a_4$ | $a_1$ | $s_9(F/x_3)$  | $a_3$ | $a_4$ | $a_1$ | $a_2$ |
| $s_{10}$ | $a_3$ | $a_4$ | $a_2$ | $s_{10}(F/x_3)$ | $a_3$ | $a_4$ | $a_2$ | $a_1$ |
| $s_{11}$ | $a_3$ | $a_4$ | $a_3$ | $s_{11}(F/x_3)$ | $a_3$ | $a_4$ | $a_3$ | $a_4$ |
| $s_{12}$ | $a_3$ | $a_4$ | $a_4$ | $s_{12}(F/x_3)$ | $a_3$ | $a_4$ | $a_4$ | $a_3$ |
| $s_{13}$ | $a_4$ | $a_3$ | $a_1$ | $s_{13}(F/x_3)$ | $a_4$ | $a_3$ | $a_1$ | $a_2$ |
| $s_{14}$ | $a_4$ | $a_3$ | $a_2$ | $s_{14}(F/x_3)$ | $a_4$ | $a_3$ | $a_2$ | $a_1$ |
| $s_{15}$ | $a_4$ | $a_3$ | $a_3$ | $s_{15}(F/x_3)$ | $a_4$ | $a_3$ | $a_3$ | $a_4$ |
| $s_{16}$ | $a_4$ | $a_3$ | $a_4$ | $s_{16}(F/x_3)$ | $a_4$ | $a_3$ | $a_4$ | $a_3$ |

if and only if the size of the structure is even. To this end, suppose first $\mathcal{M}$ is a finite structure with $M = \{a_1, \ldots, a_{2n}\}$. Let

$$X = \{\{(x_0, a_i), (x_1, f(a_i)), (x_2, a_j)\} : 1 \le i \le n, 1 \le j \le n\},$$

where $f(a_i) = a_{i+1}$ if $i$ is odd and $f(a_i) = a_{i-1}$ if $i$ is even (see Table 4.1). Let $F : X \to M$ such that

$$F(\{(x_0, a_i), (x_1, f(a_i)), (x_2, a_j)\}) = f(a_j).$$

Now we note that

$$\mathcal{M} \models_{X(F/x_3)} (=(x_2, x_3) \land \neg(x_0 = x_1)$$
$$\land (x_0 = x_2 \to x_1 = x_3)$$
$$\land (x_1 = x_2 \to x_3 = x_0)),$$

and therefore

$$\mathcal{M} \models_X \exists x_3 (=(x_2, x_3) \land \neg(x_0 = x_1)$$
$$\land (x_0 = x_2 \to x_1 = x_3)$$
$$\land (x_1 = x_2 \to x_3 = x_0)).$$

Table 4.2. *The teams $Y$ and $Y(G/x_1)$*

|       | $x_0$ |                | $x_0$ | $x_1$ |
|-------|-------|----------------|-------|-------|
| $s_1$ | $a_1$ | $s_1(G/x_1)$   | $a_1$ | $a_2$ |
| $s_5$ | $a_2$ | $s_5(G/x_1)$   | $a_2$ | $a_1$ |
| $s_9$ | $a_3$ | $s_9(G/x_1)$   | $a_3$ | $a_4$ |
| $s_{13}$ | $a_4$ | $s_{13}(G/x_1)$ | $a_4$ | $a_3$ |

Let $Y = \{\{(x_0, a_i)\} : 1 \le i \le n\}$ and let $G : X \to M$ be such that $G(\{(x_0, a_i)\}) = f(a_i)$.

$$\mathcal{M} \models_{Y(G/x_1)} \forall x_2 \exists x_3 (= (x_2, x_3) \wedge \neg(x_0 = x_1)$$
$$\wedge (x_0 = x_2 \to x_1 = x_3)$$
$$\wedge (x_1 = x_2 \to x_3 = x_0)),$$

whence

$$\mathcal{M} \models_Y \exists x_1 \forall x_2 \exists x_3 (= (x_2, x_3) \wedge \neg(x_0 = x_1)$$
$$\wedge (x_0 = x_2 \to x_1 = x_3)$$
$$\wedge (x_1 = x_2 \to x_3 = x_0)),$$

and finally $\mathcal{M} \models_{\{\emptyset\}} \Phi_{\text{even}}$. See Table 4.2.

For the converse, suppose $X = \{\emptyset\}$ and $\mathcal{M} \models_X \Phi_{\text{even}}$. Thus there are $F : X(M/x_0) \to M$ and[1] $G : X(M/x_0, F/x_1, M/x_2) \to M$ such that

$$\mathcal{M} \models_{X(M/x_0, F/x_1, M/x_2, G/x_3)} (= (x_2, x_3) \wedge \neg(x_0 = x_1)$$
$$\wedge (x_0 = x_2 \to x_1 = x_3)$$
$$\wedge (x_1 = x_2 \to x_3 = x_0)).$$

Let $F(\{(x_0, a)\}) = f(a)$ for each $a \in M$. Then $G(\{(x_0, a), (x_1, f(a)), (2, b)\}) = f(b)$. It follows that $f$ is one-to-one, $f(a) \ne a$ and $f(f(a)) = a$ for all $a$. Thus, if $M$ is finite, its cardinality must be even.

Note that even cardinality is not expressible in first order logic, as a simple application of the Compactness Theorem[2] (or Ehrenfeucht–Fraïssé games) shows.

---

[1] We compose $X(M/x_0)(F/x_1)(M/x_2)$ into $X(M/x_0, F/x_1, M/x_2)$.
[2] Suppose a first order $\phi$ expresses even cardinality in finite models of the empty vocabulary. By the Compactness Theorem (see Section 6.2), both $\phi$ and $\neg\phi$ have an infinite model. By the Downward Löwenheim–Skolem Theorem (see Section 6.2), $\phi$ has a countably infinite model $\mathcal{M}$, and similarly $\neg\phi$ has a countably infinite model $\mathcal{N}$. But since the vocabulary is empty, $\mathcal{M} \cong \mathcal{N}$. This contradicts $\mathcal{M} \models \phi$ and $\mathcal{N} \models \neg\phi$.

**Exercise 4.1** *Give a sentence of dependence logic which is true in a finite structure if and only if the size of the structure is odd. Note that* $\neg \Phi_{\text{even}}$ *would not do.*

**Exercise 4.2** *Give a sentence of dependence logic which is true in a finite structure if and only if the size of the structure is divisible by three.*

## 4.2 Cardinality

The domain of a structure is infinite if and only if there is a one-to-one function that maps the domain into a proper subset. For example, if the domain contains an infinite set $A = \{a_0, a_1, \ldots\}$, we can map $A$ onto the proper subset $\{a_1, a_2, \ldots\}$ with the mapping $a_n \mapsto a_{n+1}$ and the outside of $A$ onto itself by the identity mapping. On the other hand, if $f : M \to M$ is a one-to-one function that does not have $a$ in its range, then $\{a, f(a), f(f(a)), \ldots\}$ is an infinite subset:

In the following sentence we think of $f(x_0)$ as $x_1$ and of $g(x_2)$ as $x_3$. So $x_1$ depends only on $x_0$, and $x_3$ depends only on $x_2$, which is guaranteed by $=(x_2, x_3)$. The condition $\neg(x_1 = x_4)$ says $x_4$ is outside the range of the function $f$. To make sure that $f = g$, we stipulate $(x_0 = x_2 \to x_1 = x_3)$. The condition $(x_1 = x_3 \to x_0 = x_2)$ says $f$ is one-to-one. Let

$$\Phi_\infty : \exists x_4 \forall x_0 \exists x_1 \forall x_2 \exists x_3 (=(x_2, x_3) \wedge \neg(x_1 = x_4)$$

$$\wedge \, (x_0 = x_2 \leftrightarrow x_1 = x_3)).$$

We conclude that $\Phi_\infty$ is true in a structure if and only if the domain of the structure is infinite.

**Exercise 4.3** *A graph is a pair $\mathcal{M} = (G, E)$, where $G$ is a set of elements called* vertices *and $E$ is an anti-reflexive symmetric binary relation on $G$ called the* edge-relation. *The degree of a vertex is the number of vertices that are connected by a (single) edge to v. The degree of v is said to be infinite if the set of vertices that are connected by an edge to v is infinite. Give a sentence of dependence logic which is true in a graph if and only if every vertex has infinite degree.*

**Exercise 4.4** *Give a sentence of dependence logic which is true in a graph if and only if the graph has infinitely many isolated vertices. (A vertex is isolated if it has no neighbors.)*

**Exercise 4.5** *Give a sentence of dependence logic which is true in a graph if and only if the graph has infinitely many vertices of infinite degree. (The degree of a vertex is the cardinality of the set of neighbors of the vertex.)*

A more general question about cardinality is *equicardinality*. In this case we have two unary predicates $P$ and $Q$ on a set $M$ and we want to know whether they have the same cardinality; that is, whether there is a bijection $f$ from $P$ to $Q$. In the following sentence $\Phi_=$ we think of $f(x_0)$ as $x_1$ and of $f(x_2)$ as $x_3$:

$$\Phi_= : \forall x_0 \exists x_1 \forall x_2 \exists x_3 (=(x_2, x_3) \wedge ((Px_0 \wedge Qx_2) \rightarrow$$
$$(Qx_1 \wedge Px_3$$
$$\wedge (x_0 = x_3 \leftrightarrow x_1 = x_2)))).$$

Suppose we want to test whether a unary predicate $Q$ has at least as many elements as another unary predicate $P$. Here we can use a simplification of $\Phi_=$:

$$\Phi_\leq : \forall x_0 \exists x_1 \forall x_2 \exists x_3 (=(x_2, x_3) \wedge (Px_0 \rightarrow (Qx_1$$
$$\wedge (x_0 = x_2 \leftrightarrow x_1 = x_3)))).$$

On the other hand, the following variant of $\Phi_=$ clearly expresses the isomorphism of two linear orders $(P, <_P)$ and $(Q, <_Q)$:

$$\Phi_\cong : \forall x_0 \exists x_1 \forall x_2 \exists x_3 (=(x_2, x_3) \wedge ((Px_0 \wedge Qx_2) \rightarrow$$
$$\wedge (Qx_1 \wedge Px_3 \wedge$$
$$\wedge (x_0 <_P x_3 \leftrightarrow x_1 <_Q x_2)))).$$

An isomorphism $\mathcal{M} \rightarrow \mathcal{M}$ is called an *automorphism*. The identity mapping is, of course, always an automorphism. An automorphism is *non-trivial* if it is not the identity mapping. Below is a picture of a finite structure with a non-trivial automorphism:

A structure is *rigid* if it has only one automorphism, namely the identity. Finite linear orders and, for example, $(\mathbb{N}, <)$ are rigid,[3] but, for example, $(\mathbb{Z}, <)$ is

---

[3] Any automorphism has to map the first element to the first element, the second element to the second element, the third element to the third element, etc.

non-rigid. We can express the non-rigidity of a linear order with the following sentence:

$$\Phi_{nr} : \exists x_4 \forall x_0 \exists x_1 \forall x_2 \exists x_3 (= (x_2, x_3) \wedge (x_0 = x_4 \rightarrow \neg(x_1 = x_4))$$

$$\wedge (x_0 < x_3 \leftrightarrow x_1 < x_2))$$

**Exercise 4.6** *Write down a sentence of $\mathcal{D}$ which is true in a group[4] if and only if the group is non-rigid.*

**Exercise 4.7** *Write down a sentence of $\mathcal{D}$ which is true in a finite structure $\mathcal{M}$ if and only if the unary predicate $P$ contains in $\mathcal{M}$ at least half of the elements of $\mathcal{M}$.*

**Exercise 4.8** *A natural number $n$ is a prime power if and only if there is a finite field of $n$ elements. Use this fact to write down a sentence of $\mathcal{D}$ in the empty vocabulary which has finite models of exactly prime power cardinalities.*

**Exercise 4.9** *A group $(G, \circ, e)$ is right orderable if there is a partial order $\leq$ in the set $G$ such that $x \leq y$ implies $x \circ z \leq y \circ z$ for all $x, y, z$ in $G$. Write down a sentence of $\mathcal{D}$ which is true in a group if and only if the group is right orderable.*

**Exercise 4.10** *An abelian group $(G, +, 0)$ is the additive group of a field if there is a binary operation $\cdot$ on $G$ and an element $1$ in $G$ such that $(G, +, \cdot, 0, 1)$ is a field. Write down a sentence of $\mathcal{D}$ which is true in an abelian group if and only if the group is the additive group of a field.*

## 4.3 Completeness

Suppose we want to test whether a linear order $<$ on a set $M$ is *complete* or not, i.e. whether every non-empty $A \subseteq M$ with an upper bound has a least upper bound. Since we have to talk about arbitrary subsets $A$ of a domain $M$, we use a technique called *guessing*. This is simply fixing an element $a$ of $M$ and then taking an arbitrary function from $M$ to $M$. We call $a$ the "head" as if we were tossing a coin. The set $A$ corresponds to the set of elements of $M$ mapped to the head. For simplicity, we take the head to be an upper bound of $A$, which we assume to exist anyway.

---

[4] A group is a structure $(G, \circ, e)$ with a binary function $\circ$ and a constant $e$ such that (1) for all $a, b, c \in G$: $(a \circ b) \circ c = a \circ (b \circ c)$, (2) for all $a \in G$: $e \circ a = a \circ e = a$, (3) for all $a \in G$ there is $b \in G$ such that $a \circ b = b \circ a = e$. The group is *abelian* if, in addition, (4) for all $a, b \in G$: $a \circ b = b \circ a$.

A linear order is incomplete if and only if there is a non-empty initial segment $A$ without a last point but with an upper bound such that for every element not in $A$ there is a smaller element not in $A$. To express this we use the following sentence:

$$\Phi_{cmpl} : \exists x_6 \exists x_7 \forall x_0 \exists x_1 \forall x_2 \exists x_3$$
$$\forall x_4 \exists x_5 \forall x_8 \exists x_9 \ (=(x_2, x_3) \wedge =(x_4, x_5) \wedge =(x_8, x_9)$$
$$\wedge \ (x_0 = x_2 \to x_1 = x_3)$$
$$\wedge \ (x_1 = x_6 \to x_0 < x_6)$$
$$\wedge \ (x_0 = x_7 \to x_1 = x_6)$$
$$\wedge \ ((x_0 < x_2 \wedge x_3 = x_6)$$
$$\to x_1 = x_6)$$
$$\wedge \ ((\neg(x_1 = x_6) \wedge x_0 = x_4 \wedge x_2 = x_5)$$
$$\to (x_5 < x_0 \wedge \neg(x_3 = x_6)))$$
$$\wedge \ ((x_0 = x_8 \wedge x_1 = x_6 \wedge x_2 = x_9)$$
$$\to (x_8 < x_9 \wedge x_3 = x_6)))$$

The sentence $\Phi_{cmpl}$ is true in a linear order if and only if the linear order is incomplete. ($\Phi_{cmpl}$ is not necessarily the simplest one with this property.)

Explanation of $\Phi_{cmpl}$: The mapping $x_0 \mapsto x_1$ is the guessing function and $x_2 \mapsto x_3$ is a copy of it, as witnessed by $(x_0 = x_2 \to x_1 = x_3)$. $x_6$ is the head, therefore we have $(x_1 = x_6 \to x_0 < x_6)$. $x_7$ manifests non-emptiness of the guessed initial segment as witnessed by $(x_0 = x_7 \to x_1 = x_6)$. The clause $((x_0 < x_2 \wedge x_3 = x_6) \to x_1 = x_6)$ guarantees the guessed set is really an initial segment. Finally we need to say that if an element $x_0$ is above the initial segment $(x_1 \neq x_6)$ then there is a smaller element $x_5$ also above the initial segment. The mapping $x_8 \mapsto x_9$ makes sure the initial segment does not have a maximal element.

The sentence $\Phi_{cmpl}$ has many quantifier alternations but that is not really essential as we could equivalently use the universal-existential sentence:

$$\Phi'_{cmpl} : \forall x_0 \forall x_2 \forall x_4 \forall x_8 \exists x_1 \exists x_3$$
$$\exists x_5 \exists x_6 \exists x_7 \exists x_9 \ (=(x_0, x_1) \wedge =(x_2, x_3)$$
$$\wedge =(x_4, x_5) \wedge =(x_6) \wedge =(x_7)$$
$$\wedge =(x_8, x_9)$$

$$\wedge\ (x_0 = x_2 \rightarrow x_1 = x_3)$$
$$\wedge\ (x_1 = x_6 \rightarrow x_0 < x_6)$$
$$\wedge\ (x_0 = x_7 \rightarrow x_1 = x_6)$$
$$\wedge\ ((x_0 < x_2 \wedge x_3 = x_6) \rightarrow x_1 = x_6)$$
$$\wedge\ ((\neg(x_1 = x_6) \wedge x_0 = x_4 \wedge x_2 = x_5)$$
$$\rightarrow (x_5 < x_0 \wedge \neg(x_3 = x_6)))$$
$$\wedge\ ((x_0 = x_8 \wedge x_1 = x_6 \wedge x_2 = x_9)$$
$$\rightarrow (x_8 < x_9 \wedge x_3 = x_6))).$$

**Exercise 4.11** *Give a sentence of $\mathcal{D}$ which is true in a linear order if and only if the linear order is isomorphic to a proper initial segment of itself.*

## 4.4 Well-foundedness

A binary relation $R$ on a set $M$ is *well-founded* if and only if there is no sequence $a_0, a_1, \ldots$ in $M$ such that $a_{n+1} R a_n$ for all $n$; otherwise it is *ill-founded*. An equivalent definition of well-foundedness is that there is no non-empty subset $X$ of $M$ such that for every element $a$ in $X$ there is an element $b$ of $X$ such that $b R a$. To express ill-foundedness, we use the following sentence:

$$\Phi_{\mathrm{wf}} : \exists x_6 \exists x_7 \forall x_0 \exists x_1 \forall x_2 \exists x_3 \forall x_4 \exists x_5\ (=(x_2, x_3)$$
$$\wedge =(x_4, x_5)$$
$$\wedge\ (x_0 = x_2 \rightarrow x_1 = x_3)$$
$$\wedge\ (x_0 = x_7 \rightarrow x_1 = x_6)$$
$$\wedge\ ((x_1 = x_6 \wedge x_0 = x_4 \wedge x_2 = x_5)$$
$$\rightarrow (x_3 = x_6 \wedge R x_5 x_4))).$$

The sentence $\Phi_{\mathrm{wf}}$ is true in a binary structure $(M, R)$ if and only if $R$ is ill-founded.

The explanation is as follows. The mapping $x_0 \mapsto x_1$ guesses the set $X$ as the pre-image of $x_6$. The mapping $x_2 \mapsto x_3$ is a copy of the mapping $x_0 \mapsto x_1$, as witnessed by $(x_0 = x_2 \rightarrow x_1 = x_3)$; $x_7$ manifests non-emptiness of the guessed initial segment as witnessed by $(x_0 = x_7 \rightarrow x_1 = x_6)$. The clause $((x_1 = x_6 \wedge x_0 = x_4 \wedge x_2 = x_5) \rightarrow (x_3 = x_6 \wedge R x_5 x_4))$ guarantees the guessed set has no $R$-smallest element.

**Exercise 4.12** *A partially ordered set is an $L$-structure $\mathcal{M} = (M, \leq^M)$ for the vocabulary $L = \{\leq\}$, where $\leq^M$ is assumed to be reflexive ($x \leq x$), transitive*

$(x \leq y \leq z \Rightarrow x \leq z)$, *and anti-symmetric* $(x \leq y \leq x \Rightarrow x = y)$. *We shorten* $(x \leq^M y \,\&\, x \neq y)$ *to* $x <^M y$. *A chain of a partial order is a subset of M which is linearly ordered by* $\leq^M$. *Give a sentence of* $\mathcal{D}$ *which is true in a partially ordered set if and only if the partial order has an infinite chain.*

**Exercise 4.13** *A* tree *is a partially ordered set* $\mathcal{M}$ *such that the set* $\{x \in M : x <^M t\}$ *of predecessors of any* $t \in M$ *is well ordered by* $\leq^M$ *and there is a unique smallest element in* $\mathcal{M}$, *called the* root *of the tree. Thus for any* $t <^M s$ *in* $\mathcal{M}$ *there is an immediate successor* $r$ *of* $t$ *such that* $t <^M r \leq^M s$. *A* subtree *of a tree is a substructure which is a tree. A tree is* binary *if every element has at most two immediate successors, and is a* full binary tree *if every element has exactly two immediate successors. Give a sentence of* $\mathcal{D}$ *which is true in a tree if and only if the tree has a full binary subtree.*

**Exercise 4.14** *The* cofinality *of a linear order is the smallest cardinal* $\kappa$ *such that the order has an unbounded subset of cardinality* $\kappa$. *In particular, a linear order has cofinality* $\omega$ *if the linear order has a cofinal increasing sequence* $a_0, a_1, \dots$ *Give a sentence of* $\mathcal{D}$ *which is true in a linear order if and only if the order is either ill-founded or else well-founded and of cofinality* $\omega$.

## 4.5 Connectedness

A graph $(G, E)$ is *connected* if for every two vertices $a \neq b$ there is a path $a_1 = a, a_2, \dots, a_n = b$ in the graph so that $a_i E a_{i+1}$ for each $i = 1, \dots, n-1$. Otherwise the graph is called *disconnected*. Thus the graph is disconnected if and only if there is a proper subset $A$ of $G$ such that $A$ is closed under $E$. A graph is disconnected if and only if it satisfies the following sentence:

$$\Phi_{\text{conn}} : \exists x_4 \exists x_5 \forall x_0 \exists x_1 \forall x_2 \exists x_3 ( \; =(x_2, x_3) \wedge$$
$$(x_0 = x_2 \to x_1 = x_3) \wedge$$
$$(x_0 = x_4 \to x_1 = x_4) \wedge$$
$$(x_0 = x_5 \to \neg(x_1 = x_4)) \wedge$$
$$((x_1 = x_4 \wedge x_2 E x_0) \to x_3 = x_4)$$

The explanation is as follows (see Fig. 4.1). The mapping $x_0 \mapsto x_1$ guesses the set $A$ as the pre-image of $x_4$. The mapping $x_2 \mapsto x_3$ is a copy of the mapping $x_0 \mapsto x_1$, as witnessed by $(x_0 = x_2 \to x_1 = x_3)$; $x_4$ itself manifests non-emptiness of the guessed set $A$ as witnessed by $(x_0 = x_4 \to x_1 = x_4)$; $x_5$ manifests non-emptiness of the complement of the $A$ as witnessed by $(x_0 = x_5 \to \neg(x_1 = x_4))$. The clause $((x_1 = x_4 \wedge x_2 E x_0) \to x_3 E x_4)$ guarantees the guessed set is closed under $E$.

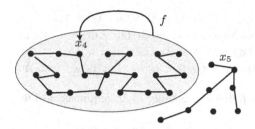

Fig. 4.1. Disconnected graph.

**Exercise 4.15** *Give a sentence of $\mathcal{D}$ which is true in a graph if and only if the graph has an infinite clique. (A* clique *is a subset in which there is an edge between any two distinct vertices.)*

**Exercise 4.16** *A* cycle *of a graph is a finite connected subgraph in which every vertex has degree 2. Give a sentence of $\mathcal{D}$ which is true in a finite graph if and only if the graph has a cycle.*

**Exercise 4.17** *A* path *in a graph $\mathcal{G}$ is a finite sequence $v_0, \ldots, v_n$ such that for every $i = 0, \ldots, n-1$ there is an edge from $v_i$ to $v_{i+1}$. A* Hamiltonian *path is a path which passes through every vertex exactly once. Give a sentence of $\mathcal{D}$ which is true in a finite graph if and only if the graph has a Hamiltonian path.*

**Exercise 4.18** *A graph is* 3-colorable *if it can be divided into three disjoint parts so that all edges are between elements from different parts. Give a sentence of $\mathcal{D}$ which is true in a graph if and only if the graph is 3-colorable.*

**Exercise 4.19** *An* equivalence relation *is an L-structure $\mathcal{M}$ for $L = \{\sim\}$ such that $\sim^{\mathcal{M}}$ is symmetric $(x \sim y \Rightarrow y \sim x)$, transitive $(x \sim y \sim z \Rightarrow x \sim z)$, and reflexive $(x \sim x)$. Give a sentence of $\mathcal{D}$ which is true in an equivalence relation if and only if the equivalence relation has infinitely many equivalence classes.*

## 4.6  Natural numbers

Let $P^-$ be the following first order sentence:

$$\forall x_0 (x_0 + 0 = 0 + x_0 = x_0) \wedge$$
$$\forall x_0 \forall x_1 (x_0 + (x_1 + 1) = (x_0 + x_1) + 1) \wedge$$
$$\forall x_0 (x_0 \cdot 0 = 0 \cdot x_0 = 0) \wedge$$

$$\forall x_0 \forall x_1 (x_0 \cdot (x_1 + 1) = (x_0 \cdot x_1) + x_0) \wedge$$
$$\forall x_0 \forall x_1 (x_0 < x_1 \leftrightarrow \exists x_2 (x_0 + (x_2 + 1) = x_1))$$
$$\forall x_0 (x_0 > 0 \to \exists x_1 (x_1 + 1 = x_0)) \wedge$$
$$0 < 1 \wedge \forall x_0 (0 < x_0 \to (1 < x_0 \vee 1 = x_0))$$

and let $\Phi_{\mathbb{N}}$ be the following sentence of $\mathcal{D}$, reminiscent of $\Phi_\infty$:

$$\neg P^- \vee \exists x_5 \exists x_4 \forall x_0 \exists x_1 \forall x_2 \exists x_3 (=(x_2, x_3) \wedge x_4 < x_5$$
$$\wedge ((x_0 = x_2 \wedge x_0 < x_5) \leftrightarrow$$
$$(x_1 = x_3 \wedge x_1 < x_4))).$$

The sentence $\Phi_{\mathbb{N}}$ is of course true in models that do not satisfy the axiom $P^-$. However, in models of $\Phi_{\mathbb{N}}$ where $P^-$ does hold, something interesting happens: the initial segments determined by $x_4$ and $x_5$ are mapped onto each other by the bijection $x_0 \mapsto x_1$. Thus such models cannot be isomorphic to $(\mathbb{N}, +, \cdot, 0, 1, <)$, in which all initial segments are finite and of different finite cardinality.

**Lemma 4.1** *If $\phi$ is a sentence of dependence logic in the vocabulary of arithmetic, then the following are equivalent:*

(i) *$\phi$ is true in $(\mathbb{N}, +, \cdot, 0, 1, <)$;*
(ii) *$\Phi_{\mathbb{N}} \vee \phi$ is valid in $\mathcal{D}$.*

*Proof* Suppose first $\models \Phi_{\mathbb{N}} \vee \phi$. Since $(\mathbb{N}, +, \cdot, 0, 1, <) \not\models \Phi_{\mathbb{N}}$, we have necessarily $(\mathbb{N}, +, \cdot, 0, 1, <) \models \phi$. Conversely, suppose $(\mathbb{N}, +, \cdot, 0, 1, <) \models \phi$ and let $\mathcal{M}$ be arbitrary. If $\mathcal{M} \models \Phi_{\mathbb{N}}$, then trivially $\models \Phi_{\mathbb{N}} \vee \phi$. Suppose then $\mathcal{M} \not\models \Phi_{\mathbb{N}}$. Necessarily $\mathcal{M} \models P^-$. If $\mathcal{M} \not\cong (\mathbb{N}, +, \cdot, 0, 1, <)$, we get $\mathcal{M} \models \Phi_{\mathbb{N}}$, contrary to our assumption. Thus $\mathcal{M} \cong (\mathbb{N}, +, \cdot, 0, 1, <)$, whence $\mathcal{M} \models \phi$. $\square$

If Lemma 4.1 is combined with Tarski's Undefinability of Truth (see Theorem 6.20), we obtain, using $\ulcorner \phi \urcorner$ to denote the Gödel number of $\phi$ according to some obvious Gödel numbering of sentences of $\mathcal{D}$, the following.

**Corollary 4.2** *The set $\{\ulcorner \phi \urcorner : \phi$ is valid in $\mathcal{D}\}$ is non-arithmetical.*

In particular, there cannot be any effective axiomatization of dependence logic, for then $\{\ulcorner \phi \urcorner : \phi$ is valid in $\mathcal{D}\}$ would be recursively enumerable and therefore arithmetical. We return to this important issue in Chapter 7.

## 4.7 Real numbers

Let $RF$ be the first order axiomatization of ordered fields:

$$\forall x_0(x_0 + 0 = x_0)\wedge$$
$$\forall x_0\forall x_1(x_0 + x_1 = x_1 + x_0)\wedge$$
$$\forall x_0\forall x_1\forall x_2((x_0 + x_1) + x_2 = x_0 + (x_1 + x_2))\wedge$$
$$\forall x_0\exists x_1(x_0 + x_1 = 0)\wedge$$
$$\forall x_0(x_0 \cdot 1 = x_0)\wedge$$
$$\forall x_0\forall x_1(x_0 \cdot x_1 = x_0 \cdot x_1)\wedge$$
$$\forall x_0\forall x_1\forall x_2((x_0 \cdot x_1) \cdot x_2 = x_0 \cdot (x_1 \cdot x_2))\wedge$$
$$\forall x_0(x_0 = 0 \vee \exists x_1(x_0 \cdot x_1 = 1))\wedge$$
$$\forall x_0\forall x_1(x_0 \cdot (x_1 + x_2) = x_0 \cdot x_1 + x_0 \cdot x_2)\wedge$$
$$\forall x_0\forall x_1(x_0 < x_1 \rightarrow x_0 + x_2 < x_1 + x_2)\wedge$$
$$\forall x_0\forall x_1((x_0 > 0 \wedge x_1 > 0) \rightarrow x_0 \cdot x_1 > 0).$$

The ordered field $(\mathbb{R}, +, \cdot, 0, 1, <)$ of real numbers is the unique ordered field in which the order is a complete order. The proof of this can be found in standard textbooks on real analysis. Accordingly, let $\Phi_\mathbb{R}$ be the sentence $\neg RF \vee \Phi_{cmpl}$ of $\mathcal{D}$. Exactly as in Lemma 4.1, we have the following.

**Lemma 4.3** *If $\phi$ is a sentence of $\mathcal{D}$ in the vocabulary of ordered fields, then the following are equivalent:*

(i) *$\phi$ is true in $(\mathbb{R}, +, \cdot, 0, 1, <)$;*
(ii) *$\Phi_\mathbb{R} \vee \phi$ is valid in $\mathcal{D}$.*

This is not as noteworthy as in the case of natural numbers, as the truth of a first order sentence in the ordered field of reals is actually effectively decidable. This is a consequence of the fact, due to Tarski, that this structure admits elimination of quantifiers (see, e.g., ref. [27]). What is noteworthy is that we can add integers to the structure $(\mathbb{R}, +, \cdot, 0, 1, <)$, obtaining the structure $(\mathbb{R}, +, \cdot, 0, 1, <, \mathbb{N})$ with a unary predicate $N$ for the set of natural numbers, making the first order theory of the structure undecidable, and still get a reduction as in Lemma 4.3. To this end, let $\Phi_N$ be

$$N0 \;\wedge\; \forall x_0(Nx_0 \rightarrow Nx_0 + 1)$$
$$\wedge\; \forall x_0(Nx_0 \rightarrow (0 = x_0 \vee 0 < x_0))$$
$$\wedge\; \forall x_0\forall x_1((Nx_0 \wedge Nx_1 \wedge x_0 < x_1)$$
$$\rightarrow (x_0 + 1 = x_1 \vee x_0 + 1 < x_1)).$$

Let $\Phi_{\mathbb{R},\mathbb{N}}$ be the sentence $\neg RF \vee \Phi_{\text{cmpl}} \vee \neg \Phi_{\mathbb{N}}$ of $\mathcal{D}$. Then any structure that is not a model of $\Phi_{\mathbb{R},\mathbb{N}}$ is isomorphic to $(\mathbb{R}, +, \cdot, 0, 1, <, \mathbb{N})$. Thus we obtain the following lemma easily.

**Lemma 4.4** *If $\phi$ is a sentence of $\mathcal{D}$ in the vocabulary of ordered fields supplemented by the unary predicate $N$, then the following are equivalent:*

(i) *$\phi$ is true in $(\mathbb{R}, +, \cdot, 0, 1, <, \mathbb{N})$;*
(ii) *$\Phi_{\mathbb{R},\mathbb{N}} \vee \phi$ is valid in $\mathcal{D}$.*

## 4.8 Set theory

The vocabulary of set theory consists of just one binary predicate symbol $E$. As a precursor to real set theory, let us consider the following simpler situation. We have, in addition to $E$, two unary predicates $R$ and $S$. Let $\theta$ be the conjunction of the first order sentence:

$$\forall x_0 \forall x_1 (x_0 E x_1 \to (R x_0 \wedge S x_1)) \wedge$$
$$\forall x_0 (S x_0 \to \neg R x_0),$$

and the *axiom of extensionality*:

$$\forall x_0 \forall x_1 (\forall x_2 (x_2 E x_0 \leftrightarrow x_2 E x_1) \to x_0 = x_1).$$

Canonical examples of models of $\theta$ are models of the form $(M, \in, X, \mathcal{P}(X))$. Indeed, $\mathcal{M} \models \theta$ if and only if $\mathcal{M} \cong \mathcal{N}$ for some $\mathcal{N}$ such that $E^{\mathcal{N}} = \{(a, b) \in N^2 : a \in R^{\mathcal{N}}, b \in S^{\mathcal{N}}, a \in b\}$ and $S^{\mathcal{N}} \subseteq \mathcal{P}(R^{\mathcal{N}})$. Let

$$\Phi_{\text{ext}} : \neg\theta \vee \exists x_6 \forall x_0 \exists x_1 \forall x_2 \exists x_3 \forall x_4 \exists x_5 \ (=(x_2, x_3)$$
$$\wedge =(x_4, x_5)$$
$$\wedge (x_0 = x_2 \to x_1 = x_3)$$
$$\wedge (x_1 = x_6 \to R x_0)$$
$$\wedge ((S x_4 \wedge x_0 = x_5) \to$$
$$(x_5 E x_4 \leftrightarrow x_1 = x_6))).$$

The sentence $\Phi_{\text{ext}}$ is true in a structure $\mathcal{M}$ if and only if $\mathcal{M} \cong \mathcal{N}$ for some $\mathcal{N}$ such that $E^{\mathcal{N}} = \{(a, b) \in N^2 : a \in R^{\mathcal{N}}, b \in S^{\mathcal{N}}, a \in b\}$ and $S^{\mathcal{N}} \neq \mathcal{P}(R^{\mathcal{N}})$.

The explanation is as follows (see Fig. 4.2). The mapping $x_0 \mapsto x_1$ guesses a set $X$ as the pre-image of $x_6$. The mapping $x_2 \mapsto x_3$ is a copy of the mapping $x_0 \mapsto x_1$, as witnessed by $(x_0 = x_2 \to x_1 = x_3)$. The clause $(x_1 = x_6 \to R x_0)$ makes sure $X$ is a subset of $R$. The clause $((S x_4 \wedge x_0 = x_5) \to (x_5 E x_4 \leftrightarrow x_1 = x_6))$ guarantees the guessed set is not in the set $S$.

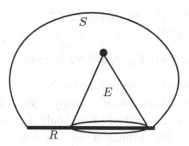

Fig. 4.2. An element coding a subset.

**Lemma 4.5** *If $\phi$ is a sentence of $\mathcal{D}$ in the vocabulary $\{E, R, S\}$, then the following are equivalent:*

(i) $\phi$ *is true in every model of the form* $(M, \in, X, \mathcal{P}(X))$;
(ii) $\Phi_{\text{ext}} \vee \phi$ *is valid in* $\mathcal{D}$.

The *cumulative hierarchy* of sets is defined as follows:

$$\begin{aligned} V_0 &= \emptyset, \\ V_{\alpha+1} &= \mathcal{P}(V_\alpha), \\ V_\nu &= \bigcup_{\beta < \alpha} V_\beta \text{ for limit } \nu. \end{aligned} \qquad (4.1)$$

Thus, $V_1$ is the powerset of $\emptyset$, i.e. $\{\emptyset\}$; $V_2$ is the powerset of $\{\emptyset\}$, i.e. $\{\emptyset, \{\emptyset\}\}$; etc. The sets $V_n$, $n \in \mathbb{N}$, are all finite, $V_\omega$ is countable, and for $\alpha > \omega$ the set $V_\alpha$ is uncountable. For more on the cumulative hierarchy, see Section 7.2.

Let $ZFC^*$ be a large but finite part of the Zermelo–Fraenkel axioms for set theory (see, e.g., ref. [26] for the axioms). It follows from the axioms that every set is in some $V_\alpha$. Models of the form $(V_\alpha, \in)$, where $\alpha$ is a limit ordinal, are canonical examples of models of $ZFC^*$.

Let

$$\begin{aligned} \Phi_{\text{set}} : \neg ZFC^* &\vee \exists x_6 \exists x_7 \forall x_0 \exists x_1 \forall x_2 \exists x_3 \; \forall x_4 \exists x_5 (=(x_2, x_3) \wedge \\ &\wedge =(x_4, x_5) \\ &\wedge (x_0 = x_2 \rightarrow x_1 = x_3) \\ &\wedge (x_1 = x_6 \rightarrow x_0 E x_7) \\ &\wedge (x_0 = x_5 \rightarrow \\ &\qquad (x_5 E x_4 \leftrightarrow x_1 = x_6))). \end{aligned}$$

**Lemma 4.6** *If $\phi$ is a sentence of $\mathcal{D}$ in the vocabulary $\{E\}$, then the following are equivalent:*

(i) $(V_\alpha, \in) \models \phi$ *for all models* $(V_\alpha, \in)$ *of* $ZFC^*$;
(ii) $\Phi_{\text{set}} \vee \phi$ *is valid in* $\mathcal{D}$.

In consequence, there are $\Psi_1$, $\Psi_2$, $\Psi_3$ and $\Psi_4$ in dependence logic such that

  (i) the Continuum Hypothesis holds if and only if $\Psi_1$ is valid in $\mathcal{D}$;
 (ii) the Continuum Hypothesis fails if and only if $\Psi_2$ is valid in $\mathcal{D}$;
(iii) there are no inaccessible cardinals if and only if $\Psi_3$ is valid in $\mathcal{D}$;
(iv) there are no measurable cardinals if and only if $\Psi_4$ is valid in $\mathcal{D}$.

These examples show that to decide whether a sentence of $\mathcal{D}$ is valid or not is tremendously difficult. One may have to search through the whole set theoretic universe. This is in sharp contrast to first order logic, where to decide whether a sentence is valid or not it suffices to search through finite proofs, i.e. essentially just through natural numbers.

By means of the sentence $\Phi_{\text{set}}$ it is easy to show that for any first order structure $\mathcal{M}$ definable in the set theoretical structure $(V_{\omega \cdot 3}, \in)$, which includes virtually all commonly used mathematical structures, and any first order $\phi$, there is a sentence $\Phi_{\mathcal{M},\phi}$ such that the following are equivalent:

  (i) $\phi$ is true in $\mathcal{M}$;
 (ii) $\Phi_{\mathcal{M},\phi}$ is valid in $\mathcal{D}$.

Moreover, $\Phi_{\mathcal{M},\phi}$ can be found effectively on the basis of $\phi$ and the defining formula of $\mathcal{M}$.

**Exercise 4.20** *Prove Lemma 4.5.*

**Exercise 4.21** *Prove Lemma 4.6.*

**Exercise 4.22** *Give a sentence $\phi$ of $\mathcal{D}$ such that $\phi$ has models of all infinite cardinalities, and for all $\kappa \geq \omega$, $\phi$ has a unique model (up to isomorphism) of cardinality $\kappa$ if and only if $\kappa$ is a strong limit cardinal (i.e. $\lambda < \kappa$ implies $2^\lambda < \kappa$).*

# 5

# Game theoretic semantics

We begin with a review of the well known game theoretic semantics of first order logic (see, e.g., ref. [17]). This is the topic of Section 5.1. There are two ways of extending the first order game to dependence logic. The first, presented in Section 5.2, corresponds to the transition in semantics from assignments to teams. The second game theoretic semantics for dependence logic is closer to the original semantics of independence friendly logic presented in refs. [16] and [19]. In the second game theoretic formulation, the dependence relation $=(x_0, \ldots, x_n)$ does not come up as an atomic formula but as the possibility to incorporate *imperfect information* into the game. A player who aims at securing $=(x_0, \ldots, x_n)$ when the game ends has to be able to choose a value for $x_n$ only on the basis of what the values of $x_0, \ldots, x_{n-1}$ are. In this sense the player's *information set* is restricted to $x_0, \ldots, x_{n-1}$ when he or she chooses $x_n$.

## 5.1 Semantic game of first order logic

The game theoretic semantics of first order logic has a long history. The basic idea is that if a sentence is true, its truth, asserted by us, can be defended against a doubter. A doubter can question the truth of a conjunction $\phi \wedge \psi$ by doubting the truth of, say, $\psi$. He can doubt the truth of a disjunction $\phi \vee \psi$ by asking which of $\phi$ and $\psi$ is the one that is true. He can doubt the truth of a negation $\neg\phi$ by claiming that $\phi$ is true instead of $\neg\phi$. At this point we become the doubter and start questioning why $\phi$ is true. This interaction can be formulated in terms of a simple game between two players, **I** and **II**. We call them opponents of each other. The opponent of player $\alpha$ is denoted by $\alpha^*$. In the literature these are sometimes called Abelard and Eloise. We go along to the extent that we refer to player **I** as "he" and to player **II** as "she." The players observe a formula $\phi$ and an assignment $s$ in the context of a given model $\mathcal{M}$. At the beginning

63

of the game, player **II** claims that assignment $s$ satisfies $\phi$ in $\mathcal{M}$, and player **I** doubts this. During the game their roles may change, as we just saw in the case of negation. To keep track of who is claiming what, we use the notation $(\phi, s, \alpha)$ for a position in the game. Here $\alpha$ is either **I** or **II**. The idea is that $\alpha$ indicates which player is claiming that $s$ satisfies $\phi$ in $\mathcal{M}$.

**Definition 5.1** *The semantic game $H(\phi)$ of first order logic in a model $\mathcal{M}$ is the following game. There are two players, **I** and **II**. A position of the game is a triple $(\psi, s, \alpha)$, where $\psi$ is a subformula of $\phi$, $s$ is an assignment, the domain of which contains the free variables of $\psi$, and $\alpha \in \{\mathbf{I}, \mathbf{II}\}$. At the beginning of the game, the position is $(\phi, \emptyset, \mathbf{II})$. The rules of the game are as follows.*

(i) *The position is $(t_1 = t_2, s, \alpha)$: if $t_1^{\mathcal{M}}\langle s \rangle = t_2^{\mathcal{M}}\langle s \rangle$, then player $\alpha$ wins and otherwise the opponent wins.*

(ii) *The position is $(Rt_1 \ldots t_n, s, \alpha)$: if $s$ satisfies $Rt_1 \ldots t_n$ in $\mathcal{M}$, then player $\alpha$ wins, otherwise the opponent wins.*

(iii) *The position is $(\neg\psi, s, \alpha)$: the game switches to the position $(\psi, s, \alpha^*)$, where $\alpha^*$ is the opponent of $\alpha$.*

(iv) *The position is $(\psi \vee \theta, s, \alpha)$: the next position is $(\psi, s, \alpha)$ or $(\theta, s, \alpha)$, and $\alpha$ decides which.*

(v) *The position is $(\exists x_n \psi, s, \alpha)$: player $\alpha$ chooses $a \in M$ and the next position is $(\psi, s(a/x_n), \alpha)$.*

The above game is a *zero-sum* game, i.e. one player's loss is the other player's victory. It is also a game of *perfect information* in the sense that the strategies of both players are allowed to depend on the whole sequence of previous positions. By the *Gale–Stewart Theorem* [10] all finite *zero-sum games of perfect information are determined*.[1] All the possible positions of this game form, in a canonical way, a tree, which we call the *game tree*. The game tree for $H(\phi)$ starts from the position $(\phi, \emptyset, \mathbf{II})$. Depending on $\phi$, it continues in different ways, as displayed in Figs. 5.1, 5.2, and 5.3. Any (maximal) branch of this tree represents a possible *play* of the game. We call proper initial segments of plays *partial* plays.

**Example 5.2** *Consider the game $H(\forall x_0 \exists x_1 (x_0 E x_1))$ in the graph of Fig. 5.4. The tree first splits according to the first move of player **I**. Since player **I** can pick any of the four elements, the tree splits into four subtrees. Then*

---

[1] The idea of the proof is the following. Suppose **I** does not have a winning strategy. Player **II** plays so that always after she has moved player **I** still does not have a winning strategy. If at some point she has only moves after which player **I** has a winning strategy, player **I** had a winning strategy already after the previous move of **II**; a contradiction.

$$(\neg\phi, s, \alpha)$$
$$\downarrow$$
$$(\phi, s, \alpha^*)$$

Fig. 5.1. The game tree at negation node.

$$(\phi \vee \psi, s, \alpha)$$
$$\swarrow \qquad \searrow$$
$$(\phi, s, \alpha) \qquad (\psi, s, \alpha)$$

Fig. 5.2. The game tree at disjunction node.

$$(\exists x_n \phi, s, \alpha)$$
$$\swarrow \qquad \downarrow \qquad \searrow$$
$$\ldots (\phi, s(a/x_n), \alpha) \quad (\phi, s(a'/x_n), \alpha) \quad (\phi, s(a''/x_n), \alpha) \ldots$$

Fig. 5.3. The game tree at quantifier node.

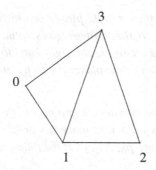

Fig. 5.4. A graph for Example 5.2.

*player II picks any of the four elements and the tree splits again into four subtrees. Altogether we end up with 16 branches (see Fig. 5.5). The figure shows who wins each play. One can see that whatever player I chooses, player II has a move that guarantees her victory.*

Inspection of the game tree is vital for success in a game. It is clear that in order to be able to declare victory a player has to have a clear picture in his or her mind what to play in each position. The following concept of strategy is the heart of game theory. It is a mathematically exact concept which tries to capture the idea of a player knowing what to play in each position.

**Definition 5.3** *A strategy of player $\alpha$ in $H(\phi)$ is any sequence $\tau$ of functions $\tau_i$ defined on the set of all partial plays $(p_0, \ldots, p_{i-1})$ satisfying the following.*

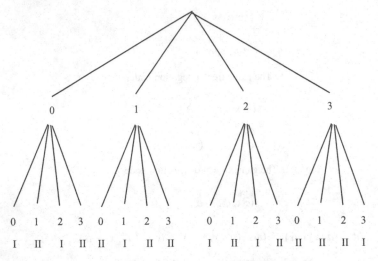

Fig. 5.5. A game tree for Example 5.2.

- *If $p_{i-1} = (\phi \lor \psi, s, \alpha)$, then $\tau$ tells player $\alpha$ which formula to pick, i.e. $\tau_i(p_0, \ldots, p_{i-1}) \in \{0, 1\}$. If the strategy gives value $0$, player $\alpha$ picks the left-hand[2] formula $\phi$, and otherwise the right-hand formula $\psi$.*
- *If $p_{i-1} = (\exists x_n \phi, s, \alpha)$, then $\tau$ tells player $\alpha$ which element $a \in M$ to pick, i.e. $\tau_i(p_0, \ldots, p_{i-1}) \in M$.*

*We say that player $\alpha$ has used strategy $\tau$ in a play of the game $H(\phi)$ if in each relevant case player $\alpha$ has used $\tau$ to make his or her choice. More exactly, player $\alpha$ has used $\tau$ in a play $p_0, \ldots, p_n$ if the following two conditions hold for all $i < n$:*

- *if $p_{i-1} = (\phi \lor \psi, s, \alpha)$ and $\tau_i(p_0, \ldots, p_{i-1}) = 0$, then $p_i = (\phi, s, \alpha)$, while if $\tau_i(p_0, \ldots, p_{i-1}) = 1$, then $p_i = (\psi, s, \alpha)$;*
- *if $p_{i-1} = (\exists x_m \phi, s, \alpha)$ and $\tau_i(p_0, \ldots, p_{i-1}) = a$, then $p_i = (\phi, s(a/x_m), \alpha)$.*

*A strategy $\tau$ of player $\alpha$ in the game $H(\phi)$ is a* winning *strategy if player $\alpha$ wins every play in which he or she has used $\tau$.*

Note that the property of a strategy $\tau$ being a winning strategy is defined without any reference to the actual playing of the game. This is not an oversight but an essential feature of the mathematical theory of games. We have reduced

---

[2] One may ask why the values of the strategy are numbers $0$ and $1$ rather than the formulas themselves. The reason is that the formulas may be one and the same. It is a delicate point whether it then makes any difference which formula is picked. For first order logic there is no difference, but there are extensions of first order logic where a difference emerges.

the intuitive act of players choosing their moves to combinatorial properties of some functions. One has to be rather careful in such a reduction. It is possible that the act of playing and handling formulas may use some property of formulas that is intuitively evident but not coded by the mathematical model. One such potential property is the place of a subformula in a formula. We will return to this point in Example 5.9 and the following discussion.

**Theorem 5.4** *Suppose $\phi$ is a sentence of first order logic. Then $\mathcal{M} \models_\emptyset \phi$ in first order logic if and only if player **II** has a winning strategy in the semantic game $H(\phi)$.*

*Proof* Suppose $\mathcal{M} \models_\emptyset \phi$ in first order logic. Consider the following strategy of player **II**. She maintains the following condition.

($\star$) If the position is $(\phi, s, \mathbf{II})$, then $\mathcal{M} \models_s \phi$. If the position is $(\phi, s, \mathbf{I})$, then $\mathcal{M} \models_s \neg\phi$.

It is completely routine to check that **II** can actually follow this strategy and win. Note that in the beginning $\mathcal{M} \models_\emptyset \phi$, so ($\star$) holds.

For the other direction, suppose player **II** has a winning strategy $\tau$ in the semantic game starting from $(\phi, \emptyset, \mathbf{II})$. It is again completely routine to show the following

($\star\star$) If a position $(\psi, s, \alpha)$ is reached in the game, player **II** using $\tau$, then $\mathcal{M} \models_s \psi$ if $\alpha = \mathbf{II}$ and $\mathcal{M} \models_s \neg\psi$ if $\alpha = \mathbf{I}$.

Since the initial position $(\phi, \emptyset, \mathbf{II})$ is reached at the beginning of the game, we obtain from ($\star\star$) the desired conclusion $\mathcal{M} \models_\emptyset \phi$.                    $\square$

**Exercise 5.1** *Consider the game $H(\exists x_0 \forall x_1 (x_0 = x_1 \lor x_0 E x_1))$ in the following graph:*

$$\bullet \quad\quad \bullet \quad\quad \bullet$$
$$1 \quad\quad 2 \quad\quad 3$$

*Draw the game tree and use it to describe the winning strategy of the player who has it.*

**Exercise 5.2** *Consider the game $H(\forall x_0 \exists x_1 (x_1 E x_0 \land x_0 = x_0))$ in the graph of Fig. 5.4. Draw the game tree and use it to describe the winning strategy of the player who has it.*

**Exercise 5.3** *Draw the game tree for $H(\phi)$, when $\phi$ is given by*

(i) $\neg \exists x_0 P x_0 \land \neg \exists x_0 R x_0$;
(ii) $\exists x_0 (P x_0 \land R x_0)$;
(iii) $\forall x_0 \exists x_1 R x_0 x_1$.

**Exercise 5.4** *Sketch the game tree for* $H(\phi)$, *when* $\phi$ *is given by*

(i) $\forall x_0(Px_0 \vee Rx_0)$;
(ii) $\forall x_0 \exists x_1(Px_0 \wedge \exists x_2 Rx_2 x_1)$.

**Exercise 5.5** *Sketch the game tree for* $H(\phi)$, *when* $\phi$ *is given by*

$$\exists x_0 \neg Px_0 \rightarrow \forall x_0(Px_0 \vee Rx_0).$$

**Exercise 5.6** *Let L consist of two unary predicates P and R. Let* $\mathcal{M}$ *be an L-structure such that* $M = \{0, 1, 2, 3\}$, $P^{\mathcal{M}} = \{0, 1, 2\}$, *and* $R^{\mathcal{M}} = \{1, 2, 3\}$. *Who has a winning strategy in* $H(\phi)$ *if* $\phi$ *is given by the following:*

(i) $\exists x_0(Px_0 \wedge Rx_0)$;
(ii) $\forall x_0 \exists x_1 \neg(x_0 = x_1)$?

*Describe the winning strategy.*

**Exercise 5.7** *Let L consist of two unary predicates P and R. Let* $\mathcal{M}$ *be an L-structure such that* $M = \{0, 1, 2, 3\}$, $P^{\mathcal{M}} = \{0, 1, 2\}$, *and* $R^{\mathcal{M}} = \{1, 2, 3\}$. *Who has a winning strategy in* $H(\phi)$ *if* $\phi$ *is given by the following:*

$$\forall x_0(Px_0 \rightarrow \exists x_1(\neg(x_0 = x_1) \wedge Px_0 \wedge Rx_1))?$$

*Describe the winning strategy.*

**Exercise 5.8** *Let L consist of two unary predicates P and R. Let* $\mathcal{M}$ *be an L-structure such that* $M = \{0, 1, 2, 3\}$, $P^{\mathcal{M}} = \{0, 1, 2\}$, *and* $R^{\mathcal{M}} = \{1, 2, 3\}$. *Who has a winning strategy in* $H(\phi)$ *if* $\phi$ *is given by the following:*

$$\exists x_0(Px_0 \wedge \forall x_1((x_0 = x_1) \vee Px_0 \vee Rx_1))?$$

*Describe the winning strategy.*

**Exercise 5.9** *Suppose* $\mathcal{M}$ *is the binary structure* $(\{0, 1, 2\}, R)$, *where R is as in Fig. 5.6. Who has a winning strategy in* $H(\phi)$ *if* $\phi$ *is given by the following:*

$$\forall x_0 \exists x_1(\neg Rx_0 x_1 \wedge \forall x_2 \exists x_3(\neg Rx_2 x_3))?$$

*Describe the winning strategy.*

**Exercise 5.10** *Suppose* $\mathcal{M}$ *is the binary structure* $(\{0, 1, 2\}, R)$, *where R is as in Fig. 5.6. Who has a winning strategy in* $H(\phi)$ *if* $\phi$ *is given by the following:*

$$\exists x_0 \forall x_1(Rx_0 x_1 \vee \exists x_2 \forall x_3(Rx_2 x_3))?$$

*Describe the winning strategy.*

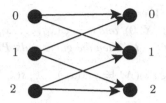

Fig. 5.6. Binary structure.

**Exercise 5.11** *Show that if $\tau$ is a strategy of player **II** in $H(\phi)$ and $\sigma$ is a strategy of player **I** in $H(\phi)$, then there is one and only one play of $H(\phi)$ in which player **II** has used $\tau$ and player **I** has used $\sigma$. (We denote this play by $[\tau, \sigma]$.)*

**Exercise 5.12** *Show that a strategy $\tau$ of player **II** in $G(\phi)$ is a winning strategy if and only if player **II** wins the play $[\tau, \sigma]$ for every strategy $\sigma$ of player **I**.*

## 5.2 Perfect information game for dependence logic

In this section we define a game of perfect information, introduced in ref. [41]. This game is very close to our definition of the semantics of dependence logic. In this game the moves are triples $(\phi, X, d)$, where $X$ is a team, $\phi$ is a formula and $d \in \{0, 1\}$. If $\phi$ is a conjunction and $d = 0$, we may have an ordered pair of teams. This game has less symmetry than $H(\phi)$. On the other hand, the game has formulas of dependence logic as arguments, and $\mathcal{D}$ does not enjoy the same kind of symmetry as first order logic. In particular, we cannot let negation correspond to exchanging the roles of the players, as in the case of $H(\phi)$, and at the same time have sentences which are neither true nor false.

The game we are going to define has two players, **I** and **II**. A *position* in the game is a triple $p = (\phi, X, d)$, where $\phi$ is a formula, $X$ is a team, the free variables of $\phi$ are in dom$(X)$, and $d \in \{0, 1\}$.

**Definition 5.5** *Let $\mathcal{M}$ be a structure. The game*

$$G(\phi)$$

*is defined by the following inductive definition for all sentences $\phi$ of dependence logic. The type of move of each player is determined by the position as follows.*

(M1) *The position is $(\phi, X, 1)$, and $\phi = \phi(x_{i_1}, \dots, x_{i_k})$ is of the form $t_1 = t_2$ or of the form $Rt_1 \dots t_n$. Then the game ends. Player **II** wins if*

$$(\forall s \in X)(\mathcal{M} \models \phi(s(x_{i_1}), \dots, s(x_{i_k}))).$$

*Otherwise player* **I** *wins.*

(M2) *The position is* $(\phi, X, 0)$, *and* $\phi = \phi(x_{i_1}, \ldots, x_{i_k})$ *is of the form* $t_1 = t_2$ *or of the form* $Rt_1 \ldots t_n$. *Then the game ends. Player* **II** *wins if*

$$(\forall s \in X)(\mathcal{M} \not\models \phi(s(x_{i_1}), \ldots, s(x_{i_k}))).$$

*Otherwise player* **I** *wins.*

(M3) *The position is* $(=(t_1, \ldots, t_n), X, 1)$. *Then the game ends. Player* **II** *wins if* $\mathcal{M} \models_X =(t_1, \ldots, t_n)$. *Otherwise player* **I** *wins.*

(M4) *The position is* $(=(t_1, \ldots, t_n), X, 0)$. *Then the game ends. Player* **II** *wins if* $X = \emptyset$. *Otherwise player* **I** *wins.*

(M5) *The position is* $(\neg\phi, X, 1)$. *The game continues from the position* $(\phi, X, 0)$.

(M6) *The position is* $(\neg\phi, X, 0)$. *The game continues from the position* $(\phi, X, 1)$.

(M7) *The position is* $(\phi \vee \psi, X, 1)$. *Now player* **II** *chooses* $Y$ *and* $Z$ *such that* $X = Y \cup Z$, *and we move to position* $(\phi \vee \psi, (Y, Z), 1)$. *Then player* **I** *chooses whether the game continues from position* $(\phi, Y, 1)$ *or* $(\psi, Z, 1)$.

(M8) *The position is* $(\phi \vee \psi, X, 0)$. *Now player* **I** *chooses whether the game continues from position* $(\phi, X, 0)$ *or* $(\psi, X, 0)$.

(M9) *The position is* $(\exists x_n \phi, X, 1)$. *Now player* **II** *chooses* $F : X \to M$, *and then the game continues from the position* $(\phi, X(F/x_n), 1)$.

(M10) *The position is* $(\exists x_n \phi, X, 0)$. *Now the game continues from the position* $(\phi, X(M/x_n), 0)$.

*At the beginning of the game, the position is* $(\phi, \{\emptyset\}, 1)$.

Note that case (M8) generates two rounds for the game: during the first round, player **II** makes a choice for $Y$ and $Z$. We call this round a *half-round*. During the next round, player **I** makes a choice between them. Thus, after the position $(\phi \vee \psi, X, 1)$ there is the position $(\phi \vee \psi, (Y, Z), 1)$, from which the game then proceeds to either $(\phi, Y, 1)$ or $(\psi, Z, 1)$. Note also that player **II** has something to do only in the cases (M7) and (M9). Likewise, player **I** has something to do only in cases (M7) and (M8). Otherwise the game goes on in a determined way with no interaction from the players.

All the possible positions of this game form a tree, just as in the case of the game $H(\phi)$. The tree for $G(\phi)$ starts from the position $(\phi, \{\emptyset\}, 1)$. Depending on $\phi$ it continues in different ways as displayed in Figs. 5.7, 5.8 and 5.9. Note that player **II** has always a winning strategy in a position of the form $(\phi, \emptyset, d)$.

**Example 5.6** *Suppose* $\mathcal{M}$ *has at least two elements. Player* **I** *has a winning strategy in* $G(\forall x_0 \exists x_1(=(x_1) \wedge x_0 = x_1))$. *The game tree is in Fig 5.10. Note*

$$(\neg\phi, X, 1) \qquad (\neg\phi, X, 0) \qquad (\exists x_n\phi, X, 0)$$
$$\downarrow \qquad\qquad \downarrow \qquad\qquad \downarrow$$
$$(\phi, X, 0) \qquad (\phi, X, 1) \qquad (\phi, X(M/x_n), 1)$$

Fig. 5.7. Game trees do not split.

$$(\phi \vee \psi, X, 0)$$
$$\swarrow \qquad \searrow$$
$$(\phi, X, 0) \qquad (\psi, X, 0)$$

Fig. 5.8. The game tree splits into two.

$$(\phi \vee \psi, X, 1)$$
$$\swarrow \qquad \downarrow \qquad \searrow$$
$$\ldots(\phi \vee \psi, (Y, Z), 1) \quad (\phi \vee \psi, (Y', Z'), 1) \quad (\phi \vee \psi, (Y'', Z''), 1)\ldots$$

$$(\exists x_n\phi, X, 1)$$
$$\swarrow \qquad \downarrow \qquad \searrow$$
$$\ldots(\phi, X(F/x_n), 1) \quad (\phi, X(F'/x_n), 1) \quad (\phi, X(F''/x_n), 1)\ldots$$

Fig. 5.9. The game tree splits into many trees.

*that when (M9) is applied, the tree splits into as many branches as there are functions F. At the end of the game, the winner is decided on the basis of (M1)–(M4).*

We define now what we mean by a strategy in the game $G(\phi)$.

**Definition 5.7** *A strategy of player* **II** *in* $G(\phi)$ *is any sequence $\tau$ of functions $\tau_i$ defined on the set of all partial plays* $(p_0, \ldots, p_{i-1})$ *satisfying the following:*

- *if $p_{i-1} = (\phi \vee \psi, X, 1)$, then $\tau$ tells player* **II** *how to cover X with two sets, one corresponding to $\phi$ and the other to $\psi$, i.e. $\tau_i(p_0, \ldots, p_{i-1}) = (Y, Z)$ such that $X = Y \cup Z$;*
- *if $p_{i-1} = (\exists x_n\phi, X, 1)$, then $\tau$ tells player* **II** *how to supplement X, i.e. $\tau_i(p_0, \ldots, p_{i-1})$ is a function $F : X \to M$.*

*We say that player* **II** *has used strategy $\tau$ in a play of the game $G(\phi)$ if in the cases (M7) and (M9) player* **II** *has used $\tau$ to make the choice. More exactly, player* **II** *has used $\tau$ in a play $p_0, \ldots, p_n$ if the following two conditions hold for all $i < n$:*

$$(\neg\exists x_0 \neg\exists x_1(=(x_1) \wedge x_0 = x_1), X, 1)$$

$$\downarrow (M5)$$

$$(\exists x_0 \neg\exists x_1(=(x_1) \wedge x_0 = x_1), X, 0)$$

$$\downarrow (M10)$$

$$(\neg\exists x_1(=(x_1) \wedge x_0 = x_1), X(M/x_0), 0)$$

$$\downarrow (M5)$$

$$(\exists x_1(=(x_1) \wedge x_0 = x_1), X(M/x_0), 1)$$

$$\swarrow \qquad\qquad \downarrow (M9) \qquad\qquad \searrow$$

$$\cdots \qquad ((=(x_1) \wedge x_0 = x_1), X(M/x_0)(F/x_1), 1) \qquad \cdots$$

$$\downarrow (M5)$$

$$\cdots \qquad ((\neg =(x_1) \vee \neg x_0 = x_1), X(M/x_0)(F/x_1), 0) \qquad \cdots$$

$$\swarrow \qquad\qquad\qquad\qquad \searrow (M8)$$

$$(\neg =(x_1), X(M/x_0)(F/x_1), 0) \qquad\qquad (\neg x_0 = x_1, X(M/x_0)(F/x_1), 0)$$

$$\downarrow \qquad\qquad\qquad\qquad\qquad \downarrow (M6)$$

$$(=(x_1), X(M/x_0)(F/x_1), 1) \qquad\qquad (x_0 = x_1, X(M/x_0)(F/x_1), 1)$$

Fig. 5.10. The game tree in Example 5.6.

- if $p_{i-1} = (\phi \vee \psi, X, 1)$ and $\tau_i(p_0, \ldots, p_{i-1}) = (Y, Z)$, then $p_i = (\phi \vee \psi, (Y, Z), 1)$;
- if $p_{i-1} = (\exists x_n\phi, X, 1)$ and $\tau_i(p_0, \ldots, p_{i-1}) = F$, then $p_i = (\phi, X(F/x_n), 1)$.

*A strategy of player **I** in $G(\phi)$ is any sequence $\sigma$ of functions $\sigma_i$ defined on the set of all partial plays $p_0, \ldots, p_{i-1}$ satisfying the following.*

- *If $p_{i-1} = (\phi \vee \psi, X, 0)$, then $\sigma$ tells player **I** which formula to pick, i.e. $\sigma_i(p_0, \ldots, p_{i-1}) \in \{0, 1\}$. If the strategy gives value 0, player **I** picks the left-hand formula $\phi$, and otherwise he chooses the right-hand formula $\psi$.*
- *If $p_{i-1} = (\phi \vee \psi, (Y, Z), 1)$, then $\sigma_i(p_0, \ldots, p_{i-1}) \in \{0, 1\}$.*

*We say that player **I** has used strategy $\sigma$ in a play of the game $G(\phi)$ if in the cases (M7) and (M8) player **I** has used $\sigma$ to make the choice. More exactly, **I** has used $\sigma$ in a play $p_0, \ldots, p_n$ if the following two conditions hold:*

- *if $p_{i-1} = (\phi \vee \psi, X, 0)$ and $\sigma_i(p_0, \ldots, p_{i-1}) = 0$, then $p_i = (\phi, X, 0)$, and if $\sigma_i(p_0, \ldots, p_{i-1}) = 1$, then $p_i = (\psi, X, 0)$.*

- if $p_{i-1} = (\phi \vee \psi, (Y, Z), 1)$, then $p_i = (\phi, Y, 0)$ if $\sigma_i(p_0, \ldots, p_{i-1}) = 0$ and $p_i = (\psi, Z, 0)$ if $\sigma_i(p_0, \ldots, p_{i-1}) = 1$.

*A strategy of player $\alpha$ in the game $G(\phi)$ is a* winning strategy *if player $\alpha$ wins every play in which he or she has used the strategy.*

**Theorem 5.8** $\mathcal{M} \models \phi$ *if and only if player* **II** *has a winning strategy in $G(\phi)$.*

*Proof* Assume first $\mathcal{M} \models \phi$. We describe a winning strategy of player **II** in $G(\phi)$. Player **II** maintains in $G(\phi)$ the condition that if the position (omitting half-rounds) is $(\psi, X, d)$, then $(\psi, X, d) \in \mathcal{T}$. We prove this by induction on $\phi$.

(S1) Position $(\phi, X, 1)$, where $\phi$ is $t = t'$ or $Rt_1 \ldots t_n$. Since $(\phi, X, 1) \in \mathcal{T}$, player **II** wins, by Definition 5.5 (M1).

(S2) Position $(\phi, X, 0)$, where $\phi$ is $t = t'$ or $Rt_1 \ldots t_n$. Since $(\phi, X, 0) \in \mathcal{T}$, player **II** wins, by Definition 5.5 (M2).

(S3) Position $(=(t_1, \ldots, t_n), X, 1)$. Since $(=(t_1, \ldots, t_n), X, 1) \in \mathcal{T}$, player **II** wins, by definition.

(S4) Position $(=(t_1, \ldots, t_n), X, 0)$. Since $(=(t_1, \ldots, t_n), X, 0) \in \mathcal{T}$, we have $X = \emptyset$, and therefore **II** wins by definition.

(S5) Position $(\neg\phi, X, 1)$. Since $(\neg\phi, X, 1) \in \mathcal{T}$, we have $(\phi, X, 0) \in \mathcal{T}$. Thus **II** can play this move according to her plan.

(S6) Position $(\neg\phi, X, 0)$. Since $(\neg\phi, X, 0) \in \mathcal{T}$, we have $(\phi, X, 1) \in \mathcal{T}$. Thus **II** can play this move according to her plan.

(S7) Position $(\phi \vee \psi, X, 0)$. We know $(\phi \vee \psi, X, 0) \in \mathcal{T}$ and therefore both $(\phi, X, 0) \in \mathcal{T}$ and $(\psi, X, 0) \in \mathcal{T}$. Thus, whether the game proceeds to $(\phi, X, 0)$ or $(\psi, X, 0)$, player **II** maintains her plan.

(S8) Position $(\phi \vee \psi, X, 1)$. We know $(\phi \vee \psi, X, 1) \in \mathcal{T}$ and hence $(\phi, Y, 1) \in \mathcal{T}$ and $(\psi, Z, 1) \in \mathcal{T}$ for some $Y$ and $Z$ with $X = Y \cup Z$. Thus we can let player **II** play the ordered pair $(Y, Z)$. After this half-round player **I** wants the game to proceed either to $(\phi, Y, 1)$ or to $(\psi, Z, 1)$. In either case player **II** can fulfil her plan, as both $(\phi, Y, 1) \in \mathcal{T}$ and $(\psi, Z, 1) \in \mathcal{T}$.

(S9) Position $(\exists x_n\phi, X, 1)$. Thus there is $F : X \to M$ such that the triple $(\phi, X(F/x_n), 1)$ is in $\mathcal{T}$. Player **II** can now play the function $F$, for in the resulting position $(\phi, X(F/x_n), 1)$ she can maintain the condition $(\phi, X(F/x_n), 1) \in \mathcal{T}$.

(S10) Position $(\exists x_n\phi, X, 0)$. We know $(\phi, X(M/x_n), 0) \in \mathcal{T}$. But the triple $(\phi, X(M/x_n), 0)$ is the next position, so **II** can maintain her plan.

For the other direction, we assume player **II** has a winning strategy $\tau$ in $G(\phi)$ and use this to show $\mathcal{M} \models \phi$. We prove by induction on $\phi$ that if **II** is using $\tau$ and a position $(\psi, X, d)$ is reached, then $(\psi, X, d) \in \mathcal{T}$. This gives the desired conclusion as the initial position $(\phi, \{\emptyset\}, 1)$ is trivially reached.

(S1′) Position $(\phi, X, 1)$, where $\phi$ is $t = t'$ or $Rt_1 \ldots t_n$. Since **II** has been playing her winning strategy, $(\phi, X, 1) \in \mathcal{T}$ by Definition 5.5 (M1).

(S2′) Position $(\phi, X, 0)$, where $\phi$ is $t = t'$ or $Rt_1 \ldots t_n$. Since **II** has been playing her winning strategy, $(\phi, X, 0) \in \mathcal{T}$ by Definition 5.5 (M2).

(S3′) Position $(=(t_1, \ldots, t_n), X, 1)$. Since **II** has been playing her winning strategy, $(=(t_1, \ldots, t_n), X, 1) \in \mathcal{T}$, by definition.

(S4′) Position $(=(t_1, \ldots, t_n), X, 0)$. Since **II** has been playing her winning strategy, $X = \emptyset$, and therefore $(=(t_1, \ldots, t_n), X, 0) \in \mathcal{T}$ by definition.

(S5′) Position $(\neg\phi, X, 1)$. The game continues, still following $\tau$, to the position $(\phi, X, 0)$. By the induction hypothesis, $(\phi, X, 0) \in \mathcal{T}$, and therefore $(\neg\phi, X, 1) \in \mathcal{T}$.

(S6′) Position $(\neg\phi, X, 0)$. The game continues, still following $\tau$, to the position $(\phi, X, 1)$. By the induction hypothesis, $(\phi, X, 1) \in \mathcal{T}$, and therefore $(\neg\phi, X, 0) \in \mathcal{T}$.

(S7′) Position $(\phi \vee \psi, X, 0)$. To prove $(\phi \vee \psi, X, 0) \in \mathcal{T}$, we need both $(\phi, X, 0) \in \mathcal{T}$ and $(\psi, X, 0) \in \mathcal{T}$. Let us try to prove $(\phi, X, 0) \in \mathcal{T}$. Since **II** is following a winning strategy, we can let the game proceed to position $(\phi, X, 0)$. By the induction hypothesis, $(\phi, X, 0) \in \mathcal{T}$. The same argument gives $(\psi, X, 0) \in \mathcal{T}$. Thus we have proved $(\phi \vee \psi, X, 0) \in \mathcal{T}$.

(S8′) Position $(\phi \vee \psi, X, 1)$. To get $(\phi \vee \psi, X, 1) \in \mathcal{T}$ we need $(\phi, Y, 1) \in \mathcal{T}$ and $(\psi, Z, 1) \in \mathcal{T}$ for some $Y$ and $Z$ with $X = Y \cup Z$. Indeed, the winning strategy $\tau$ gives an ordered pair $(Y, Z)$ with $X = Y \cup Z$. Let us try to prove $(\phi, Y, 1) \in \mathcal{T}$. Since **II** is following a winning strategy, we can let the game proceed to position $(\phi, Y, 1)$. By the induction hypothesis, $(\phi, Y, 1) \in \mathcal{T}$. The same argument gives $(\psi, Z, 1) \in \mathcal{T}$. Thus we have proved $(\phi \vee \psi, X, 1) \in \mathcal{T}$.

(S9′) Position $(\exists x_n \phi, X, 1)$. The strategy $\tau$ gives $F : X \to M$ such that **II** has a winning strategy in position $(\phi, X(F/x_n), 1)$. By the induction hypothesis, $(\phi, X(F/x_n), 1) \in \mathcal{T}$. Hence $(\exists x_n \phi, X, 1) \in \mathcal{T}$.

(S10′) Position $(\exists x_n \phi, X, 0)$. The game continues, still following $\tau$, to the position $(\phi, X(M/x_n), 0)$. By the induction hypothesis the triple $(\phi, X(M/x_n), 0)$ is in $\mathcal{T}$, and therefore $(\exists x_n \phi, X, 0) \in \mathcal{T}$.               $\square$

Theo Janssen [24] has pointed out the following example.

Table 5.1.

| | | $(= (x_0) \vee \neg x_0 = x_1)$ | | | | | | | | |
|---|---|---|---|---|---|---|---|---|---|---|
| ( | = | ( $x_0$ ) | $\vee$ | $\neg$ | $x_0$ | = | $x_1$ | ) | | |
| 1 | 2 | 3 | 4 | 5 | 6 | 7 | 8 | 9 | 10 | 11 |

**Example 5.9** *The sentence*

$$\forall x_0 \exists x_1 ((=(x_1) \wedge \neg x_0 = x_1) \vee (=(x_1) \wedge \neg x_0 = x_1)) \qquad (5.1)$$

*is true in the natural numbers. The trick is the following: for $s \in \{\emptyset\}(\mathbb{N}/x_0)$ let $F(s) \in \{0, 1\}$ be such that $F(s) \neq s(x_0)$. The team $\{\emptyset\}(\mathbb{N}/x_0)(F/x_1)$ is a subset of the union of $\{\emptyset\}(\mathbb{N}/x_0)(F_0/x_1)$ and $\{\emptyset\}(\mathbb{N}/x_0)(F_1/x_1)$, where $F_0$ is the constant function 0 and $F_1$ is the constant function 1. Both parts satisfy $(=(x_1) \wedge \neg x_0 = x_1)$.*

The above example shows that when we play a game that follows the structure of a formula, we may have to take the formula structure into the game. To accomplish this, and in order to be sufficiently precise, we identify formulas with finite strings of symbols. Variables $x_n$ are treated as separate symbols. The other symbols are the symbols of the vocabulary, $=$, ), (, $\neg$, $\wedge$, $\exists$, and the comma. Each string $S$ has a length, which we denote by len($S$). We number the symbols in a formula with positive integers starting from the left, as in Table 5.1 If the $n$th symbol of $\phi$ starts a string which is a subformula of $\phi$, we denote the subformula by $\Lambda(\phi, n)$. Thus, every subformula of $\phi$ is of the form $\Lambda(\phi, n)$ for some $n$ and some may occur with several $n$. In the case that $\phi$ is the formula in Eq. (5.1), the subformula $(=(x_1) \wedge \neg x_0 = x_1)$ occurs as both $\Lambda(\phi, 6)$ and $\Lambda(\phi, 18)$. Note that

(i) if $\Lambda(\phi, n) = \neg \psi$, then $\Lambda(\phi, n+1) = \psi$;
(ii) if $\Lambda(\phi, n) = (\psi \vee \theta)$, then $\Lambda(\phi, n+1) = \psi$ and $\Lambda(\phi, n+2+\text{len}(\psi)) = \theta$;
(iii) if $\Lambda(\phi, n) = \exists x_m \psi$, then $\Lambda(\phi, n+2) = \psi$.

We let $G_{\text{place}}(\phi)$ be the elaboration of the game $G(\phi)$ in which the rules are the same as in $G(\phi)$ but the positions are of the form $(\psi, n, X, d)$, and it is assumed all the time that $\Lambda(\phi, n) = \psi$. Thus the formula $\psi$ could be computed from the number $n$, and it is mentioned in the position only for the sake of clarity.

**Definition 5.10** *Let $\mathcal{M}$ be a structure. The game*

$$G_{\text{place}}(\phi)$$

*is defined by the following inductive definition for all sentences $\phi$ of dependence logic. The type of the move of each player is determined by the position as follows.*

(M1′) *The position is $(\phi, n, X, 1)$ and $\phi = \phi(x_{i_1}, \ldots, x_{i_k})$ is of the form $t_1 = t_2$ or of the form $Rt_1 \ldots t_n$. Then the game ends. Player $\mathbf{II}$ wins if*

$$(\forall s \in X)(\mathcal{M} \models \phi(s(x_{i_1}), \ldots, s(x_{i_k}))).$$

*Otherwise player $\mathbf{I}$ wins.*

(M2′) *The position is $(\phi, n, X, 0)$ and $\phi = \phi(x_{i_1}, \ldots, x_{i_k})$ is of the form $t_1 = t_2$ or of the form $Rt_1 \ldots t_n$. Then the game ends. Player $\mathbf{II}$ wins if*

$$(\forall s \in X)(\mathcal{M} \not\models \phi(s(x_{i_1}), \ldots, s(x_{i_k}))).$$

*Otherwise player $\mathbf{I}$ wins.*

(M3′) *The position is $(=(t_1, \ldots, t_n), m, X, 1)$. Then the game ends. Player $\mathbf{II}$ wins if $\mathcal{M} \models_X =(t_1, \ldots, t_n)$. Otherwise player $\mathbf{I}$ wins.*

(M4′) *The position is $(=(t_1, \ldots, t_n), m, X, 0)$. Then the game ends. Player $\mathbf{II}$ wins if $X = \emptyset$. Otherwise player $\mathbf{I}$ wins.*

(M5′) *The position is $(\neg\phi, n, X, 1)$. The game continues from the position $(\phi, n+1, X, 0)$.*

(M6′) *The position is $(\neg\phi, n, X, 0)$. The game continues from the position $(\phi, n+1, X, 1)$.*

(M7′) *The position is $(\phi \vee \psi, n, X, 0)$. Now player $\mathbf{I}$ chooses whether the game continues from position $(\phi, n+1, X, 0)$ or $(\psi, n+2+\text{len}(\phi), X, 0)$.*

(M8′) *The position is $(\phi \vee \psi, n, X, 1)$. Now player $\mathbf{II}$ chooses $X_0$ and $X_1$ such that $X = X_0 \cup X_1$, and we move to position $(\phi \vee \psi, n, (X_0, X_1), 1)$. Then player $\mathbf{I}$ chooses whether the game continues from position $(\phi, n+1, X_0, 1)$ or $(\psi, n+2+\text{len}(\phi), X_1, 1)$.*

(M9′) *The position is $(\exists x_n\phi, m, X, 1)$. Now player $\mathbf{II}$ chooses $F : X \to M$ and then the game continues from the position $(\phi, m+2, X(F/x_n), 1)$.*

(M10′) *The position is $(\exists x_n\phi, m, X, 0)$. Now the game continues from the position $(\phi, m+2, X(M/x_n), 0)$.*

*At the beginning of the game, the position is $(\phi, 1, \{\emptyset\}, 1)$.*

The following easy observation shows that coding the location of the sub-formula into the game makes no difference. However, we shall use $G_{\text{place}}(\phi)$ in the proof of Theorem 5.16.

**Proposition 5.11** *If player* **II** *has a winning strategy in* $G(\phi)$, *she has a winning strategy in* $G_{\text{place}}(\phi)$, *and vice versa.*

*Proof* Assume **II** has a winning strategy $\tau$ in $G(\phi)$. We describe her winning strategy in $G_{\text{place}}(\phi)$. While playing $G_{\text{place}}(\phi)$, she also plays $G(\phi)$, maintaining the condition that if the position in $G_{\text{place}}(\phi)$ is $(\phi, n, X, d)$, then the position in $G(\phi)$ is $(\phi, X, d)$ while she uses $\tau$ in $G(\phi)$.

  (i) Position is $(\phi, n, X, d)$, where $\phi$ is $t = t'$ or $Rt_1 \ldots t_n$. Player **II** wins since, by assumption, she wins $G(\phi)$ in position $(\phi, X, d)$.
 (ii) Position is $(=(t_1, \ldots, t_n), n, X, d)$. Player **II** wins since, by assumption, she wins $G(\phi)$ in position $(=(t_1, \ldots, t_n), X, d)$.
(iii) Position is $(\neg\phi, n, X, d)$. Player **II** moves to the position $(\phi, n + 1, X, 1 - d)$ in $G_{\text{place}}(\phi)$ and to the position $(\phi, X, 1 - d)$ in $G(\phi)$. Her strategy is still valid.
 (iv) Position is $((\phi \vee \psi), n, X, 0)$. Player **II** proceeds, according to the choice of player **I**, to the position $(\phi, n + 1, X, 0)$ or to the position $(\psi, n + 2 + \text{len}(\phi), X, 0)$ in $G_{\text{place}}(\phi)$ and, respectively, to the position $(\phi, X, 0)$ or to the position $(\psi, X, 0)$ in $G(\phi)$. Whichever way the game proceeds, her strategy is still valid.
  (v) Position is $((\phi \vee \psi), n, X, 1)$. We know that **II** is using $\tau$ in $G(\phi)$, so in the position $((\phi \vee \psi), X, 1) \in T$ she has $Y$ and $Z$ with $X = Y \cup Z$, and she can still win with $\tau$ from positions $(\phi, Y, 1)$ and $(\psi, Z, 1)$. Thus we let player **II** play the ordered pair $(Y, Z)$. After this half-round, player **I** wants the game to proceed either to $(\phi, n + 1, Y, 1)$ or to $(\psi, n + 2 + \text{len}(\phi), Z, 1)$. In either case player **II** can maintain her plan.
 (vi) Position is $(\exists x_n \phi, m, X, 1)$. The strategy $\tau$ gives player **II** a function $F : X \to M$ and the game $G(\phi)$ proceeds to the position $(\phi, X(F/x_n), 1)$. We let **II** play this function $F$. The game $G_{\text{place}}(\phi)$ proceeds to the position $(\phi, m + 2, X(F/x_n), 1)$ and **II** maintains her plan.
(vii) Position is $(\exists x_n \phi, m, X, 0)$. Player **II** proceeds to the position $(\phi, \yen m + 2, X(M/x_n), 0)$ in $G_{\text{place}}(\phi)$ and to the respective position $(\phi, X(M/x_n), d)$ in $G(\phi)$. Player **II** maintains her plan.

For the other direction, assume **II** has a winning strategy $\tau$ in $G_{\text{place}}(\phi)$. We describe her winning strategy in $G(\phi)$. While playing $G(\phi)$ she also plays $G_{\text{place}}(\phi)$, maintaining the condition that if the position in $G(\phi)$ is $(\phi, X, d)$, then the position in $G(\phi)$ is $(\phi, n, X, d)$ for some $n$ while she uses $\tau$ in $G_{\text{place}}(\phi)$.

(i) Position is $(\psi, X, d)$, where $\psi$ is atomic. Player **II** wins since, by assumption she wins $G_{\text{place}}(\phi)$ in position $(\psi, n, X, d)$ for some $n$.

(ii) Position in $G_{\text{place}}(\phi)$ is $(\neg\phi, X, d)$ and in $G_{\text{place}}(\phi)$ it is $(\neg\phi, n, X, d)$. Player **II** moves to the position $(\phi, n + 1, X, 1 - d)$ in $G_{\text{place}}(\phi)$ and to the position $(\phi, X, 1 - d)$ in $G(\phi)$. Her strategy is still valid.

(iii) Position is $((\phi \vee \psi), X, 0)$ and in $G_{\text{place}}(\phi)$ it is $((\phi \vee \psi), n, X, 0)$. Player **II** moves to the position $(\phi, X, 0)$ or to the position $(\psi, X, 0)$ in $G_{\text{place}}(\phi)$ and, respectively, to the position $(\phi, n + 1, X, 0)$ or to the position $(\psi, n + 2 + \text{len}(\phi), X, 0)$ in $G(\phi)$. Whichever way the game proceeds, her strategy is still valid.

(iv) Position is $((\phi \vee \psi), X, 1)$ and in $G_{\text{place}}(\phi)$ it is $((\phi \wedge \psi), n, X, 1)$. We know that **II** is using $\tau$ in $G_{\text{place}}(\phi)$ so in the position $((\phi \vee \psi), n, X, 1) \in \mathcal{T}$ she has $Y$ and $Z$ with $X = Y \cup Z$, and she can still win with $\tau$ from positions $(\phi, n + 1, Y, 1)$ and $(\psi, n + 2 + \text{len}(\phi), Z, 1)$. Thus we let player **II** play the ordered pair $(Y, Z)$. After this half-round, player **I** wants the game to proceed either to $(\phi, Y, 1)$ or to $(\psi, Z, 1)$. Player **II** moves in $G_{\text{place}}(\phi)$, respectively, to $(\phi, n + 1, Y, 1)$ or to $(\psi, n + 2 + \text{len}(\phi), Z, 1)$. In either case, player **II** can maintain her plan.

(v) Position is $(\exists x_n\phi, X, 1)$ and in $G_{\text{place}}(\phi)$ it is $(\exists x_n\phi, m, X, 1)$. The strategy $\tau$ gives player **II** a function $F : X \to M$ and the game $G_{\text{place}}(\phi)$ proceeds to the position $(\phi, m + 2, X(F/x_n), 1)$. We let **II** play this function $F$. The game $G(\phi)$ proceeds to the position $(\phi, X(F/x_n), 1)$, and **II** maintains her plan.

(vi) Position is $(\exists x_n\phi, X, 0)$ and in $G_{\text{place}}(\phi)$ it is $(\exists x_n\phi, m, X, 0)$. Player **II** proceeds to the position $(\phi, ¥m + 2, X(M/x_n), 0)$ in $G_{\text{place}}(\phi)$ and to the position $(\phi, X(M/x_n), d)$ in $G(\phi)$. Player **II** maintains her plan.   $\square$

**Lemma 5.12** *Suppose player* **II** *uses the strategy* $\tau$ *in* $G_{\text{place}}(\phi)$ *and the game reaches a position* $(\psi, n, X, d)$. *Then $X$ and $d$ are uniquely determined by* $\tau$ *and* $n$.

*Proof* Let $k_0, \ldots, k_m$ be the unique sequence of numbers such that if we denote $\Lambda(\phi, k_i)$ by $\phi_i$, then $\phi_0 = \phi$, $\phi_{i+1}$ is an immediate subformula of $\phi_i$, and $\phi_m = \psi$. This sequence is uniquely determined by the number $n$.

During the game that ended in $(\psi, n, X, d)$, the positions (omitting half-rounds) were $(\phi_i, k_i, X_i, d_i)$, $i = 0, \ldots, m$. We know that $\phi_0 = \phi$, $X_0 = \{\emptyset\}$, and $d_0 = 1$. If $\phi_i = \neg\phi_{i+1}$, then necessarily $X_{i+1} = X_i$ and $d_{i+1} = 1 - d_i$. If $\phi_{i+1}$ is a conjunct of $\phi_i$, and $d_i = 1$, then $X_{i+1} = X_i$ and $d_{i+1} = 1$. If $\phi_{i+1}$ is a conjunct of $\phi_i = \psi \wedge \theta$, and $d_i = 0$, then $\tau$ determines, on the basis of $(\phi_j, k_j, X_j, d_j)$, $j = 0, \ldots, i$, two sets $Y$ and $Z$ such that $X_i = Y \cup Z$. Player **I** chooses whether $X_{i+1} = Y$ or $X_{i+1} = Z$. The result is completely determined by whether $k_{i+1} = k_i + 1$ or $k_{i+1} = k_i + 2 + \text{len}(\psi)$. If $\phi_i = \exists x_n \phi_{i+1}$ and $d_i = 0$, then $X_{i+1} = X_i(M/x_n)$ and $d_{i+1} = d_i$, both uniquely determined by $X_i$ and $d_i$. If $\phi_i = \exists x_n \phi_{i+1}$ and $d_i = 1$, then $\tau$ determines $F : X \to M$ on the basis of $(\phi_j, k_j, X_j, d_j)$, $j = 0, \ldots, i$. Then $X_{i+1} = X_i(F/x_n)$ and $d_i = 1$, again uniquely determined by $\tau$ and $X_i$.

$\square$

**Exercise 5.13** *Draw the game tree for $G(\phi)$, when $\phi$ is given by*

(i) $\neg\exists x_0 P x_0 \wedge \neg\exists x_0 R x_0$,
(ii) $\exists x_0 (P x_0 \wedge R x_0)$.

**Exercise 5.14** *Draw the game tree for $G(\phi)$, when $\phi$ is given by*

(i) $\forall x_0 (P x_0 \vee R x_0)$,
(ii) $\forall x_0 \exists x_1 (P x_0 \wedge \exists x_2 R x_2 x_1)$.

**Exercise 5.15** *Draw the game tree for $G(\phi)$, when $\phi$ is given by*

$$\exists x_0 \neg P x_0 \to \forall x_0 (P x_0 \vee R x_0).$$

**Exercise 5.16** *Let L consist of two unary predicates P and R. Let M be an L-structure such that $M = \{0, 1, 2, 3\}$, $P^M = \{0, 1, 2\}$, and $R^M = \{1, 2, 3\}$. Who has a winning strategy in $G(\phi)$ if $\phi$ is given by the following:*

(i) $\exists x_0 (P x_0 \wedge R x_0)$,
(ii) $\forall x_0 \exists x_1 \neg(x_0 = x_1)$?

*Describe the winning strategy.*

**Exercise 5.17** *Let L consist of two unary predicates P and R. Let M be an L-structure such that $M = \{0, 1, 2, 3\}$, $P^M = \{0, 1, 2\}$, and $R^M = \{1, 2, 3\}$. Who has a winning strategy in $G(\phi)$ if $\phi$ is given by the following*

$$\forall x_0 (P x_0 \to \exists x_1 (\neg(x_0 = x_1) \wedge P x_0 \wedge R x_1))?$$

*Describe the winning strategy.*

**Exercise 5.18** *Let L consist of two unary predicates P and R. Let M be an L-structure such that $M = \{0, 1, 2, 3\}$, $P^M = \{0, 1, 2\}$, and $R^M = \{1, 2, 3\}$. Who has a winning strategy in $G(\phi)$ if $\phi$ is given by the following:*

$$\exists x_0 (P x_0 \wedge \forall x_1 ((x_0 = x_1) \vee P x_0 \vee R x_1))?$$

*Describe the winning strategy.*

**Exercise 5.19** *Use the game tree to analyze the formula in Eq. (5.1).*

**Exercise 5.20** *Suppose $\mathcal{M}$ is the binary structure $(\{0, 1, 2\}, R)$, where $R$ is as in Fig. 5.6. Who has a winning strategy in $G(\phi)$ if $\phi$ is given by the following:*

$$\forall x_0 \exists x_1 (\neg R x_0 x_1 \wedge \forall x_2 \exists x_3 (= (x_2, x_3) \wedge \neg R x_0 x_3))?$$

*Describe the winning strategy.*

**Exercise 5.21** *Suppose $\mathcal{M}$ is the binary structure $(\{0, 1, 2\}, R)$, where $R$ is as in Fig. 5.6. Who has a winning strategy in $G(\phi)$ if $\phi$ is given j by the following:*

$$\exists x_0 \forall x_1 (R x_0 x_1 \vee \exists x_2 \forall x_3 (= (x_2, x_3) \rightarrow R x_2 x_3))?$$

*Describe the winning strategy.*

**Exercise 5.22** *Show that if $\tau$ is a strategy of player II in $G(\phi)$ and $\sigma$ is a strategy of player I in $G(\phi)$, then there is one and only one play of $G(\phi)$ in which player II has used $\tau$ and player I has used $\sigma$. (We denote this play by $[\tau, \sigma]$.)*

**Exercise 5.23** *Show that a strategy $\tau$ of player II in $G(\phi)$ is a winning strategy if and only if player II wins the play $[\tau, \sigma]$ for every strategy $\sigma$ of player I.*

## 5.3 Imperfect information game for dependence logic

The semantics of dependence logic can also be defined by means of a simpler game. In this case, however, we have to put a uniformity restriction on strategies in order to get the correct truth definition. The restriction has the effect of making the game a game of partial information.

As in Section 5.2, we pay attention to where a subformula occurs in a formula. This is taken care of by the parameter $n$ in Definition 5.13. It should be borne in mind that disjunctions are assumed to have brackets around them, as in $(\psi \vee \theta)$. Then if this formula is $\Lambda(\phi, n)$, we can infer that $\psi$ is $\Lambda(\phi, n + 1)$ and $\theta$ is $\Lambda(\phi, n + 2 + \text{len}(\psi))$.

**Definition 5.13** *Let $\phi$ be a sentence of dependence logic. The semantic game $H(\phi)$ in a model $\mathcal{M}$ is the following game: there are two players, I and II. A position of the game is a quadruple $(\psi, n, s, \alpha)$, where $\psi$ is $\Lambda(\phi, n)$, $s$ is an assignment, the domain of which contains the free variables of $\psi$, and $\alpha \in \{I, II\}$. At the beginning of the game the position is $(\phi, 1, \emptyset, II)$. The rules of the game are as follows.*

(i) *The position is $(t_1 = t_2, n, s, \alpha)$: if $t_1^{\mathcal{M}} \langle s \rangle = t_2^{\mathcal{M}} \langle s \rangle$, then player $\alpha$ wins and otherwise the opponent wins.*

$$(\neg\phi, n, s, \alpha)$$
$$\downarrow$$
$$(\phi, n + 1, s, \alpha^*)$$

Fig. 5.11. The game tree of $H(\phi)$ at negation node.

$$(\phi \vee \psi, n, s, \alpha)$$

$(\phi, n + 1, s, \alpha)$ $\qquad$ $(\psi, n + 2 + \mathrm{len}(\phi), s, \alpha)$

Fig. 5.12. The game tree of $H(\phi)$ at disjunction node.

(ii) *The position is* $(Rt_1 \ldots t_m, n, s, \alpha)$: *if* $s$ *satisfies* $Rt_1 \ldots t_m$ *in* $\mathcal{M}$, *then player* $\alpha$ *wins, otherwise the opponent wins.*

(iii) *The position is* $(=(t_1, \ldots, t_m), n, s, \alpha)$: *player* $\alpha$ *wins.*

(iv) *The position is* $(\neg\phi, n, s, \alpha)$: *the game switches to the position* $(\phi, n + 1, s, \alpha^*)$, *where* $\alpha^*$ *is the opponent of* $\alpha$.

(v) *The position is* $((\psi \vee \theta), n, s, \alpha)$: *the next position is* $(\psi, n + 1, s, \alpha)$ *or* $(\theta, n + 2 + \mathrm{len}(\psi), s, \alpha)$, *and* $\alpha$ *decides which.*

(vi) *The position is* $(\exists x_m\phi, n, s, \alpha)$: *player* $\alpha$ *chooses* $a \in M$ *and the next position is* $(\phi, n + 2, s(a/x_m), \alpha)$.

Thus $(=(t_1, \ldots, t_n), n, s, \alpha)$ is a safe haven for $\alpha$. Note that the game is a determined zero-sum game of perfect information. However, we are not really interested in who has a winning strategy in this determined game, but in who has a winning strategy with extra uniformity, as defined below. The uniformity requirement in effect makes the game into a non-determined game of imperfect information.

The concepts of a game tree, play, and partial play are defined for this game exactly as for the game $G(\phi)$.

**Definition 5.14** *A* strategy *of player* $\alpha$ *in* $H(\phi)$ *is any sequence* $\tau$ *of functions* $\tau_i$ *defined on the set of all partial plays* $(p_0, \ldots, p_{i-1})$ *satisfying the following.*

- *If* $p_{i-1} = ((\phi \vee \psi), n, s, \alpha)$, *then* $\tau$ *tells player* $\alpha$ *which formula to pick, i.e.* $\tau_i(p_0, \ldots, p_{i-1}) \in \{n + 1, n + 2 + \mathrm{len}(\phi)\}$. *If the strategy gives the lower value, player* $\alpha$ *picks the left-hand formula* $\phi$, *and otherwise chooses the right-hand formula* $\psi$.

- *If* $p_{i-1} = (\exists x_m\phi, n, s, \alpha)$, *then* $\tau$ *tells player* $\alpha$ *which element* $a \in M$ *to pick, i.e.* $\tau_i(p_0, \ldots, p_{i-1}) \in M$.

$$(\exists x_m \phi, n, s, \alpha)$$

$$\ldots (\phi, n+2, s(a/x_m), \alpha) \quad (\phi, , n+2, s(a'/x_m), \alpha) \quad (\phi, n+2, s(a''/x_m), \alpha) \ldots$$

Fig. 5.13. The game tree of $H(\phi)$ at quantifier node.

*We say that player $\alpha$ has used strategy $\tau$ in a play of the game $H(\phi)$ if in each relevant case player $\alpha$ has used $\tau$ to make his or her choice. More exactly, player $\alpha$ has used $\tau$ in a play $p_0, \ldots, p_n$ if the following two conditions hold for all $i < n$:*

- *if $p_{i-1} = ((\phi \vee \psi), m, s, \alpha)$ and $\tau_i(p_0, \ldots, p_{i-1}) = m+1$, then $p_i = (\phi, m+1, s, \alpha)$, while if $\tau_i(p_0, \ldots, p_{i-1}) = m+2+\text{len}(\phi)$, then $p_i = (\psi, m+2+\text{len}(\phi), s, \alpha)$;*
- *if $p_{i-1} = (\exists x_k \phi, m, s, \alpha)$ and $\tau_i(p_0, \ldots, p_{i-1}) = a$, then $p_i = (\phi, m+2, s(a/x_k), \alpha)$.*

*A strategy of player $\alpha$ in the game $H(\phi)$ is a winning strategy if player $\alpha$ wins every play in which she has used the strategy.*

**Definition 5.15** *We call a strategy $\tau$ of player **II** in the game $H(\phi)$ uniform if the following condition holds. Suppose $(\Lambda(\phi, m), m, s, \mathbf{II})$ and $(\Lambda(\phi, m), m, s', \mathbf{II})$ are two positions arising in the game when **II** has played according to $\tau$. Moreover we assume that $\Lambda(\phi, m)$ is $= (t_1, \ldots, t_n)$. Then if $s$ and $s'$ agree about the values of $t_1, \ldots, t_{n-1}$, they agree about the value of $t_n$.*

**Theorem 5.16** *Suppose $\phi$ is a sentence of dependence logic. Then $\mathcal{M} \models_{\{\emptyset\}} \phi$ if and only if player **II** has a uniform winning strategy in the semantic game $H(\phi)$.*

*Proof* Suppose $\mathcal{M} \models_X \phi$. Let $\tau$ be a winning strategy of **II** in $G_{\text{place}}(\phi)$. Consider the following strategy of player **II**. She keeps playing $G_{\text{place}}(\phi)$ as an auxiliary game such that if she is in a position $(\phi, n, X, d)$ in $G_{\text{place}}(\phi)$ and has just moved in the semantic game, the following holds.

($\star$) Suppose the position is $(\phi, n, s, \alpha)$. Then **II** is in a position $(\phi, n, X, d)$, playing $\tau$, in $G_{\text{place}}(\phi)$ and $s \in X$. If $\alpha = \mathbf{II}$, then $d = 1$. If $\alpha = \mathbf{I}$, then $d = 0$.

Let us check that **II** can actually follow this strategy and win. In the beginning $\mathcal{M} \models_{\{\emptyset\}} \phi$, so ($\star$) holds.

(i) $\phi$ is $t_1 = t_2$ or $Rt_1 \ldots t_n$. If $\alpha = \mathbf{II}$, $s$ satisfies $\phi$ in $\mathcal{M}$. So **II** wins. If $\alpha = \mathbf{I}$, $s$ does not satisfy $\phi$ in $\mathcal{M}$, and again **II** wins.

(ii) $\phi$ is $=(t_1, \ldots, t_n)$. If $\alpha = \textbf{II}$, then $\textbf{II}$ wins by definition. On the other hand, $s \in X$, so $X \neq \emptyset$, and we must have $\alpha = \textbf{II}$.

(iii) $\phi$ is $\neg \psi$ and the position in $G_{\text{place}}(\phi)$ is $(\neg \psi, n, X, d)$. By the rules of $G_{\text{place}}(\phi)$, the next position is $(\psi, n+1, X, 1-d)$. So the game can proceed to position $(\psi, n+1, s, \alpha^*)$ and $\textbf{II}$ maintains $(\star)$.

(iv) $\phi$ is $(\psi \vee \theta)$ and the position in $G_{\text{place}}(\phi)$ is $((\psi \vee \theta), n, X, d)$. Suppose $\alpha = \textbf{I}$ and $d = 0$. Then both $(\psi, n+1, X, d)$ and $(\theta, n+2+\text{len}(\psi), X, d)$ are possible positions in $G_{\text{place}}(\phi)$ while $\textbf{II}$ uses $\tau$. The next position is $(\psi, n+1, s, 0)$ or $(\theta, n+2+\text{len}(\psi_0), s, 0)$, and $\textbf{I}$ chooses which. Condition $(\star)$ remains valid, whichever he chooses. Suppose then $\alpha = \textbf{II}$ and $d = 1$. Strategy $\tau$ gives $X_0$ and $X_1$ such that $X = X_0 \cup X_1$ and $\textbf{II}$ wins with $\tau$ both in the position $(\psi, n+1, X_0, 1)$ and in $(\theta, n+2+\text{len}(\psi_0), X_1, 1)$. Since $s \in X$, we have either $s \in X_0$ or $s \in X_1$. Let us say $s \in X_0$. We let $\textbf{I}$ play $\psi$ in $G_{\text{place}}(\phi)$. The game $G_{\text{place}}(\phi)$ proceeds to $(\psi, n+1, X_0, 1)$. We let $\textbf{II}$ play in $H(\phi)$ the sentence $\psi$. Condition $(\star)$ remains valid. The situation is similar if $s \in X_1$.

(v) $\phi$ is $\exists x \psi$. We leave this as an exercise.

We claim that the strategy is uniform. Suppose $s$ and $s'$ are assignments arising from the game when $\textbf{II}$ plays the above strategy and the game ends in the same dependence formula $=(t_1, \ldots, t_n)$. Let the ending positions be $(\Lambda(\phi, n), s, \alpha)$ and $(\Lambda(\phi, n), s', \alpha)$. Since $\textbf{II}$ wins, $\alpha = \textbf{II}$. When the game ended, player $\textbf{II}$ had reached the position $(=(t_1, \ldots, t_n), n, X, 1)$ on one hand and the position $(=(t_1, \ldots, t_n), n, X', 1)$ on the other hand in $G_{\text{place}}(\phi)$ playing, $\tau$. By Lemma 5.12, $X = X'$. Suppose $s$ and $s'$ agree about the values of $t_1, \ldots, t_{n-1}$. Since $\textbf{II}$ wins in the position $(=(t_1, \ldots, t_n), n, X, 1)$ and $s, s' \in X$, it follows that $s$ and $s'$ agree about the value of $t_n$. This strategy gives one direction of the theorem.

For the other direction, suppose player $\textbf{II}$ has a uniform winning strategy $\tau$ in the semantic game starting from $(\phi, 1, \emptyset, \textbf{II})$. Let $X_n$ be the set of $s$ such that $(\Lambda(\phi, n), n, s, \alpha)$ is the position in some play where $\textbf{II}$ used $\tau$. Note that $\alpha$ depends only on $n$, so we can denote it by $\alpha_n$. We show by induction on subformulas $\Lambda(\phi, n)$ of $\phi$ that $(\Lambda(\phi, n), X_n, d_n) \in \mathcal{T}$, where $d_n = 1$ if and only if $\alpha_n = \textbf{II}$. Putting $n = 1$, we obtain $\alpha_1 = \textbf{II}$ and we get the desired result.

(i) Suppose $\Lambda(\phi, n)$ is $t_1 = t_2$ or $R(t_1, \ldots, t_n)$. We show that the quadruple $(\Lambda(\phi, n), n, X_n, d)$ is in $\mathcal{T}$. Let $s \in X_n$. Let the quadruple $(\Lambda(\phi, n), n, s, \alpha_n)$ be a position in some play where $\textbf{II}$ used $\tau$. Since $\textbf{II}$ wins with $\tau$, $(\Lambda(\phi, n), X_n, d) \in \mathcal{T}$.

(ii) Suppose $\Lambda(\phi, n)$ is $=(t_1, \ldots, t_n)$. Suppose first $\alpha_n = \textbf{II}$. Suppose $s$ and $s'$ are in $X_n$ and agree about the values of $t_1, \ldots, t_{n-1}$. By the definition of

$X_n$, $(\Lambda(\phi, n), n, s, \mathbf{II})$ and $(\Lambda(\phi, n), n, s', \mathbf{II})$ are positions in some plays where $\mathbf{II}$ used $\tau$. Since $\tau$ is uniform, $s$ and $s'$ agree about the value of $t_n$. The case $\alpha_n = \mathbf{I}$ cannot occur since $\tau$ is a winning strategy.

(iii) Suppose $\Lambda(\phi, n)$ is $\neg\psi$. Note that $X_n = X_{n+1}$. By the induction hypothesis, $(\psi, X_n, 1 - d) \in \mathcal{T}$, hence $(\neg\psi, X_n, d) \in \mathcal{T}$.

(iv) Suppose $\Lambda(\phi, n)$ is $(\psi \vee \theta)$. Suppose first $\alpha_n = \mathbf{I}$. Then both $(\Lambda(\phi, n + 1), n + 1, s, \mathbf{I})$ and $(\Lambda(\phi, n + 2 + \text{len}(\psi)), n + 2 + \text{len}(\psi), s, \mathbf{I})$ can be positions in some plays where $\mathbf{II}$ has used $\tau$. By the induction hypothesis, $(\psi, X_{n+1}, 0) \in \mathcal{T}$ and $(\theta, X_{n+2+\text{len}(\psi)}, 0) \in \mathcal{T}$. Note that $X_n \subseteq X_{n+1} \cap X_{n+2+\text{len}(\psi)}$. Hence $(\psi \vee \theta, X_n, 0) \in \mathcal{T}$. Suppose then $\alpha_n = \mathbf{II}$. Now $X = Y \cup Z$, where $Y$ is the set of $s \in X_n$ such that $(\Lambda(\phi, n + 1), n + 1, s, \alpha_n)$ and $Z$ is the set of $s \in X_n$ such that $(\Lambda(\phi, n + 2 + \text{len}(\psi)), n + 2 + \text{len}(\psi), s, \alpha_n)$, By the induction hypothesis, $(\psi, X_{n+1}, 1) \in \mathcal{T}$ and $(\theta, X_{n+2+\text{len}(\psi)}, 1) \in \mathcal{T}$. Hence $(\psi \wedge \theta, X_n, 1) \in \mathcal{T}$.

(v) Suppose $\Lambda(\phi, n)$ is $\exists x\psi$. We leave this as an exercise.     $\square$

**Exercise 5.24** *Draw the game tree for $H(\phi)$, when $\phi$ is given by*

(i) $(\neg\exists x_0 =(x_0) \vee \neg\exists x_0 Rx_0)$,
(ii) $\exists x_0 \exists x_1 (=(x_1) \vee Rx_0)$.

**Exercise 5.25** *Draw the game tree for $H(\phi)$, when $\phi$ is given by*

(i) $\forall x_0 \exists x_1 (=(x_1) \vee Rx_0)$,
(ii) $\forall x_0 \exists x_1 (Px_0 \vee \exists x_2 (=(x_0, x_2) \vee Rx_2 x_1))$.

**Exercise 5.26** *Draw the game tree for $H(\phi)$, when $\phi$ is given by*

$$\exists x_0(\neg Px_0 \rightarrow \forall x_1(=(x_0, x_1) \wedge Rx_1)).$$

**Exercise 5.27** *Let $L$ consist of two unary predicates $P$ and $R$. Let $\mathcal{M}$ be an $L$-structure such that $M = \{0, 1, 2, 3\}$, $P^{\mathcal{M}} = \{0, 1, 2\}$, and $R^{\mathcal{M}} = \{1, 2, 3\}$. Who has a winning strategy in $H(\phi)$ if $\phi$ is given by the following:*

(i) $\exists x_0 (=(x_0) \wedge Rx_0)$,
(ii) $\forall x_0 \exists x_1 (=(x_1) \wedge \neg(x_0 = x_1))$?

*Describe the winning strategy.*

**Exercise 5.28** *Let $\mathcal{M}$ be as in Exercise 5.27. Does $\mathbf{II}$ have a uniform winning strategy in $H(\phi)$ if $\phi$ is given by the following:*

$$\forall x_0(Px_0 \rightarrow \exists x_1(=(x_0, x_1) \wedge \neg(x_0 = x_1) \wedge Px_0 \wedge Rx_1))?$$

Table 5.2.

| $x_0$ | $x_1$ | $\vee$ | $x_0$ | $x_1$ | $\vee$ |
|-------|-------|--------|-------|-------|--------|
| 0 | 0 | left | 0 | 0 | left |
| 1 | 1 | left | 1 | 1 | right |
| 2 | 2 | right | 2 | 2 | right |

Table 5.3.

| $x_0$ | $x_1$ | $\vee$ | $x_0$ | $x_1$ | $\vee$ |
|-------|-------|--------|-------|-------|--------|
| 0 | 2 | left | 0 | 1 | left |
| 1 | 2 |  | 1 | 2 |  |
| 2 | 0 | right | 2 | 0 | right |

**Exercise 5.29** *Let $\mathcal{M}$ be as in Exercise 5.27. Does II have a uniform winning strategy in $H(\phi)$ if $\phi$ is given by following:*

$$\exists x_0(Px_0 \wedge \forall x_1(=(x_0, x_1) \vee (x_0 = x_1) \vee Rx_1))?$$

**Exercise 5.30** *Suppose $\mathcal{M}$ is the binary structure $(\{0, 1, 2\}, R)$, where $R$ is as in Fig. 5.6. Does $\mathbf{II}$ have a uniform winning strategy in $H(\phi)$ if $\phi$ is given by the following:*

$$\forall x_0 \exists x_1(\neg Rx_0x_1 \wedge \forall x_2 \exists x_3(=(x_2, x_3) \wedge \neg Rx_0x_3))?$$

**Exercise 5.31** *Suppose $\mathcal{M}$ is the binary structure $(\{0, 1, 2\}, R)$, where $R$ is as in Fig. 5.6. Does $\mathbf{II}$ have a uniform winning strategy in $H(\phi)$ if $\phi$ is given by the following:*

$$\exists x_0 \forall x_1(Rx_0x_1 \vee \exists x_2 \forall x_3(=(x_2, x_3) \rightarrow Rx_2x_3))?$$

**Exercise 5.32** *Show that neither of the winning strategies (shown in Table 5.2) of player $\mathbf{II}$ in $H(\phi)$ is uniform, when $\phi$ is the sentence $\forall x_0 \exists x_1((=(x_1) \wedge x_0 = x_1) \vee (=(x_1) \wedge x_0 = x_1))$ and the universe is $\{0, 1, 2\}$.*

**Exercise 5.33** *Which of the strategies (in Table 5.3) of player $\mathbf{II}$ in $H(\phi)$ can be completed so that the strategy becomes a uniform winning strategy of $\mathbf{II}$? Here $\phi$ is the sentence $\forall x_0 \exists x_1((=(x_1) \wedge \neg x_0 = x_1) \vee (=(x_1) \wedge \neg x_0 = x_1)))$ and the universe is $\{0, 1, 2\}$.*

# 6

# Model theory

Many model theoretic results for dependence logic can be proved by means of a reduction to existential second order logic. We establish this reduction in the first section of this chapter. This immediately gives such results as the Compactness Theorem, the Löwenheim–Skolem Theorem, and the Craig Interpolation Theorem.

## 6.1 From $\mathcal{D}$ to $\Sigma_1^1$

We associate with every formula $\phi$ of dependence logic a second order sentence which is in a sense equivalent to $\phi$. This is in fact nothing more than a formalization of the truth definition of $\phi$ (Definition 3.5). What is interesting is that the second order sentence, which we denote by $\tau_{1,\phi}(S)$, is not just any second order sentence but a particularly simple second order existential sentence, called a $\Sigma_1^1$-sentence. Such sentences have a close relationship with first order logic, especially on countable models. It turns out that their relationship with dependence logic is even closer. In a sense they are one and the same thing. It is the main purpose of this section to explain exactly what is this sense in which they are one and the same thing.

**Definition 6.1** Let $L$ be a vocabulary. The class of $\Sigma_1^1$-formulas of $L$ is defined as follows.

(i) Any first order formula of $L$ is a $\Sigma_1^1$-formula of $L$.
(ii) If $\phi$ is a $\Sigma_1^1$-formula of $L \cup \{R\}$, then $\exists R\phi$ is a $\Sigma_1^1$-formula of $L$. $\mathcal{M} \models_s \exists R\phi$ if and only if there is an expansion $\mathcal{M}'$ of $\mathcal{M}$ to an $L \cup \{R\}$-structure such that $\mathcal{M}' \models_s \phi$.

(iii) *If $\phi$ is a $\Sigma_1^1$-formula of $L \cup \{f\}$, then $\exists f\phi$ is a $\Sigma_1^1$-formula of $L$. $\mathcal{M} \models_s$ $\exists f\phi$ if and only if there is an expansion $\mathcal{M}'$ of $\mathcal{M}$ to an $L \cup \{f\}$-structure such that $\mathcal{M}' \models_s \phi$.*

An equivalent concept is PC-definability. A property $\mathcal{O}(\mathcal{M})$ of $L$-structures is *PC-definable* if there is a vocabulary $L' \supseteq L$ and a first order $\phi$ in the vocabulary $L'$ such that an $L$-structure $\mathcal{M}$ has the property $\mathcal{O}(\mathcal{M})$ if and only if it is a reduct of an $L'$-structure satisfying $\phi$. Clearly, PC-definable properties are exactly the $\Sigma_1^1$-definable properties of models.

The logic $\Sigma_1^1$ is closed under conjunction in the sense that if $\phi$ and $\psi$ are $\Sigma_1^1$-formulas then there is a $\Sigma_1^1$-formula $\theta$ such that

$$\mathcal{M} \models_s \theta \iff \mathcal{M} \models_s \phi \text{ and } \mathcal{M} \models_s \psi.$$

For example, if $\phi = \exists R \exists R' \phi'$ and $\psi = \exists R \exists f \psi'$, then we first change the relation symbol $R$ in $\psi'$ to a new relation symbol $R''$, obtaining $\psi''$, and then take $\exists R \exists R' \exists R'' \exists f(\phi' \wedge \psi'')$ as $\theta$. Similarly, $\Sigma_1^1$ is closed under disjunction in the sense that if $\phi$ and $\psi$ are $\Sigma_1^1$-formulas then there is a $\Sigma_1^1$-formula $\theta$ such that

$$\mathcal{M} \models_s \theta \iff \mathcal{M} \models_s \phi \text{ or } \mathcal{M} \models_s \psi.$$

It is fairly obvious that $\Sigma_1^1$ is closed under first order existential quantification in the sense that if $\phi$ is an $\Sigma_1^1$-formula then there is a $\Sigma_1^1$-formula $\theta$ such that

$$\mathcal{M} \models_s \theta \iff \text{ there is } a \in M \text{ such that } \mathcal{M} \models_{s(a/x_n)} \phi.$$

For example, if $\phi = \exists R \exists R' \psi$, then we can take $\exists R \exists R' \exists x_n \psi$ as $\theta$. It is a little more tricky to see that $\Sigma_1^1$ is also closed under first order universal quantification in the sense that if $\phi$ is an $\Sigma_1^1$-formula then there is a $\Sigma_1^1$-formula $\theta$ such that

$$\mathcal{M} \models_s \theta \iff \text{ for all } a \in M \text{ we have } \mathcal{M} \models_{s(a/x_n)} \phi.$$

To see why this is so, let us consider a simple $\Sigma_1^1$-formula $\exists R\psi$, where $\psi$ is first order. Let $\phi'$ be obtained from $\phi$ by replacing everywhere $Rt_1 \ldots t_m$ by $R'x_n t_1 \ldots t_n$, where $R'$ is a new predicate symbol of arity $\#_L(R) + 1$. Now,

$$\mathcal{M} \models_s \exists R' \forall x_n \phi' \iff \text{ for all } a \in M \text{ we have } \mathcal{M} \models_{s(a/x_n)} \exists R\phi \qquad . (6.1)$$

(see Exercise 6.7).

We adopt the convention of writing a formula $\phi$ of dependence logic with free variables $x_{i_1}, \ldots, x_{i_n}$ as $\phi(x_{i_1}, \ldots, x_{i_n})$, where it is always assumed that $i_1 < \cdots < i_n$. This notation includes the case that $n = 0$ which corresponds to the case that $\phi$ is a sentence. Similarly we write $t(x_{i_1}, \ldots, x_{i_n})$ for a term built up from the variables $x_{i_1}, \ldots, x_{i_n}$. In the following theorem we refer to $S$ as an $n$-ary predicate symbol. If $n = 0$, $S$ is a $\top$ or $\neg\top$. In the following theorem, $\mathrm{rel}(X) = \{(s(x_{i_1}), \ldots, s(x_{i_n})) : s \in X\}$.

**Theorem 6.2** *We can associate with every formula $\phi(x_{i_1}, \ldots, x_{i_n})$ of $\mathcal{D}$ in vocabulary $L$ and every $d \in \{0, 1\}$ a $\Sigma_1^1$-sentence $\tau_{d,\phi}(S)$, where $S$ is $n$-ary, such that for all $L$-structures $\mathcal{M}$ and teams $X$ with $\mathrm{dom}(X) = \{x_{i_1}, \ldots, x_{i_n}\}$ the following are equivalent:*

(i) $(\phi, X, d) \in \mathcal{T}$,
(ii) $(\mathcal{M}, \mathrm{rel}(X)) \models \tau_{d,\phi}(S)$.

*Proof* We modify Hodges' approach (see ref. [22], sect. 3) to fit our setup. The sentence $\tau_{d,\phi}(S)$ is simply Definition 3.5 written in another way. There is nothing new in $\tau_{d,\phi}(S)$, and in each case the proof of the claimed equivalence is straightforward (see Exercises 6.4 and 6.5).

Case (1)  Suppose $\phi(x_{i_1}, \ldots, x_{i_n})$ is $t_1 = t_2$ or $Rt_1 \ldots t_n$. We rewrite (D1), (D2), (D5) and (D6) of Definition 3.5 by letting $\tau_{1,\phi}(S)$ be

$$\forall x_{i_1} \ldots \forall x_{i_n} (S x_{i_1} \ldots x_{i_n} \to \phi(x_{i_1}, \ldots, x_{i_n}))$$

and by letting $\tau_{0,\phi}(S)$ be

$$\forall x_{i_1} \ldots \forall x_{i_n} (S x_{i_1} \ldots x_{i_n} \to \neg\phi(x_{i_1}, \ldots, x_{i_n})).$$

Case (2)  Suppose $\phi(x_{i_1}, \ldots, x_{i_n})$ is the dependence formula

$$= (t_1(x_{i_1}, \ldots, x_{i_n}), \ldots, t_m(x_{i_1}, \ldots, x_{i_n})),$$

where $i_1 < \cdots < i_n$. Recall conditions (D3) and (D4) of Definition 3.5. Following these conditions, we define $\tau_{1,\phi}(S)$ as follows.
Subcase (2.1)  $m = 0$. We let $\tau_{1,\phi}(S) = \top$ and $\tau_{0,\phi}(S) = \neg\top$.
Subcase (2.2)  $m = 1$. Now $\phi(x_{i_1}, \ldots, x_{i_n})$ is the dependence formula $= (t_1(x_{i_1}, \ldots, x_{i_n}))$. We let $\tau_{1,\phi}(S)$ be the formula

$$\forall x_{i_1} \ldots \forall x_{i_n} \forall x_{i_n+1} \ldots \forall x_{i_n+n}((S x_{i_1} \ldots x_{i_n} \wedge S x_{i_n+1} \ldots x_{i_n+n})$$
$$\to t_1(x_{i_1}, \ldots, x_{i_n}) = t_1(x_{i_n+1}, \ldots, x_{i_n+n}))$$

and we further let $\tau_{0,\phi}(S)$ be the formula $\forall x_{i_1} \ldots \forall x_{i_n} \neg S x_{i_1} \ldots x_{i_n}$.

**Subcase (2.3)** If $m > 1$ we let $\tau_{1,\phi}(S)$ be the formula

$$\forall x_{i_1} \ldots \forall x_{i_n} \forall x_{i_n+1} \ldots \forall x_{i_n+n}((Sx_{i_1} \ldots x_{i_n} \wedge Sx_{i_n+1} \ldots x_{i_n+n}$$
$$\wedge\, t_1(x_{i_1}, \ldots, x_{i_n}) = t_1(x_{i_n+1}, \ldots, x_{i_n+n})$$
$$\wedge \ldots$$
$$t_{m-1}(x_{i_1}, \ldots, x_{i_n}) = t_{m-1}(x_{i_n+1}, \ldots, x_{i_n+n}))$$
$$\rightarrow t_m(x_{i_1}, \ldots, x_{i_n}) = t_m(x_{i_n+1}, \ldots, x_{i_n+n})),$$

and we further let $\tau_{0,\phi}(S)$ be the formula $\forall x_{i_1} \ldots \forall x_{i_n} \neg\, Sx_{i_1} \ldots x_{i_n}$.

**Case (3)** Suppose $\phi(x_{i_1}, \ldots, x_{i_n})$ is the disjunction

$$(\psi(x_{j_1}, \ldots, x_{j_p}) \vee \theta(x_{k_1}, \ldots, x_{k_q})),$$

where $\{i_1, \ldots, i_n\} = \{j_1, \ldots, j_p\} \cup \{k_1, \ldots, k_q\}$. We let the sentence $\tau_{1,\phi}(S)$ be

$$\exists R \exists T (\tau_{1,\psi}(R) \wedge \tau_{1,\theta}(T) \wedge$$
$$\forall x_{i_1} \ldots \forall x_{i_n}(Sx_{i_1} \ldots x_{i_n} \rightarrow (Rx_{j_1} \ldots x_{j_p} \vee Tx_{k_1} \ldots x_{k_q})))$$

and we let the sentence $\tau_{0,\phi}(S)$ be

$$\exists R \exists T (\tau_{0,\psi}(R) \wedge \tau_{0,\theta}(T) \wedge$$
$$\forall x_{i_1} \ldots \forall x_{i_n}(Sx_{i_1} \ldots x_{i_n} \rightarrow (Rx_{j_1} \ldots x_{j_p} \wedge Tx_{k_1} \ldots x_{k_q}))).$$

**Case (4)** $\phi$ is $\neg\psi$. $\tau_{d,\phi}(S)$ is the formula $\tau_{1-d,\psi}(S)$.

**Case (5)** Suppose $\phi(x_{i_1}, \ldots, x_{i_n})$ is the formula $\exists x_{i_{n+1}} \psi(x_{i_1}, \ldots, x_{i_{n+1}})$. $\tau_{1,\phi}(S)$ is the formula

$$\exists R(\tau_{1,\psi}(R) \wedge \forall x_{i_1} \ldots \forall x_{i_n}(Sx_{i_1} \ldots x_{i_n} \rightarrow \exists x_{i_{n+1}} Rx_{i_1} \ldots x_{i_{n+1}}))$$

and $\tau_{0,\phi}(S)$ is the formula

$$\exists R(\tau_{0,\psi}(R) \wedge \forall x_{i_1} \ldots \forall x_{i_n}(Sx_{i_1} \ldots x_{i_n} \rightarrow \forall x_{i_{n+1}} Rx_{i_1} \ldots x_{i_{n+1}})).$$

$\square$

**Corollary 6.3** *For every sentence* $\phi$ *of* $\mathcal{D}$ *there are* $\Sigma_1^1$-*sentences* $\tau_{1,\phi}$ *and* $\tau_{0,\phi}$ *such that for all models* $\mathcal{M}$ *we have*

$$\mathcal{M} \models \phi \text{ if and only if } \mathcal{M} \models \tau_{1,\phi};$$
$$\mathcal{M} \models \neg\phi \text{ if and only if } \mathcal{M} \models \tau_{0,\phi}.$$

*Proof* Let $\tau_{d,\phi}$ be the result of replacing in $\tau_{d,\phi}(S)$ every occurrence of the 0-ary relation symbol $S$ by $\top$. Now the claim follows from Theorem 6.2. $\square$

**Exercise 6.1** *Write down* $\tau_{1,\exists x_1 =(x_1)}(S)$ *and* $\tau_{0,\exists x_1 =(x_1)}(S)$.

**Exercise 6.2** *What is* $\tau_{1,\phi}(S)$ *if* $\phi$ *is* $\exists x_1(=(x_1) \vee x_1 = x_0)$?

**Exercise 6.3** *What is $\tau_{1,\phi}(S)$ if $\phi$ is given by the following:*

$$\forall x_0 \exists x_1 \forall x_2 \exists x_3 (=(x_2, x_3) \wedge \neg(x_1 = x_4) \wedge (x_0 = x_2 \leftrightarrow x_1 = x_3))?$$

**Exercise 6.4** *Fill in the details of Case (3) of the proof of Theorem 6.2.*

**Exercise 6.5** *Fill in the details of Case (5) of the proof of Theorem 6.2.*

**Exercise 6.6** *Show that if $\phi$ is $\Sigma_1^1$, $\mathcal{M} \models \phi$ and $\mathcal{M} \cong \mathcal{N}$, then $\mathcal{N} \models \phi$.*

**Exercise 6.7** *Prove the claim given in Eq. (6.1).*

# 6.2 Applications of $\Sigma_1^1$

The $\Sigma_1^1$-representation of $\mathcal{D}$-formulas yields some immediate but nevertheless very important applications. They are all based on model theoretic properties of first order logic, which we now review.

### Compactness Theorem of first order logic

Suppose $T$ is an arbitrary set of sentences of first order logic such that every finite subset of $T$ has a model. Then $T$ itself has a model. There are many different proofs of this classical result due to Gödel and Mal'cev. The proof is somewhat easier in the case that $T$ is countable. One main line of proof [15] constructs a sufficiently complete extension $T^*$ of $T$ in a vocabulary which has infinitely many new constant symbols. After this a syntactical (term-) model is constructed by means of $T^*$. In another line of proof a model of $T$ is constructed by gluing together the models of the finite parts of $T$ into one structure which models all of $T$. This is the ultraproduct approach [9].

### Löwenheim–Skolem Theorem of first order logic

Suppose $\phi$ is a sentence of first order logic such that $\phi$ has an infinite model or arbitrarily large finite models. Then $\phi$ has models of all infinite cardinalities [36], [37], [39]. This is one of the oldest results of model theory of first order logic. The proof combines the method of Skolem functions and compactness.

### Craig Interpolation Theorem of first order logic

Suppose $\phi$ and $\psi$ are sentences of first order logic such that $\models \phi \rightarrow \psi$. Suppose the vocabulary of $\phi$ is $L_\phi$ and that of $\psi$ is $L_\psi$. Then there is a first order sentence $\theta$ of vocabulary $L_\phi \cap L_\psi$ such that $\models \phi \rightarrow \theta$ and $\models \theta \rightarrow \psi$. There are several different proofs of this result. The original proof of William Craig [6] was proof theoretic, based on cut-elimination. Most model theoretic proofs, starting with ref. [34], use compactness in one form or another.

We can now easily derive similar results for dependence logic by appealing to the $\Sigma_1^1$-representation of $\mathcal{D}$-sentences.

**Theorem 6.4 (Compactness Theorem of $\mathcal{D}$)** *Suppose $\Gamma$ is an arbitrary set of sentences of dependence logic such that every finite subset of $\Gamma$ has a model. Then $\Gamma$ itself has a model.*

*Proof* Let $\Gamma = \{\phi_i : i \in I\}$ and let $L$ be the vocabulary of $\Gamma$. Let $\tau_{1,\phi_i} = \exists S_1^i \ldots \exists S_{n_i}^i \psi_i$, where $\psi_i$ is first order. By changing symbols we can assume that all $S_j^i$ are different symbols. Let $T$ be the first order theory $\{\psi_i : i \in I\}$ in the vocabulary $L' = L \cup \{S_j^i : i \in I, 1 \le j \le n_i\}$. Every finite subset of $T$ has a model. By the Compactness Theorem of first order logic, there is an $L'$-structure $\mathcal{M}'$ that is a model of the theory $T$ itself. The reduction $\mathcal{M} = \mathcal{M}'{\restriction}L$ of $\mathcal{M}'$ to the original vocabulary $L$ is, by definition, a model of $\Gamma$. $\qquad\square$

**Theorem 6.5 (Löwenheim–Skolem Theorem of $\mathcal{D}$)** *Suppose $\phi$ is a sentence of dependence logic, such that $\phi$ either has an infinite model or has arbitrarily large finite models. Then $\phi$ has models of all infinite cardinalities, in particular $\phi$ has a countable model and an uncountable model.*

*Proof* Let $\tau_{1,\phi} = \exists S_1 \ldots \exists S_n \psi$, where $\psi$ is first order in the vocabulary $L' = L \cup \{S_1, \ldots, S_n\}$. Suppose $\kappa$ is an arbitrary infinite cardinal number. By the Löwenheim–Skolem Theorem of first order logic, there is an $L'$-model $\mathcal{M}'$ of $\psi$ of cardinality $\kappa$. The reduction $\mathcal{M} = \mathcal{M}'{\restriction}L$ of $\mathcal{M}'$ to the original vocabulary $L$ is a model of $\phi$ of cardinality $\kappa$. $\qquad\square$

**Corollary 6.6** (ref. [31]) *A sentence of dependence logic in the empty vocabulary is true in one infinite model (or arbitrarily large finite ones) if and only it is true in every infinite model.*

*Proof* All models of the empty vocabulary of the same cardinality are isomorphic. Thus the claim follows from Theorem 6.5. $\qquad\square$

We shall address the Craig Interpolation Theorem in Corollary 6.17 and derive first a Separation Theorem which is an equivalent formulation in the case of first order logic.

**Theorem 6.7 (Separation Theorem)** *Suppose $\phi$ and $\psi$ are sentences of dependence logic such that $\phi$ and $\psi$ have no models in common. Let the vocabulary of $\phi$ be $L$ and the vocabulary of $\psi$ be $L'$. Then there is a sentence $\theta$ of $\mathcal{D}$ in the vocabulary $L \cap L'$ such that every model of $\phi$ is a model of $\theta$, but $\theta$ and $\psi$ have no models in common. In fact, $\theta$ can be chosen to be first order.*

*Proof* Let $\tau_{1,\phi} = \exists S_1 \ldots \exists S_n \phi_0$, where $\phi_0$ is first order in the vocabulary $L_0$. Let $\tau_{1,\psi} = \exists S'_1 \ldots \exists S'_m \psi_0$, where $\psi_0$ is first order in the vocabulary $L'_0$. Without loss of generality, $\{S_1, \ldots, S_n\} \cap \{S'_1, \ldots, S'_m\} = \emptyset$. Note that $\models \phi_0 \rightarrow \neg\psi_0$ for if $\mathcal{M}$ is a model of $\phi_0 \wedge \psi_0$, then $\mathcal{M}{\restriction}L \models \phi \wedge \psi$, contrary to the assumption that $\phi$ and $\psi$ have no models in common. By the Craig Interpolation Theorem for first order logic, there is a first order sentence $\theta$ of vocabulary $L \cap L'$ such that $\models \phi_0 \rightarrow \theta$ and $\models \theta \rightarrow \neg\psi_0$. Every model of $\phi$ is a model of $\theta$, but $\theta$ and $\psi$ have no models in common.                                                                    □

A particularly striking application of Theorem 6.7 is the following special case in which $\phi$ and $\psi$ not only have no models in common but, furthermore, every model satisfies one of them.

**Theorem 6.8 (Failure of the Law of Excluded Middle)** *Suppose $\phi$ and $\psi$ are sentences of dependence logic such that for all models $\mathcal{M}$ we have*

$$\mathcal{M} \models \phi \text{ if and only if } \mathcal{M} \not\models \psi.$$

*Then $\phi$ is logically equivalent to a first order sentence $\theta$ such that $\psi$ is logically equivalent to $\neg\theta$.*

*Proof* The first order $\theta$ obtained in the proof of Theorem 6.7 is the $\theta$ we seek.                                                                    □

Note that it is perfectly possible to have for all *finite* models $\mathcal{M}$

$$\mathcal{M} \models \phi \text{ if and only if } \mathcal{M} \not\models \psi$$

without $\phi$ or $\psi$ being logically equivalent to a first order sentence. For example, in the empty vocabulary $\phi$ can say the size of the universe is even while $\psi$ says it is odd.

**Definition 6.9** *A sentence $\phi$ of dependence logic is called* determined in $\mathcal{M}$ *if $\mathcal{M} \models \phi$ or $\mathcal{M} \models \neg\phi$. Otherwise $\phi$ is called* non-determined in $\mathcal{M}$. *We say that $\phi$ is* determined *if $\phi$ is determined in every structure.*

A typical non-determined sentence (from ref. [2]) is given by

$$\forall x_0 \exists x_1 (=(x_1) \wedge x_0 = x_1),$$

which is non-determined in every structure with at least two elements. The following corollary shows that it is not at all difficult to find other non-determined sentences.

**Corollary 6.10** *Every determined sentence of dependence logic is strongly logically equivalent to a first order sentence.*

*Proof* Suppose $\phi$ is determined. Thus for all $\mathcal{M}$ we have $\mathcal{M} \models \phi$ or $\mathcal{M} \models \neg\phi$. It follows that for all $\mathcal{M}$ we have $\mathcal{M} \models \phi$ if and only if $\mathcal{M} \not\models \neg\phi$. By Theorem 6.8 there is a first order $\theta$ such that $\phi$ is logically equivalent to $\theta$ and $\neg\phi$ is logically equivalent to $\neg\theta$. Thus $\phi$ is strongly logically equivalent to $\theta$.      □

Thus we can take any sentence of dependence logic, which is not strongly equivalent to a first order sentence, and we know that there are models in which the sentence is non-determined.

**Example 6.11** *The sentence* $\Phi_{\mathrm{wf}}$ *is non-determined in every infinite well ordered structure. This can be seen either by a direct argument based on the truth definition, or by the following indirect argument. Suppose* $\mathcal{M}$ *is an infinite well ordered linear order in which* $\Phi_{\mathrm{wf}}$ *is determined. Thus* $\mathcal{M} \models \neg\Phi_{\mathrm{wf}}$. *Let* $\Gamma$ *be the following set of sentences of dependence logic:*

$$\neg\Phi_{\mathrm{wf}},$$
$$c_1 < c_0,$$
$$c_2 < c_1,$$
$$\dots$$
$$c_{n+1} < c_n,$$
$$\dots$$

*It is evident that every finite subset of* $\Gamma$ *is true in an expansion of* $\mathcal{M}$. *By the Compactness Theorem, there is a model* $\mathcal{M}'$ *of the whole* $\Gamma$. *Then* $\mathcal{M}'$ *is ill-founded and therefore satisfies* $\Phi_{\mathrm{wf}}$. *This contradicts the fact that* $\mathcal{M}'$ *also satisfies* $\neg\Phi_{\mathrm{wf}}$.

For more examples of non-determinacy, see ref. [42].

**Exercise 6.8** *Show that the sentence* $\Phi_{\mathrm{wf}}$ *is non-determined in all sufficiently big finite linear orders.*

**Exercise 6.9** *Show that* $\Phi_{\mathrm{even}}$ *is non-determined in every sufficiently large finite model of odd size.*

**Exercise 6.10** *Show that* $\Phi_\infty$ *is non-determined in every sufficiently big finite model.*

**Exercise 6.11** *Show that* $\Phi_{\mathrm{cmpl}}$ *is non-determined in every complete dense linear order.*

**Exercise 6.12** *Suppose* $\phi_n$, $n \in \mathbb{N}$, *are sentences of* $\mathcal{D}$ *such that each* $\phi_n$ *is true in some model, and moreover* $\phi_{n+1} \Rightarrow \phi_n$ *for all n. Show that there is one model* $\mathcal{M}$ *in which each* $\phi_n$ *is true.*

**Exercise 6.13** *Show that if $\phi$ is a sentence of $\mathcal{D}$ and $\psi$ is a first order sentence such that every countable[1] model of $\phi$ is a model of $\psi$, then every model of $\phi$ is a model of $\psi$.*

**Exercise 6.14** *Give a sentence $\phi$ of $\mathcal{D}$ and a first order sentence $\psi$ such that every countable model of $\psi$ is a model of $\phi$ and vice versa, but some model of $\psi$ is not a model of $\phi$.*

# 6.3 From $\Sigma_1^1$ to $\mathcal{D}$

We have seen that representing formulas of dependence logic in $\Sigma_1^1$ form is a powerful method for getting model theoretic results about dependence logic. We now show that this method is in a sense the best possible. Namely, we can also translate any $\Sigma_1^1$-sentence back to dependence logic.

We prove first a fundamental property of first order and $\Sigma_1^1$-formulas. Its various formulations all carry the name of Thoralf Skolem [36] (see also ref. [37]), who proved the following result in 1920. The basic idea is that the existential second order quantifiers in front of $\Sigma_1^1$-formulas are so powerful that they subsume all other existential quantifiers.

**Theorem 6.12 (Skolem Normal Form Theorem)** *Every $\Sigma_1^1$-formula $\phi$ is logically equivalent to an existential second order formula*

$$\exists f_1 \ldots \exists f_n \forall x_1 \ldots \forall x_m \psi, \tag{6.2}$$

*where $\psi$ is quantifier free and $f_1, \ldots, f_n$ are function symbols. The formula in Eq. (6.2) is called a Skolem Normal Form of $\phi$.*

*Proof* For atomic and negated atomic $\phi$ we choose $\phi$ itself as a Skolem Normal Form. Suppose then $\phi_0$ has a Skolem Normal Form,

$$\exists f_1^0 \ldots \exists f_{n_0}^0 \forall x_1^0 \ldots \forall x_{m_0}^0 \psi_0, \tag{6.3}$$

and $\phi_1$ has a Skolem Normal Form,

$$\exists f_1^1 \ldots \exists f_{n_1}^1 \forall x_1^1 \ldots \forall x_{m_1}^1 \psi_1. \tag{6.4}$$

By changing bound variables, we may assume, w.l.o.g., that $\{f_1^0, \ldots, f_{n_0}^0\} \cap \{f_1^1, \ldots, f_{n_1}^1\} = \emptyset$ and that $\{x_1^0, \ldots, x_{m_0}^0\} \cap \{x_1^1, \ldots, x_{m_1}^1\} = \emptyset$. Now we obtain a Skolem Normal Form,

$$\exists f_1^0 \ldots \exists f_{n_0}^0 \exists f_1^1 \ldots \exists f_{n_1}^1 \forall x_1^0 \ldots \forall x_{m_0}^0 \forall x_1^1 \ldots \forall x_{m_1}^1 (\psi_0 \wedge \psi_1),$$

---

[1] i.e. countably infinite or finite.

for $\phi_0 \wedge \phi_1$ and a Skolem Normal Form,

$$\exists f_1^0 \ldots \exists f_{n_0}^0 \exists f_1^1 \ldots \exists f_{n_1}^1 \forall x_1^0 \ldots \forall x_{m_0}^0 \forall x_1^1 \ldots \forall x_{m_1}^1 (\psi_0 \vee \psi_1),$$

for $\phi_0 \vee \phi_1$. For a definition of a Skolem Normal Form for $\forall x_0 \phi$, suppose $\phi$ has a Skolem Normal Form as given in Eq. (6.2). Let $f_1', \ldots, f_n'$ be new function symbols such that $\#(f_i') = \#(f_i) + 1$ and let $x_0$ be a variable not occurring in $\psi$ so that also $x_0 \notin \{x_1, \ldots, x_n\}$. Let $\psi'$ be obtained from $\psi$ by inductively replacing everywhere $f_i(t_1, \ldots, t_{\#(f_i)})$ by $f_i'(x_0, t_1, \ldots, t_{\#(f_i)})$. We define

$$\exists f_1' \ldots \exists f_n' \forall x_0 \forall x_1 \ldots \forall x_m \psi'$$

to be the Skolem Normal Form of $\forall x_0 \phi$. Let $f_0$ be a new 0-place function variable and let $\psi''$ be obtained from $\psi$ by replacing everywhere $x_0$ by $f_0$. We let

$$\exists f_0 \exists f_1 \ldots \exists f_n \forall x_1 \ldots \forall x_m \psi''$$

be a Skolem Normal Form of $\exists x_0 \phi$. For the definition of a Skolem Normal Form for $\exists R \phi$, suppose $\phi$ has the Skolem Normal Form given in Eq. (6.2). A Skolem Normal Form of $\exists R \phi$ is simply

$$\exists f_1 \ldots \exists f_n \exists f_{n+1} \exists f_{n+2} \forall x_1 \ldots \forall x_m \theta,$$

where $\theta$ is obtained from $\psi$ by replacing everywhere $R t_1 \ldots t_k$ by the formula $f_{n+1} t_1 \ldots t_k = f_{n+2}$. Finally, the construction of a Skolem Normal Form of $\exists f \phi$ is easy. Suppose $\phi$ has the Skolem Normal Form as in Eq. (6.2). We obtain the Skolem Normal Form

$$\exists f \exists f_1 \ldots \exists f_n \forall x_1 \ldots \forall x_m \psi$$

for $\exists f \phi$. $\qquad \square$

**Example 6.13** *The sentence*

$$\forall x_1 \exists x_2 R x_1 x_2$$

*in Skolem Normal Form reads as follows:*

$$\exists f \forall x_1 R x f x_1,$$

*and*

$$\forall x_1 (\exists x_2 P x_1 x_2 \vee \forall x_3 \exists x_4 R x_1 x_3 x_4)$$

*reads as follows:*

$$\exists f_1 \exists f_2 \forall x_1 \forall x_3 (P x_1 f_1 x_1 \vee R x_1 x_3 f_2 x_1 x_3).$$

**Corollary 6.14 (Skolem Normal Form Theorem for $\mathcal{D}$)** *For every formula $\phi$ of dependence logic and every $d \in \{0, 1\}$, there is a $\Sigma_1^1$-sentence $\tau_{d,\phi}^*(S)$ of the following form:*

$$\exists f_1 \ldots \exists f_n \forall x_1 \ldots \forall x_m \psi, \tag{6.5}$$

*where $\psi$ is quantifier-free, such that the following are equivalent:*

(i) $(\phi, X, d) \in \mathcal{T}$,
(ii) $(\mathcal{M}, X) \models \tau_{d,\phi}^*(S)$.

*In particular, for every sentence $\phi$ of dependence logic there are $\Sigma_1^1$-sentences $\tau_{1,\phi}^*$ and $\tau_{0,\phi}^*$ of the form (6.5) such that for all models $\mathcal{M}$ we have*

$$\mathcal{M} \models \phi \text{ if and only if } \mathcal{M} \models \tau_{1,\phi}^*.$$

$$\mathcal{M} \models \neg\phi \text{ if and only if } \mathcal{M} \models \tau_{0,\phi}^*.$$

Corollary 6.14 gives an easy proof of the Löwenheim–Skolem Theorem for dependence logic (Theorem 6.5). Namely, suppose $\phi$ is a given sentence of dependence logic with an infinite model or arbitrarily large finite models. By compactness we may assume $\phi$ indeed has an infinite model $\mathcal{M}$. Let $\tau_{1,\phi}^*$ be of the form $\exists f_1 \ldots \exists f_n \forall x_1 \ldots \forall x_m \psi$. Thus there are interpretations $f_i^{\mathcal{M}'}$ of the function symbols $f_i$ in an expansion of $\mathcal{M}'$ of $\mathcal{M}$ such that $\mathcal{M}'$ satisfies $\forall x_1 \ldots \forall x_m \psi$. Let $N$ be a countable subset of $M$ such that $N$ is closed under all the $n$ functions $f_i^{\mathcal{M}'}$. Because $\forall x_1 \ldots \forall x_m \psi$ is universal, it is still true in the countable substructure $\mathcal{M}^*$ of $\mathcal{M}$ generated by $N$. Thus $\mathcal{M}^*$ is a countable model of $\phi$. This is in line with the original proof of Skolem. If we wanted a model of size $\kappa$ for a given infinite cardinal number $\kappa$, the argument would be similar, but we would first use compactness to get a model of size at least $\kappa$.

**Theorem 6.15** (*refs. [7] and [44]*) *For every $\Sigma_1^1$-sentence $\phi$ there is a sentence $\phi^*$ in dependence logic such that for all $\mathcal{M}$:*

$$\mathcal{M} \models \phi \iff \mathcal{M} \models \phi^*.$$

*Proof* We may assume $\phi$ is of the following form:

$$\Phi = \exists f_1 \ldots \exists f_n \forall x_1 \ldots \forall x_m \psi, \tag{6.6}$$

where $\psi$ is quantifier-free. We will perform some reductions on Eq. (6.6) in order to make it more suitable for the construction of $\phi$.

Step (1) If $\psi$ contains nesting of the function symbols $f_1, \ldots, f_n$ or of the function symbols of the vocabulary, we can remove them one by one by using the equivalence of

$$\models \psi(f_i t_1 \ldots t_m)$$

and

$$\forall x_1 \ldots \forall x_m ((t_1 = x_1 \wedge \ldots \wedge t_m = x_m) \rightarrow \psi(f_i x_1 \ldots x_m)).$$

Thus we may assume that all terms occurring in $\psi$ are of the form $x_i$ or $f_i x_{i_1} \ldots x_{i_k}$.

Step (2) If $\psi$ contains an occurrence of a function symbol $f_i x_{i_1} \ldots x_{i_k}$ with the same variable occurring twice, e.g. $i_s = i_r, 1 < r < k$, we can remove it by means of a new variable $x_l$ and the equivalence

$$\models \forall x_1 \ldots \forall x_m \psi(f_i x_{i_1} \ldots x_{i_k}) \leftrightarrow$$
$$\forall x_1 \ldots \forall x_m \forall x_l (x_l = x_r \rightarrow \psi(f_i x_{i_1} \ldots x_{i_{r-1}}, x_l, x_{i_{r+1}} \ldots x_{i_k})).$$

Thus we may assume that if a term such as $f_i x_{i_1} \ldots x_{i_k}$ occurs in $\psi$, its variables are all distinct.

Step (3) If $\psi$ contains two occurrences of the same function symbol but with different variables or with the same variables in different order, we can remove it using appropriate equivalences. If $\{i_1, \ldots, i_k\} \cap \{j_1, \ldots, j_k\} = \emptyset$, we have the equivalence

$$\models \forall x_1 \ldots \forall x_m \psi(f_i x_{i_1} \ldots x_{i_k}, f_i x_{j_1} \ldots x_{j_k})$$
$$\leftrightarrow \exists f_i' \forall x_1 \ldots \forall x_m (\psi(f_i x_{i_1} \ldots x_{i_k}, f_i' x_{j_1} \ldots x_{j_k})$$
$$\wedge ((x_{i_1} = x_{j_1} \wedge \ldots \wedge x_{i_k} = x_{j_k})$$
$$\rightarrow f_i x_{i_1} \ldots x_{i_k} = f_i' x_{j_1} \ldots x_{j_k})).$$

We can reduce the more general case, where $\{i_1, \ldots, i_k\} \cap \{j_1, \ldots, j_k\} \neq \emptyset$, to this case by introducing new variables, as in Step (2). (We are grateful to Ville Nurmi for pointing out the necessity of this.) Thus we may assume that for each function symbol $f_i$ occurring in $\psi$, there are $j_1^i, \ldots, j_{n_i}^i$ such that *all* occurrences of $f_i$ are of the form $f_i x_{j_1^i} \ldots x_{j_{m_i}^i}$ and $j_1^i, \ldots, j_{m_i}^i$ are all different from each other.

In sum, we may assume the function terms that occur in $\psi$ are of the form $f_i x_{j_1^i} \ldots x_{j_{m_i}^i}$ and for each $i$ the variables $x_{j_1^i}, \ldots, x_{j_{m_i}^i}$ and their order is the same. Let $N$ be greater than all the $x_{j_k^i}$. Following the notation of Eq. (6.6), let $\phi^*$ be the sentence

$$\forall x_1 \ldots \forall x_m \exists x_{N+1} \ldots \exists x_{N+n} \ (=(x_{j_1^1}, \ldots, x_{j_{m_1}^1}, x_{N+1}) \wedge$$

$$\cdots$$

$$(=(x_{j_1^n}, \ldots, x_{j_{m_n}^n}, x_{N+n}) \wedge \psi'),$$

where $\psi'$ is obtained from $\psi$ by replacing everywhere $f_i x_{j_1^i} \ldots x_{j_{m_i}^i}$ by $x_{N+i}$. This is clearly the desired sentence.

□

It is noteworthy that the $\mathcal{D}$-representation of a given $\Sigma_1^1$-sentence given above is of universal-existential form, i.e. of the form

$$\forall x_{n_1} \ldots \forall x_{n_k} \exists x_{m_1} \ldots \exists x_{m_l} \psi,$$

where $\psi$ is quantifier-free. Moreover, $\psi$ is just a conjunction of dependence statements $=(x_1, \ldots, x_n)$ and a quantifier-free first order formula. This is a powerful *normal form* for dependence logic.

**Corollary 6.16** *For any sentence $\phi$ of dependence logic of vocabulary L and for every $L' \subseteq L$ there is a sentence $\phi'$ of dependence logic of vocabulary $L'$ such that the following are equivalent:*

(i) $\mathcal{M} \models \phi'$;
(ii) *there is an expansion $\mathcal{N}$ of $\mathcal{M}$ such that $\mathcal{N} \models \phi$.*

*Proof* Let $\psi$ be a $\Sigma_1^1$-sentence logically equivalent with $\phi$. We assume for simplicity that $L \setminus L'$ consists of just one predicate symbol R. Let $\phi'$ be a sentence of dependence logic logically equivalent with the $\Sigma_1^1$-sentence $\exists R \psi$. Then $\phi'$ is a sentence satisfying the equivalence of the conditions (i) and (ii). □

Corollary 6.16 implies in a trivial way the following strong form of the Craig Interpolation Theorem.

**Corollary 6.17 (uniform interpolation property)** *Suppose $\phi$ is a sentence of $\mathcal{D}$. Let L be the vocabulary of $\phi$. For every $L' \subseteq L$ there is a sentence $\phi'$ of $\mathcal{D}$ in the vocabulary $L'$ which is a* uniform interpolant *of $\phi$ in the following sense: $\phi \Rightarrow \phi'$, and, if $\psi$ is a sentence of $\mathcal{D}$ in a vocabulary $L''$ such that $\phi \Rightarrow \psi$ and $L \cap L'' = L'$, then $\phi' \Rightarrow \psi$.*

*Proof* Let $\phi'$ be as in Corollary 6.16. By its very definition, $\phi'$ is a logical consequence of $\phi$. Suppose then $\psi$ is a sentence of $\mathcal{D}$ in a vocabulary $L''$ such that $\phi \Rightarrow \psi$ and $L \cap L'' = L'$. If $\mathcal{M}''$ is an $L''$-structure which is a model of $\phi'$, then $\mathcal{M}'' \restriction L'$ is a model of $\phi'$, whence there is an expansion $\mathcal{M}$ of $\mathcal{M}'' \restriction L'$ to a model of $\phi$. Since $\phi \Rightarrow \psi$, $\mathcal{M}$ is a model of $\psi$. But $\mathcal{M} \restriction L'' = \mathcal{M}'' \restriction L''$. Thus $\mathcal{M}'' \models \psi$.

□

For a version of the *Beth Definability Theorem*, see Exercise 6.23.

**Exercise 6.15** *Give a Skolem Normal Form for the following first order formulas:*

(i) $\forall x_0 \exists x_1 x_0 = x_1$;

(ii) $\exists x_0 \forall x_1 \neg x_0 = x_1$;

(iii) $(\exists x_0 P x_0 \vee \forall x_0 \neg P x_0)$.

**Exercise 6.16** *Give a Skolem Normal Form for the following first order formula:*

$$\forall x_0 \exists x_1 \forall x_2 \exists x_3 ((x_0 = x_4 \rightarrow \neg(x_1 = x_4)) \wedge (x_0 < x_3 \leftrightarrow x_1 < x_2)).$$

**Exercise 6.17** *Write the following $\Sigma_1^1$-sentences in Skolem Normal Form:*

(i) $\exists x_0 \exists f \forall x_1 (\neg f x_1 = x_0 \wedge \forall x_2 (f x_0 = f x_1 \rightarrow x_0 = x_1))$;

(ii) $\exists R (\forall x_0 \forall x_1 \forall x_2 ((R x_0 x_1 \wedge R x_1 x_2) \rightarrow R x_0 x_2)$
$\wedge \forall x_0 \forall x_1 (R x_0 x_1 \vee R x_1 x_0 \vee x_0 = x_1)$
$\wedge \forall x_0 \neg R x_0 x_0 \wedge \forall x_0 \exists x_1 R x_0 x_1)$.

**Exercise 6.18** *Give a sentence of $\mathcal{D}$ which is logically equivalent to the following $\Sigma_1^1$-sentence in Skolem Normal Form:*

(i) $\exists f_0 \exists f_1 \forall x_0 \forall x_1 \phi(x_0, x_1, f_0(x_0, x_1), f_1(x_0, x_1))$;

(ii) $\exists f_0 \exists f_1 \forall x_0 \forall x_1 \phi(x_0, x_1, f_0(x_0, x_1), f_1(x_1))$;

(iii) $\exists f_0 \exists f_1 \forall x_0 \phi(x_0, f_0(x_0), f_1(x_0))$.

*In each case $\phi$ is quantifier-free and first order.*

**Exercise 6.19** *Give a sentence of $\mathcal{D}$ which is logically equivalent to the following $\Sigma_1^1$-sentence:*

$$\exists f \forall x_0 \forall x_1 \phi(x_0, x_1, f(x_0, x_1), f(x_1, x_0)),$$

*where $\phi$ is quantifier-free.*

**Exercise 6.20** *Express the Henkin quantifier*

$$\begin{pmatrix} \forall x & \exists y \\ \forall u & \exists v \end{pmatrix} R(x, y, u, v) \leftrightarrow \exists f \exists g \forall x \forall u R(x, f(x), u, g(u))$$

*in dependence logic.*

**Exercise 6.21** *Suppose $\mathcal{M}$ is an $L$-structure and $P \subseteq M^n$. We say that $P$ is $\mathcal{D}$-definable in $\mathcal{M}$ if there is a sentence $\phi(c_1, \ldots, c_n)$ of $\mathcal{D}$ with new constant symbols $c_1, \ldots, c_n$ such that the following are equivalent for all $a_1, \ldots, a_n \in \mathcal{M}$:*

(i) $(a_1, \ldots, a_n) \in P$;

(ii) $(\mathcal{M}, a_1, \ldots, a_n) \models \phi$,

where $(\mathcal{M}, a_1, \ldots, a_n)$ *denotes the expansion of* $\mathcal{M}$ *obtained by interpreting* $c_i$ *in* $(\mathcal{M}, a_1, \ldots, a_n)$ *by* $a_i$. *We then say that* $\phi$ *defines* $P$ *in* $\mathcal{M}$. *Show that if* $P$ *and* $Q$ *are* $\mathcal{D}$-*definable, then so are* $P \cap Q$ *and* $P \cup Q$, *but not necessarily* $P \setminus Q$.

**Exercise 6.22**   *Recall the definition of* $\mathcal{D}$-*definability in a model in Exercise 6.21. Let* $L$ *be a vocabulary. Suppose* $\psi$ *is a* $\mathcal{D}$-*sentence in a vocabulary* $L \cup \{R\}$, *where* $R$ *is a new n-ary predicate symbol. We say that* $R$ *is* $\mathcal{D}$-definable *in models of* $\psi$ *if there is a* $\mathcal{D}$-*sentence* $\phi$ *of vocabulary* $L \cup \{c_1, \ldots, c_n\}$ *such that* $\phi$ *defines* $R$ *in every model of* $\psi$. *Prove the following useful criterion for* $\mathcal{D}$-*undefinability: if* $\psi$ *has two models* $\mathcal{M}$ *and* $\mathcal{N}$ *such that* $\mathcal{M}{\upharpoonright}L = \mathcal{N}{\upharpoonright}L$ *but* $R^{\mathcal{M}} \neq R^{\mathcal{N}}$, *then* $R$ *is not* $\mathcal{D}$-*definable in models of* $\psi$. *(In the case of first order logic this is known as the* Padoa Principle.*)*

**Exercise 6.23**   *Recall the definition of* $\mathcal{D}$-*definability in models of a sentence in Exercise 6.22. Let* $L$ *be a vocabulary. Suppose* $\psi$ *is a* $\mathcal{D}$-*sentence in a vocabulary* $L \cup \{R\}$, *where* $R$ *is a new n-ary predicate symbol. Suppose any two models* $\mathcal{M}$ *and* $\mathcal{N}$ *of* $\psi$ *such that* $\mathcal{M}{\upharpoonright}L = \mathcal{N}{\upharpoonright}L$ *satisfy also* $R^{\mathcal{M}} = R^{\mathcal{N}}$. *Show that* $R$ *is* $\mathcal{D}$-*definable in models of* $\psi$ *(In the case of first order logic this is known as the* Beth Definability Theorem.*)*

**Exercise 6.24**   *(See ref. [3].) Suppose* $\phi$ *and* $\psi$ *are sentences of* $\mathcal{D}$ *such that no model satisfies both* $\phi$ *and* $\psi$. *Show that there is a sentence* $\theta$ *of* $\mathcal{D}$ *such that*

$$\mathcal{M} \models \phi \text{ if and only if } \mathcal{M} \models \theta$$

*and*

$$\mathcal{M} \models \psi \text{ if and only if } \mathcal{M} \models \neg\theta.$$

## 6.4  Truth definitions

In 1933, the Polish logician Alfred Tarski defined the concept of truth in a general setting (see, e.g., ref. [38]) and pointed out what is known as Tarski's Undefinability of Truth argument: no language can define its own truth, owing to the *Liar Paradox*, namely to the sentence

"This sentence is false."

This sentence is neither true nor false, contrary to the Law of Excluded Middle, which Tarski took for granted. By 1931, the Austrian logician Kurt Gödel [12], working not on arbitrary formalized languages but on first order number theory, had constructed, using a lengthy process referred to as the arithmetization of syntax, the following sentence:

"This sentence is unprovable."

This sentence cannot be provable, for then it would be true, and hence unprovable. So it is unprovable and hence true. Its negation cannot be provable either, for else the negation would be true. So it is an example of a sentence which is *independent* of first order number theory. This is known as Gödel's First Incompleteness Theorem. Gödel's technique could be used to make exact sense of undefinability of truth (see below) and to prove it exactly for first order number theory.

**Exercise 6.25** *Consider the following: "If this sentence is true, then its negation is true." Derive a contradiction.*

**Exercise 6.26** *Consider the following: "It is not true that this sentence is true." Derive a contradiction.*

**Exercise 6.27** *(See ref. [28].) Consider the following sentences.*

(1) *It is raining in Warsaw.*
(2) *It is raining in Vienna.*
(3) *Exactly one of the sentences (1)–(3) is true.*

*Under what kind of weather conditions in Europe is sentence (3) paradoxical?*

### 6.4.1 Undefinability of truth

Even to formulate the concept of definability of truth, we have to introduce a method for speaking about a formal language in the language itself. The clearest way of doing this is by means of Gödel-numbering. Each sentence $\phi$ is associated with a natural number $\ulcorner\phi\urcorner$, its *Gödel-number*, in a systematic way, described in Section 6.4.2. Moreover, we assume that our language has a name $\underline{n}$ for each natural number $n$.

**Definition 6.18** *A truth definition for any model $\mathcal{M}$ and any formal language $\mathcal{L}$, such as first order logic or dependence logic, is a formula $\tau(x_0)$ of some possibly other formal language $\mathcal{L}'$ such that for each sentence $\phi$ of $\mathcal{L}$ we have*

$$\mathcal{M} \models \phi \text{ if and only if } \mathcal{M} \models \tau(\ulcorner\phi\urcorner). \tag{6.7}$$

A stronger requirement would be $\mathcal{M} \models \phi \leftrightarrow \tau(\ulcorner\phi\urcorner)$, but this would be true only in the presence of the Law of Excluded Middle, as $(\phi \leftrightarrow \psi) \Rightarrow (\phi \vee \neg\phi)$.

An even stronger requirement would be the provability of $\phi \leftrightarrow \tau(\ulcorner \phi \urcorner)$ from some axioms, but we abandon this also in the current setup.

By the vocabulary $L_{\{+,\times\}}$ of arithmetic we mean a vocabulary appropriate for the study of number theory, with a symbol $N$ for the set of natural numbers. We specify $L_{\{+,\times\}}$ in detail below. We call an $L$-structure $\mathcal{M}_\omega$, where $L \supseteq L_{\{+,\times\}}$, "a model of Peano's axioms" if the reduct of $\mathcal{M}_\omega$ to the vocabulary $\{N, +, \times\}$ satisfies the first order Peano axioms of number theory. The results on definability of truth are relevant even if we assume that $N^{\mathcal{M}_\omega}$ is the whole universe of the model $\mathcal{M}_\omega$. Below, $\mathcal{M}_\omega$ denotes such a model of Peano's axioms.

**Theorem 6.19 (Gödel's Fixed Point Theorem)** *For any first order formula* $\phi(x_0)$, *in the vocabulary of arithmetic there is a first order sentence* $\psi$ *of the same vocabulary such that for all models* $\mathcal{M}_\omega$ *of Peano's axioms,*

$$\mathcal{M}_\omega \models \psi \text{ if and only if } \mathcal{M}_\omega \models \phi(\ulcorner \psi \urcorner).$$

*Proof* Let $Sub$ be the set of triples $\langle \ulcorner w \urcorner, \ulcorner w' \urcorner, n \rangle$, where $w'$ is obtained from $w$ by replacing $x_0$ by the term $\underline{n}$. Since recursive relations are representable in models of Peano's axioms, there is a first order formula $\sigma(x_0, x_1, x_2)$ such that

$$\langle n, m, k \rangle \in Sub \iff \mathcal{M}_\omega \models \sigma(\underline{n}, \underline{m}, \underline{k}).$$

W.l.o.g., $x_0$ is not bound in $\sigma(x_0, x_1, x_2)$ and $x_0$ and $x_1$ are not bound in $\phi(x_0)$. Let $\theta(x_0)$ be the formula $\exists x_1(\phi(x_1) \wedge \sigma(x_0, x_1, x_0))$. Let $k = \ulcorner \theta(x_0) \urcorner$ and $\psi = \theta(\underline{k})$. Then $\mathcal{M}_\omega \models \psi$ if and only if $\mathcal{M}_\omega \models \phi(\ulcorner \psi \urcorner)$. $\qquad\square$

The above result does not hold just for first order logic but for any extension of first order logic, the syntax of which is sufficiently effectively given, for example dependence logic.

**Theorem 6.20 (Tarski's Undefinability of Truth Result)** *First order logic does not have a truth definition in first order logic for any model* $\mathcal{M}_\omega$ *of Peano's axioms.*

*Proof* Let $\tau(x_0)$ be as in Definition 6.18. By Theorem 6.19 there is a sentence $\psi$ such that

$$\mathcal{M}_\omega \models \psi \text{ if and only if } \mathcal{M}_\omega \models \neg\tau(\ulcorner \psi \urcorner). \qquad (6.8)$$

If $\mathcal{M}_\omega \models \psi$, then $\mathcal{M}_\omega \models \tau(\ulcorner \psi \urcorner)$ by Eq. (6.7), and $\mathcal{M}_\omega \models \neg\tau(\ulcorner \psi \urcorner)$ by Eq. (6.8). Hence $\mathcal{M}_\omega \not\models \psi$. Now $\overline{\mathcal{M}_\omega \not\models \tau(\ulcorner \psi \urcorner)}$ by Eq. (6.7), and $\mathcal{M}_\omega \not\models \neg\tau(\ulcorner \psi \urcorner)$ by Eq. (6.8). So neither $\tau(\ulcorner \psi \urcorner)$ nor $\neg\tau(\ulcorner \psi \urcorner)$ is true in $\mathcal{M}_\omega$. This contradicts the Law of Excluded Middle, which says in this case $\mathcal{M}_\omega \models (\tau(\ulcorner \psi \urcorner) \vee \neg\tau(\ulcorner \psi \urcorner))$. $\qquad\square$

Theorem 6.20 has many stronger formulations. As the almost trivial proof above shows, no language for which the Gödel Fixed Point Theorem can be proved and which satisfies the Law of Excluded Middle for its negation can have a truth definition in the language itself. We shall not elaborate more on this point here, as our topic, dependence logic, certainly does not satisfy the Law of Excluded Middle for its negation.

**Exercise 6.28** *Prove* $\mathcal{M}_\omega \models \psi$ *if and only if* $\mathcal{M}_\omega \models \phi(\ulcorner\underline{\psi}\urcorner)$ *in the proof of Theorem 6.19.*

## 6.4.2 Definability of truth in first order logic

We turn to another important contribution of Tarski, namely that truth is *implicitly* (or even better – *inductively*) definable in first order logic. In dependence logic the implicit definition can even be turned into an explicit definition by means of Theorem 6.15, as emphasized by Hintikka [19]. So, after all, truth *is* definable, albeit only implicitly. The realization of this may be an even more important contribution of Tarski to logic than the undefinability of truth.

We shall carry out in some detail the definition of truth for first order logic. We shall omit many details, as these are well covered by the literature. For simplicity, we assume that the vocabulary $L_{\{+,\times\}}$ of arithmetic includes all the machinery needed for arithmetization. All that we really need is a pairing function, but the pursuit of such minimalism is not relevant for the main argument and belongs to other contexts. Another simplification is that we only consider truth in models $\mathcal{M}_\omega$ of Peano's axioms.

We consider a finite vocabulary $L = \{c_1, \ldots, c_n, R_1, \ldots, R_m, f_1, \ldots, f_k\}$ containing $L_{\{+,\times\}}$. When we specify in the following what $L_{\{+,\times\}}$ should contain we assume they are all among $c_1, \ldots, c_n, R_1, \ldots, R_m, f_1, \ldots, f_k$. Let $r_i = \#(R_i)$. If $w = w_0 \ldots w_k$ is a string of symbols in the following alphabet:

$$=, c_i, R_i, f_i, (,), \neg, \vee, \exists,$$

the Gödel-number $\ulcorner w \urcorner$ of $w$ is the natural number given by

$$\ulcorner w \urcorner = p_0^{\#(w_0)+1} \cdot \ldots \cdot p_k^{\#(w_k)+1},$$

where $p_0, p_1, \ldots$ are the prime numbers $2,3,5,\ldots$ in increasing order and

| | | | |
|---|---|---|---|
| $\#(=) = 0,$ | $\#(() = 1,$ | $\#()) = 2,$ | $\#(\neg) = 3,$ |
| $\#(\vee) = 4,$ | $\#(\wedge) = 5,$ | $\#(\exists) = 6,$ | $\#(\forall) = 7,$ |
| $\#(c_i) = 4 + 4i,$ | $\#(x_i) = 5 + 4i,$ | $\#(R_i) = 6 + 4i,$ | $\#(f_i) = 7 + 4i.$ |

Table 6.1.

| Symbol | Interpretation in $\mathcal{M}_\omega$ |
|---|---|
| POS-ID$x_0x_1x_2$ | $x_0$ is $\ulcorner t = t' \urcorner$, where $x_1 = \ulcorner t \urcorner$ and $x_2 = \ulcorner t' \urcorner$ |
| NEG-ID$x_0x_1x_2$ | $x_0$ is $\ulcorner \neg t = t' \urcorner$, where $x_1 = \ulcorner t \urcorner$ and $x_2 = \ulcorner t' \urcorner$ |
| POS-ATOM$_i x_0 x_1 \ldots x_{r_i}$ | $x_0$ is $\ulcorner R_i t_1 \ldots t_{r_i} \urcorner$, where $x_1 = \ulcorner t_1 \urcorner, \ldots, x_{r_i} = \ulcorner t_{r_i} \urcorner$ |
| NEG-ATOM$_i x_0 x_1 \ldots x_{r_i}$ | $x_0$ is $\ulcorner \neg R_i t_1 \ldots t_{r_i} \urcorner$, where $x_1 = \ulcorner t_1 \urcorner, \ldots, x_{r_i} = \ulcorner t_{r_i} \urcorner$ |
| CONJ$x_0x_1x_2$ | $x_0$ is $\ulcorner (\phi \wedge \psi) \urcorner$, where $x_1 = \ulcorner \phi \urcorner$ and $x_2 = \ulcorner \psi \urcorner$ |
| DISJ$x_0x_1x_2$ | $x_0$ is $\ulcorner (\phi \vee \psi) \urcorner$, where $x_1 = \ulcorner \phi \urcorner$ and $x_2 = \ulcorner \psi \urcorner$ |
| EXI$x_0x_1x_2$ | $x_0$ is $\ulcorner \exists x_n \phi \urcorner$, where $x_1 = n$ and $x_2 = \ulcorner \phi \urcorner$ |
| UNI$x_0x_1x_2$ | $x_0$ is $\ulcorner \forall x_n \phi \urcorner$, where $x_1 = n$ and $x_2 = \ulcorner \phi \urcorner$ |

The vocabulary $L_{\{+,\times\}}$ has a symbol $\underline{0}$ for zero, a symbol $\underline{1}$ for one, and the names $\underline{n}$ of the other natural numbers $n$ are defined inductively as terms $+\underline{n}\underline{1}$. We assume $L_{\{+,\times\}}$ uses the symbols listed in Table 6.1 to represent syntactic operations:

We assume that among the symbols of $L_{\{+,\times\}}$ are functions that provide a bijection between elements of the model and finite sequences of elements of the model.[2] Thus it makes sense to interpret arbitrary elements of $\mathcal{M}_\omega$ as assignments. We also assume that $L_{\{+,\times\}}$ has the symbols given in Table 6.2 (we think of $x_0$ as an assignment).

All the above symbols are easily definable in terms of $+$ and $\cdot$ in first order logic in any model $\mathcal{M}_\omega$ of Peano's axioms, if wanted. Now we take a new predicate symbol SAT, not to be included in $L_{\{+,\times\}}$ (and not to be definable in terms of $+$ and $\cdot$ in first order logic) with the following intuitive meaning:

SAT$x_0x_1$    $x_0$ is an assignment $s$ and $x_1$ is $\ulcorner \phi \urcorner$ for some
$L$-formula $\phi$ such that $\mathcal{M}_\omega \models_s \phi$.

---

[2] In the special case that $N^{\mathcal{M}_\omega}$ (i.e. $\mathbb{N}$) is the whole universe of $\mathcal{M}_\omega$, the encoding of finite sequences of elements of the model by elements of the model can be achieved by means of the unique factorization of integers, or alternatively by means of the Chinese Remainder Theorem.

Table 6.2.

| Symbol | Interpretation in $\mathcal{M}_\omega$ |
|---|---|
| TRUE-ID$x_0 x_1 x_2$ | $x_0$ satisfies the identity $t = t'$, where $x_1 = \ulcorner t \urcorner$ and $x_2 = \ulcorner t' \urcorner$ |
| FALSE-ID$x_0 x_1 x_2$ | $x_0$ satisfies the non-identity $\neg t = t'$, where $x_1 = \ulcorner t \urcorner$ and $x_2 = \ulcorner t' \urcorner$ |
| TRUE-ATOM$_i x_0 x_1 \ldots x_{r_i}$ | $x_0$ satisfies $R_i t_1 \ldots t_{r_i}$, where $x_1 = \ulcorner t_1 \urcorner, \ldots, x_{r_i} = \ulcorner t_{r_i} \urcorner$ |
| FALSE-ATOM$_i x_0 x_1 \ldots x_{r_i}$ | $x_0$ satisfies $\neg R_i t_1 \ldots t_{r_i}$, where $x_1 = \ulcorner t_1 \urcorner, \ldots, x_{r_i} = \ulcorner t_{r_i} \urcorner$ |
| AGR$x_0 x_1 x_2$ | $x_0$ and $x_2$ are assignments that agree about variables other than $x_1$ |

The point is that SAT is (implicitly) definable in terms of the others by the first order sentence $\theta_L$ as follows:

$$\forall x_0 \forall x_1 (\text{SAT}x_0 x_1$$
$$\leftrightarrow \exists x_2 \exists x_3 (\text{POS-ID}x_1 x_2 x_3 \wedge \text{TRUE-ID}x_0 x_2 x_3)$$
$$\vee \exists x_2 \exists x_3 (\text{NEG-ID}x_1 x_2 x_3 \wedge \text{FALSE-ID}x_0 x_2 x_3)$$
$$\vee \exists x_2 \ldots \exists x_{r_1+1} (\text{POS-ATOM}_1 x_1 x_2 \ldots x_{r_1+1} \wedge \text{TRUE-ATOM}_1 x_0 x_2 \ldots x_{r_1+1})$$
$$\vee \ldots$$
$$\exists x_2 \ldots \exists x_{r_m+1} (\text{POS-ATOM}_m x_1 x_2 \ldots x_{r_m+1} \wedge \text{TRUE-ATOM}_m x_0 x_2 \ldots x_{r_m+1})$$
$$\vee \exists x_2 \ldots \exists x_{r_1+1} (\text{NEG-ATOM}_1 x_1 x_2 \ldots x_{r_1+1} \wedge \text{FALSE-ATOM}_1 x_0 x_2 \ldots x_{r_1+1})$$
$$\vee \ldots$$
$$\exists x_2 \ldots \exists x_{r_m+1} (\text{NEG-ATOM}_m x_1 x_2 \ldots x_{r_m+1} \wedge \text{FALSE-ATOM}_m x_0 x_2 \ldots x_{r_m+1})$$
$$\vee \exists x_2 \exists x_3 (\text{CONJ}x_1 x_2 x_3 \wedge (\text{SAT}x_0 x_2 \wedge \text{SAT}x_0 x_3))$$
$$\vee \exists x_2 \exists x_3 (\text{DISJ}x_1 x_2 x_3 \wedge (\text{SAT}x_0 x_2 \vee \text{SAT}x_0 x_3))$$
$$\vee \exists x_2 \exists x_3 (\text{EXI}x_1 x_2 x_3 \wedge \exists x_4 (\text{AGR}x_0 x_2 x_4 \wedge \text{SAT}x_4 x_3))$$
$$\vee \exists x_2 \exists x_3 (\text{UNI}x_1 x_2 x_3 \wedge \forall x_4 (\text{AGR}x_0 x_2 x_4 \rightarrow \text{SAT}x_4 x_3))).$$

The implicit (or inductive) nature of this definition (known as Tarski's Truth Definition) manifests itself in the fact that SAT appears on both sides of the equivalence sign in $\theta_L$, and only positively in each case. There is no guarantee that $\theta$ really fixes what SAT is (see, e.g., ref. [26]). However, for the part of the actual formulas, or their representatives in $\mathcal{M}_\omega$, the set SAT is unique.

**Theorem 6.21** *If the L-structure $\mathcal{M}_\omega$ is a model of Peano's axioms, then:*

(i) $(\mathcal{M}_\omega, Sat_\mathbb{N}) \models \theta_L$, *where $Sat_\mathbb{N}$ is the set of pairs $\langle s, \ulcorner \phi \urcorner^{\mathcal{M}_\omega} \rangle$ such that $\mathcal{M}_\omega \models_s \phi$;*

(ii) *if $(\mathcal{M}_\omega, S) \models \theta_L$ and $(\mathcal{M}_\omega, S') \models \theta_L$, then $S \cap Sat_\mathbb{N} = S' \cap Sat_\mathbb{N}$.*

*Proof* Claim (i) is tedious but trivial, assuming that our concepts are correctly defined. The claim we prove is the second one. To this end, suppose $(\mathcal{M}_\omega, S) \models \theta_L$ and $(\mathcal{M}_\omega, S') \models \theta_L$. We prove

$$\langle s, \ulcorner \phi \urcorner^{\mathcal{M}_\omega} \rangle \in S \text{ if and only if } \mathcal{M}_\omega \models_s \phi \qquad (6.9)$$

for all $\phi$. Since the same holds for $S'$ by symmetry, we obtain the desired result.

Case (1) $\phi$ is of the form $t = t'$, $\neg t = t'$, $R_i t_1 \ldots t_n$ or $\neg R_i t_1 \ldots t_n$. The claim in Eq. (6.9) follows from our interpretation of POS-ID, NEG-ID, POS-ATOM, NEG-ATOM, TRUE-ID, FALSE-ID, TRUE-ATOM$_i$, and FALSE-ATOM$_i$ in $\mathcal{M}_\omega$.

Case (2) $\phi$ is of the form $(\phi_0 \wedge \phi_1)$. The claim in Eq. (6.9) follows from the conjunction part of the definition of $\theta_L$ and our interpretation of CONJ in $\mathcal{M}_\omega$.

Case (3) $\phi$ is of the form $(\phi_0 \vee \phi_1)$. The claim in Eq. (6.9) follows from the disjunction part of the definition of $\theta_L$ and our interpretation of DISJ in $\mathcal{M}_\omega$.

Case (4) $\phi$ is of the form $\exists x_n \psi$. The claim in Eq. (6.9) follows from the part of the existential quantifier of the definition of $\theta_L$ and our interpretation of AGR and EXI in $\mathcal{M}_\omega$.

Case (5) $\phi$ is of the form $\forall x_n \psi$. The claim in Eq. (6.9) follows from the part of the universal quantifier of the definition of $\theta_L$ and our interpretation of AGR and UNI in $\mathcal{M}_\omega$.                                    □

By means of the satisfaction relation SAT we can define truth by means of the following formula:

$$\text{TRUE}(x_0) = \exists x_1 \text{SAT} x_1 x_0.$$

**Corollary 6.22** *The following are equivalent for all first order sentences $\phi$ in the vocabulary $L$ and any model $\mathcal{M}_\omega$ of Peano's axiom:*

(i) $\mathcal{M}_\omega \models \phi$;

(ii) $(\mathcal{M}_\omega, S) \models (\theta_L \wedge \text{TRUE}(\ulcorner \phi \urcorner))$ *for some* $S \subseteq \mathbb{N}^2$;

(iii) $(\mathcal{M}_\omega, S) \models (\theta_L \to \text{TRUE}(\ulcorner \phi \urcorner))$ *for all* $S \subseteq \mathbb{N}^2$.

Let us call a model $\mathcal{M}_\omega$ of Peano's axioms *standard* if the interpretation of the predicate $N$ of the vocabulary of arithmetic in $\mathcal{M}_\omega$ is the set of natural numbers, the interpretation of $+$ is the addition of natural numbers, and the interpretation of $\times$ is the multiplication of natural numbers.

**Corollary 6.23** *Suppose $\mathcal{M}_\omega$ is any standard model of Peano's axioms. If $(\mathcal{M}_\omega, S) \models \theta_L$ and $(\mathcal{M}_\omega, S') \models \theta_L$, then $S = S'$.*

First order definable relations on standard $\mathcal{M}_\omega$ are called *arithmetical*. The definition of truth given by Corollary 6.22 is not first order, so we cannot say truth is arithmetical. A definition that has both existential second order and universal second order definition, as truth in the above corollary, is called *hyperarithmetical*. So we can say that first order truth on $\mathcal{M}_\omega$ is not arithmetical but hyperarithmetical. For more on hyperarithmetical definitions, see ref. [33].

**Exercise 6.29** *Prove Corollary 6.22.*

**Exercise 6.30** *Prove Corollary 6.23.*

**Exercise 6.31** *Show that we cannot remove the word "standard" from Corollary 6.23.*

### 6.4.3 Definability of truth in $\mathcal{D}$

We now move from first order logic back to dependence logic. We observed in Corollary 6.22 that the truth definition of first order logic on a model $\mathcal{M}_\omega$ of Peano's axioms can be given in first order logic if one existential second order quantifier is allowed. In dependence logic we can express the existential second order quantifier, and thus the truth definition of first order logic on $\mathcal{M}_\omega$ can be given in dependence logic. This can be extended to a truth definition of all of dependence logic, and this is our goal in this section.

The fact that $\Sigma_1^1$ has a truth definition in $\Sigma_1^1$ on a structure with enough coding is well known in descriptive set theory. It was observed in ref. [19] that, as an application of Theorem 6.15, we have a truth definition for $\mathcal{D}$ in $\mathcal{D}$ on any structure with enough coding.

We shall consider below a vocabulary $L \supseteq L_{\{+,\times\}} \cup \{c\}$, where $c$ is a new constant symbol. If $\phi$ is a sentence of $\mathcal{D}$ in this vocabulary, we indicate the inclusion of a new constant by writing $\phi$ as $\phi(c)$. Then, if $d$ is another constant symbol, $\phi(d)$ is the sentence obtained from $\phi(c)$ by replacing $c$ everywhere by $d$.

**Theorem 6.24** *(ref. [19]) Suppose $L \supseteq L_{\{+,\times\}} \cup \{c\}$ is finite. There is a sentence $\tau(c)$ of $\mathcal{D}$ in the vocabulary $L$ such that for all sentences $\phi$ of $\mathcal{D}$ in the vocabulary $L$ and all models $\mathcal{M}_\omega$ of Peano's axioms:*

$$\mathcal{M}_\omega \models \phi \text{ if and only if } \mathcal{M}_\omega \models \tau(\ulcorner\phi\urcorner).$$

*Proof* By Theorem 6.2, every sentence of $\mathcal{D}$ is logically equivalent to a $\Sigma_1^1$-sentence of following the form:

$$\exists R_1 \ldots \exists R_n \phi^*, \tag{6.10}$$

where $\phi^*$ is first order. We now replace the second order quantifiers $\exists R_1 \ldots \exists R_n$ by just one second order quantifier $\exists R_0$, where $R_0$ is a unary predicate symbol not in $L$. At the same time we replace every occurrence of $R_i t_1 \ldots t_n$ in $\phi^*$ by $R_0\langle i, t_1, \ldots, t_n\rangle$, where $(a_1, \ldots, a_n) \mapsto \langle a_1, \ldots, a_n\rangle$ is the function in $L_{\{+,\times\}}$ coding $n$-sequences. Let the result be $\phi^{**}$. Now the following are equivalent for any $\phi$ in $\mathcal{D}$ in vocabulary $L$ and any $L$-structure $\mathcal{M}_\omega$, which is a model of Peano's axioms:

(i) $\mathcal{M}_\omega \models \phi$.
(ii) $(\mathcal{M}_\omega, Z) \models \phi^{**}$ for some $Z \subseteq \mathbb{N}$.

Let $L' = L \cup \{R_0\}$. By Corollary 6.22, we obtain the equivalence of (ii) with

(iii) $(\mathcal{M}_\omega, Z, Sat) \models (\theta_{L'} \wedge \mathrm{TRUE}(\ulcorner \phi^{**} \urcorner))$ for some $Sat \subseteq \mathbb{N}^2$ and some $Z \subseteq \mathbb{N}$.

By Corollary 6.16, there is a sentence $\tau_0(c)$ of vocabulary $L$ such that (iii) is equivalent to

(iv) $\mathcal{M}_\omega \models \tau_0(\ulcorner \phi^{**} \urcorner)$.

Let $t(x_0)$ be a term of the vocabulary $L_{\{+,\times\}}$ such that for all sentences $\phi$ of $\mathcal{D}$ of vocabulary $L$ the value of $t(\ulcorner \phi \urcorner)$ in any $L_{\{+,\times\}}$-model of Peano's axioms is $\ulcorner \phi^{**} \urcorner$. Let $\tau(c)$ be the $L$-sentence $\tau_0(t(c))$. Then (i)–(iv) are equivalent to

(v) $\mathcal{M}_\omega \models \tau(\ulcorner \phi \urcorner)$. $\qquad\qquad\qquad\square$

Now that we have constructed the truth definition by recourse to $\Sigma_1^1$-sentences, it should be pointed out that we could write $\tau(c)$ also directly in $\mathcal{D}$, imitating the inductive definition of truth given in Definition 3.5. This is the approach taken in a forthcoming publication by Hodges and Väänänen.[3]

Let us now go back to the Liar Paradox. By Theorem 6.19, there is a sentence $\lambda$ of $\mathcal{D}$ in the vocabulary $L_{\{+,\times\}}$ such that for all $\mathcal{M}_\omega$

$$\mathcal{M}_\omega \models \lambda \text{ if and only if } \mathcal{M}_\omega \models \neg\tau(\ulcorner \lambda \urcorner).$$

Intuitively, $\lambda$ says "This sentence is not true." By Theorem 6.24,

$$\mathcal{M}_\omega \models \lambda \text{ if and only if } \mathcal{M}_\omega \models \tau(\ulcorner \lambda \urcorner).$$

Thus

$$\mathcal{M}_\omega \models \tau(\ulcorner \lambda \urcorner) \text{ if and only if } \mathcal{M}_\omega \models \neg\tau(\ulcorner \lambda \urcorner),$$

[3] W. Hodges and J. Väänänen, Dependence of variables construed as an atomic formula; to appear.

which is, of course, only possible if

$$\mathcal{M}_\omega \not\models \tau(\ulcorner\underline{\lambda}\urcorner) \text{ and } \mathcal{M}_\omega \not\models \neg\tau(\ulcorner\underline{\lambda}\urcorner).$$

Still another way of putting this is:

$$\mathcal{M}_\omega \not\models \tau(\ulcorner\underline{\lambda}\urcorner) \vee \neg\tau(\ulcorner\underline{\lambda}\urcorner).$$

Thus we get the pleasing result that the assertion that the Liar sentence is true (in the sense of $\tau(c)$) is non-determined. This is in harmony with the intuition that the Liar sentence does not have a truth value.

**Exercise 6.32** *Suppose*

$$\mathcal{M}_\omega \models \psi_0 \text{ if and only if } \mathcal{M}_\omega \models \tau(\ulcorner\underline{\psi_1}\urcorner)$$

*and*

$$\mathcal{M}_\omega \models \psi_1 \text{ if and only if } \mathcal{M}_\omega \models \neg\tau(\ulcorner\underline{\psi_0}\urcorner).$$

*Show that* $\tau(\ulcorner\underline{\psi_0}\urcorner)$ *and* $\tau(\ulcorner\underline{\psi_1}\urcorner)$ *are non-determined.*

**Exercise 6.33** *Contemplation of the sentence "This sentence is not true" leads immediately to the paradox ($\star$): the sentence is true if and only if it is not true. We have seen that we can write a sentence $\lambda$ with the meaning "This sentence is not true" in $\mathcal{D}$. Still we do not get the result that $(\lambda \leftrightarrow \neg\lambda)$ is true. Indeed, show that $(\phi \leftrightarrow \neg\phi)$ has no models, whatever $\phi$ in $\mathcal{D}$ is. Explain why the existence of $\lambda$ does not lead to the paradox ($\star$).*

**Exercise 6.34** *Suppose $\psi$ says "If $\psi$ is true, then $\neg\psi$ is true," i.e.*

$$\mathcal{M}_\omega \models \psi \text{ if and only if } \mathcal{M}_\omega \models (\tau(\ulcorner\underline{\psi}\urcorner) \to \tau(\ulcorner\underline{\neg\psi}\urcorner)).$$

*Show that $\psi$ is non-determined.*

**Exercise 6.35** *Suppose $\psi$ says "It is not true that $\psi$ is true," i.e.*

$$\mathcal{M}_\omega \models \psi \text{ if and only if } \mathcal{M}_\omega \models \neg\tau(\ulcorner\tau(\ulcorner\underline{\psi}\urcorner)\urcorner).$$

*Show that $\tau(\ulcorner\tau(\ulcorner\underline{\psi}\urcorner)\urcorner)$ is non-determined.*

**Exercise 6.36** *Show that there cannot be $\tau'(c)$ in $\mathcal{D}$ such that for all $\phi$ and all $\mathcal{M}_\omega$ we have*

$$\mathcal{M}_\omega \not\models \phi \text{ if and only if } \mathcal{M}_\omega \models \tau'(\ulcorner\phi\urcorner).$$

**Exercise 6.37** *(See ref. [35].) Suppose $\mathcal{M}_\omega$ is a model of Peano's axioms, as above. Let $T$ be the set of sentences of $\mathcal{D}$ that are true in $\mathcal{M}_\omega$. Show that there* *is a model $\mathcal{M}$ of $T$ such that for some $a \in M$ we have*

$$\mathcal{M} \models (\tau(a) \wedge \tau(\neg a)).$$

## 6.5 Model existence game

In this section we learn a new game associated with trying to construct a model for a sentence or a set of sentences. The new game, called the model existence game, is like $H(\phi)$ except that there is no model present. So the game is about an imagined model. In particular, player **II** claims she has a model for $\phi$ but she does not tell player **I** what it is. It turns out that if she has a winning strategy, then there actually is a model for $\phi$ and we can build it from the winning strategy.

### 6.5.1 First order case

We first treat the case of first order logic. After this we shall easily extend the method to all of dependence logic.

In the model existence game we talk about semantics without actually having any model. Let us take a countably infinite set $C$ of new constant symbols. Intuitively the new constants $c$ are elements of the imaginary model.

**Definition 6.25** *The* model existence game *MEG $(T, L)$ of the set $T$ of first order $L$-sentences is defined as follows. Let $C$ be a countably infinite set of new constant symbols. Let $L' = L \cup C$. There are two players, **I** and **II**. We denote players by $\alpha$, and the opponent of $\alpha$ by $\alpha^*$. A position of the game is a pair $(\phi, \alpha)$, where $\phi$ is an $L'$-sentence and $\alpha \in \{\mathbf{I}, \mathbf{II}\}$. At the beginning of the game the position is $(\top, \mathbf{II})$. The rules of the game are as follows.*

| | |
|---|---|
| (1) Theory move | *Player **I** chooses $\phi \in T$ and the game continues from the position $(\phi, \mathbf{II})$.* |
| (2) Identity move | *Player **I** chooses a constant $L'$-term $t$. The game continues from the position $(t = t, \mathbf{II})$.* |
| (3) Substitution move | *Player **I** chooses previous positions $(\phi(t), \alpha)$ and $(t = t', \mathbf{II})$, where $\phi(t)$ is atomic, and then the game continues from the position $(\phi(t'), \alpha)$.* |
| (4) Negation move | *Player **I** chooses a previous position $(\neg\phi, \alpha)$. Then the game continues from the position $(\phi, \alpha^*)$.* |
| (5) Disjunction move | *Player **I** chooses a previous position $(\phi_0 \vee \phi_1, \alpha)$. Then player $\alpha$ decides whether the game continues from the position $(\phi_0, \alpha)$ or from the position $(\phi_1, \alpha)$.* |

Table 6.3.

| I | II | Rule |
|---|---|---|
| | $\exists x_0 \neg (Px_0 \vee \neg Pfx_0)$ | (1) |
| | $\neg(Pc \vee \neg Pfc)$ | (6) |
| $Pc \vee \neg Pfc$ | | (4) |
| $\neg Pfc$ | | (5) |
| | $Pfc$ | (4) |
| $Pc$ | | (5) |
| | $fc = c'$ | (7) |
| | $Pc'$ | (3) |
| | $fc' = c''$ | (7) |
| | $fc'' = c'''$ | (7) |
| | ... | |
| | **II wins** | |

(6) Existential move    *Player* **I** *chooses a previous position of the form* $(\exists x_n \phi(x_n), \alpha)$. *Player* $\alpha$ *chooses* $c \in C$ *and the game continues from the position* $(\phi(c), \alpha)$.

(7) Constant move    *Player* **I** *chooses a constant $L'$-term t. Then player* **II** *chooses* $c \in C$ *and the game continues from the position* $(t = c, \textbf{II})$.

*Player* **I** *wins if during the game both* $(\phi, \textbf{II})$ *and* $(\phi, \textbf{I})$ *occur as positions for some atomic $L'$-sentence* $\phi$. *Otherwise* **II** *wins*.

**Example 6.26** *Player* **II** *has a winning strategy in* $\mathrm{MEG}(\{\exists x_0 \neg (Px_0 \vee \neg Pfx_0)\}, \{P, f\})$. *Table 6.3 shows a play of the game, when* **II** *plays her winning strategy. The rules would allow* **I** *to prolong the game by repeating previous moves but* **II** *would always give the same responses, and in the end* **II** *wins anyway*.

**Example 6.27** *In this example,* **I** *has a winning strategy in* $\mathrm{MEG}(\{\neg \exists x_0 (Pd \vee \neg Px_0)\}, \{P, d\})$ *(see Table 6.4)*.

A winning strategy of **II** in $\mathrm{MEG}(T, L)$ can be conveniently presented in the form of a set of sets.

**Definition 6.28** *Let L be a countable vocabulary, let C be a countable set of new constant symbols, and let $L' = L \cup C$. A* consistency property *is any set* $\Delta$ *of finite sets S of pairs* $(\phi, \alpha)$, *where* $\phi$ *is a first order L-formula, and* $\alpha \in \{\textbf{I}, \textbf{II}\}$, *which satisfies the following conditions*.

Table 6.4.

| I | II | Rule |
|---|---|---|
| | $\neg \exists x_0(Pd \vee \neg Px_0)$ | (1) |
| $\exists x_0(Pd \vee \neg Px_0)$ | | (4) |
| | $d = c$ | (7) |
| $Pd \vee \neg Pc$ | | (6) |
| $Pd$ | | (5) |
| $Pc$ | | (3) |
| $\neg Pc$ | | (5) |
| | $Pc$ | (4) |
| **I has won** | | |

(i) *If $S \in \Delta$, then $S \cup \{(t = t, \mathbf{II})\} \in \Delta$ for every constant $L'$-term $t$.*

(ii) *If $(\phi(t), \alpha) \in S \in \Delta$, $\phi(t)$ atomic, and $(t = t', \mathbf{II}) \in S$, then $S \cup \{(\phi(t'), \alpha)\} \in \Delta$.*

(iii) *If $(\neg\phi, \alpha) \in S \in \Delta$, then $S \cup \{(\phi, \alpha^*)\} \in \Delta$.*

(iv) *If $(\phi \vee \psi, \mathbf{II}) \in S \in \Delta$, then $S \cup \{(\phi, \mathbf{II})\} \in \Delta$ or $S \cup \{(\psi, \mathbf{II})\} \in \Delta$.*

(v) *If $(\phi \vee \psi, \mathbf{I}) \in S \in \Delta$, then $S \cup \{(\phi, \mathbf{I})\} \in \Delta$ and $S \cup \{(\psi, \mathbf{I})\} \in \Delta$.*

(vi) *If $(\exists x_n \phi(x_n), \mathbf{II}) \in S \in \Delta$, then $S \cup \{(\phi(c), \mathbf{II})\} \in \Delta$ for some $c \in C$.*

(vii) *If $(\exists x_n \phi(x_n), \mathbf{I}) \in S \in \Delta$, then $S \cup \{(\phi(c), \mathbf{I})\} \in \Delta$ for all $c \in C$.*

(viii) *For every constant $L'$-term $t$ there is $c \in C$ such that $S \cup \{(t = c, \mathbf{II})\} \in \Delta$.*

(ix) *There is no atomic formula $\phi$ such that $(\phi, \mathbf{II}) \in S$ and $(\phi, \mathbf{I}) \in S$.*

*The consistency property $\Delta$ is a consistency property* for *a set $T$ of first order $L$-sentences if for all $S \in \Delta$ and all $\phi \in T$ we have $S \cup \{(\phi, \mathbf{II})\} \in \Delta$.*

**Lemma 6.29** *The following are equivalent.*

(i) *Player $\mathbf{II}$ has a winning strategy in* MEG$(T, L)$.

(ii) *There is a consistency property $\Delta$ for $T$.*

*Proof* Suppose $\sigma$ is a winning strategy of $\mathbf{II}$ in MEG$(T, L)$. Suppose the game MEG$(T, L)$ is played for $n$ rounds and $\mathbf{II}$ plays $\sigma$. A certain set $S$ of positions $(\phi, \alpha)$ is generated. Let $\Delta$ be the set of all possible such sets $S$, when $n \in \mathbb{N}$. Clearly $\Delta$ is a consistency property and for all $S \in \Delta$ and all $\phi \in T$ we have $S \cup \{(\phi, \mathbf{II})\} \in \Delta$. Conversely, if such a consistency property $\Delta$ existed, player $\mathbf{II}$ would win MEG$(T, L)$ by making sure the set of positions played is in $\Delta$. $\square$

An extreme case of a consistency property $\Delta$ is the situation in which all the finite sets in $\Delta$ are actually subsets of one and the same set $H$, and this set $H$ has nice closure properties (see below). Such sets are called Hintikka sets and they were introduced in ref. [18]. In many applications of the model existence game such a set is all we need. More refined proofs use a consistency property.

**Definition 6.30** *Let L be a let countable vocabulary, let C be a countable set of new constant symbols, and let $L' = L \cup C$. A* Hintikka set *is any set H of pairs $(\phi, \alpha)$, where $\phi$ is a first order L'-sentence and $\alpha \in \{I, II\}$, which satisfies the following conditions.*

   (i) $(t = t, II) \in H$ *for every constant L'-term t.*
  (ii) *If $(\phi(t), \alpha) \in H$, $\phi(t)$ atomic, and $(t = t', II) \in H$, then $(\phi(t'), \alpha) \in H$.*
 (iii) *If $(\neg\phi, \alpha) \in H$, then $(\phi, \alpha^*) \in H$.*
  (iv) *If $(\phi \vee \psi, II) \in H$, then $(\phi, II) \in H$ or $(\psi, II) \in H$.*
   (v) *If $(\phi \vee \psi, I) \in H$, then $(\phi, I) \in H$ and $(\psi, I) \in H$.*
  (vi) *If $(\exists x_n \phi(x_n), II) \in H$, then $(\phi(c), II) \in H$ for some $c \in C$*
 (vii) *If $(\exists x_n \phi, I) \in H$, then $(\phi(c), I) \in H$ for all $c \in C$*
(viii) *For every constant L'-term t and every $(\phi, \alpha) \in H$ there is $c \in C$ such that $(t = c, II) \in H$.*
  (ix) *There is no atomic sentence $\phi$ such that $(\phi, II) \in H$ and $(\phi, I) \in H$.*

*The Hintikka set H is a* Hintikka set for *a set T of first order L-sentences if $(\phi, II) \in H$ for all $\phi \in T$.*

Lemma 6.31 establishes the basic connection between consistency properties and Hintikka sets. Roughly speaking, Hintikka sets are unions of increasing chains in a consistency property, but this is not quite true: the increasing chains have to be constructed carefully.

**Lemma 6.31** *Let T be a set of first order L-sentences.*

 (i) *Suppose $\Delta$ is a consistency property for T. Then there is an increasing sequence of $S_n \in \Delta$ such that $H = \bigcup_n S_n$ is a Hintikka set for T.*
(ii) *If H is a Hintikka set for T, then $\Delta = \{S \subseteq H : S \text{ is finite}\}$ is a consistency property for T.*

*Proof* To prove the first claim, let $Trm$ be the set of all constant $L'$-terms. Let

$$T = \{\phi_n : n \in \mathbb{N}\};$$
$$C = \{c_n : n \in \mathbb{N}\};$$
$$Trm = \{t_n : n \in \mathbb{N}\}.$$

Let $\{p_n : n \in \mathbb{N}\}$ be a list of all elements of $\Delta$. We let $S_0 \in \Delta$ be arbitrary and define an increasing sequence $S_n \in \Delta$ as follows:

(i) If $n = 3^i$, then $S_{n+1}$ is $S_n \cup \{(\phi_i, \mathbf{II})\} \in \Delta$.

(ii) If $n = 2 \cdot 3^i$, then $S_{n+1}$ is $S_n \cup \{(t_i = t_i, \mathbf{II})\} \in \Delta$.

(iii) If $n = 4 \cdot 3^i \cdot 5^j$, $p_i = (t = t', \mathbf{II}) \in S_n$, and $p_j = (\phi(t), \alpha) \in S_n$ with $\phi(t)$ atomic, then $S_{n+1}$ is $S_n \cup \{(\phi(t'), \alpha)\} \in \Delta$.

(iv) If $n = 8 \cdot 3^i$ and $p_i = (\neg\psi, \alpha) \in S_n$, then $S_{n+1}$ is $S_n \cup \{(\psi, \alpha^*)\}$.

(v) If $n = 16 \cdot 3^i$ and $p_i = (\theta_0 \vee \theta_1, \mathbf{II}) \in S_n$, then $S_{n+1}$ is $S_n \cup \{(\theta_1, \mathbf{II})\}$ or $S_n \cup \{(\theta_1, \mathbf{II})\}$, whichever is in $\Delta$.

(vi) If $n = 32 \cdot 3^i \cdot 5^j$, $j \in \{0, 1\}$ and $p_i = (\theta_0 \vee \theta_1, \mathbf{I}) \in S_n$, then $S_{n+1}$ is $S_n \cup \{(\theta_j, \mathbf{I})\} \in \Delta$.

(vii) If $n = 64 \cdot 3^i$ and $p_i = (\exists x_k \phi, \mathbf{II}) \in S_n$, then $S_{n+1}$ is $S_n \cup \{(\phi(c), \mathbf{II})\}$ for such $c \in C$ that $S_{n+1} \in \Delta$.

(viii) If $n = 128 \cdot 3^i \cdot 5^j$ and $p_i = (\exists x_k \phi, \mathbf{I}) \in S_n$, then $S_{n+1}$ is $S_n \cup \{(\phi(c_j), \mathbf{I})\} \in \Delta$.

(ix) If $n = 256 \cdot 3^i$, then $S_{n+1}$ is $S_n \cup \{(t_i = c, \mathbf{II})\}$ for such $c \in C$ that $S_{n+1} \in \Delta$.

Clearly $\bigcup_n S_n$ is a Hintikka set for $T$.

The second claim is trivial.                                                  $\square$

Combining Lemmas 6.29 and 6.31 immediately yields the following.

**Corollary 6.32** *The following are equivalent.*

(i) *There is a Hintikka set H for T.*

(ii) *Player $\mathbf{II}$ has a winning strategy in* $\mathrm{MEG}(T, L)$

**Theorem 6.33 (Model Existence Theorem for first order logic)** *Suppose L is a countable vocabulary and T is a set of L-sentences. The following are equivalent.*

(i) *There is an L-structure $\mathcal{M}$ such that $\mathcal{M} \models T$.*

(ii) *Player II has a winning strategy in* $\mathrm{MEG}(T, L)$.

*Proof* Let us first assume such an $\mathcal{M}$ exists. The winning strategy of player II in $\mathrm{MEG}(T, L)$ is to maintain the condition $(\star)$ that if the position is $(\phi, \alpha)$, then $(\phi, \emptyset, \alpha) \in \mathcal{T}$. Player $\mathbf{II}$ defines interpretations of the constants $c \in C$ in $\mathcal{M}$ during the game, step by step, as soon as they appear.

(1) Position $(\phi, \emptyset, \mathbf{II})$, where $\phi \in T$. Then $(\phi, \emptyset, \mathbf{II}) \in \mathcal{T}$, since $\mathcal{M}$ is a model of $T$.

(2) Position $(t = t, \mathbf{II})$. Of course, $\mathcal{M} \models t = t$, if the constants of $t$ are interpreted in $\mathcal{M}$. If they are not, we can interpret them in an arbitrary way.

(3) Player **I** has chosen previous positions $(\phi(t), \alpha)$ and $(t = t', \mathbf{II})$, and then the game has continued to the position $(\phi(t'), \alpha)$. By the induction hypothesis, $\mathcal{M} \models t = t'$. Thus $(\phi(t), \emptyset, \alpha) \in \mathcal{T}$ implies $(\phi(t'), \emptyset, \alpha) \in \mathcal{T}$.

(4) Position $(\neg\phi, \alpha)$. Since $(\neg\phi, \emptyset, \alpha) \in \mathcal{T}$, we have $(\phi, \emptyset, \alpha^*) \in \mathcal{T}$. Thus **II** can play this move according to her plan.

(5) Position $(\phi \vee \psi, \alpha)$. We know $(\phi \vee \psi, \emptyset, \alpha) \in \mathcal{T}$. Let us first assume $\alpha = \mathbf{II}$. By the definition of $\mathcal{T}$, $(\phi, \emptyset, \mathbf{II})$ or $(\psi, \emptyset, \mathbf{II})$ is in $\mathcal{T}$. Thus player **II** can choose one of $(\phi, \mathbf{II})$, $(\psi, \mathbf{II})$, and the respective triple $(\phi, \emptyset, \mathbf{II})$ or $(\psi, \emptyset, \mathbf{II})$ is in $\mathcal{T}$. Condition $(\star)$ remains valid. Let us then assume $\alpha = \mathbf{I}$. Whichever choice player **I** makes in position $(\phi \vee \psi, \mathbf{I})$, condition $(\star)$ remains valid, as both $(\phi, \emptyset, \mathbf{I})$ and $(\psi, \emptyset, \mathbf{I})$ are in $\mathcal{T}$.

(6) Position $(\exists x_n \phi(x_n), \alpha)$. We know $(\exists x_n \phi(x_n), \emptyset, \alpha) \in \mathcal{T}$. Let us first assume $\alpha = \mathbf{II}$. By the definition of $\mathcal{T}$, there is $a \in M$ such that $(\phi, \{(n, a)\}, \mathbf{II})$ is in $\mathcal{T}$. Let us choose a new constant symbol $c$ and interpret it in $\mathcal{M}$ as $a$. Now the game can proceed to $(\phi(c), \mathbf{II})$ and condition $(\star)$ remains valid. Suppose then $\alpha = \mathbf{I}$ and **I** chooses $c \in C$. Now $(\phi, \{(n, c^{\mathcal{M}})\}, \alpha) \in \mathcal{T}$, whatever $c^{\mathcal{M}} \in M$ is. If $c^{\mathcal{M}}$ is defined already, we are done, otherwise we let $c^{\mathcal{M}}$ be an arbitrary[4] element of $M$. Thus condition $(\star)$ remains valid.

(7) Player **I** chooses a constant $L'$-term $t$. If some constants of $t$ are not interpreted yet in $\mathcal{M}$, we interpret them in an arbitrary way (it does not matter which value they are given). Then $a = t^{\mathcal{M}}$ is defined. Now player **II** chooses any new $c \in C$ and the game continues to the position $(t = c, \mathbf{II})$. We let $c^{\mathcal{M}} = a$, and condition $(\star)$ remains valid.

Let us then observe that if **II** plays according to this plan, she wins, for if she ends up playing both $(\psi, \mathbf{II})$ and $(\psi, \mathbf{I})$ for some atomic $L'$-sentence $\psi$, then $\{(\psi, \emptyset, \mathbf{II}) \in \mathcal{T}$ and $(\psi, \emptyset, \mathbf{I})\} \in \mathcal{T}$, a contradiction.

On the other hand, suppose player **II** has a winning strategy in the game $\mathrm{MEG}(T, L)$. By Corollary 6.32, there is a Hintikka set $H$ such that $(\phi, \mathbf{II}) \in H$ for all $\phi \in T$. Define $c \sim c'$ if $(c = c', \mathbf{II}) \in H$. The relation $\sim$ is an equivalence relation on $C$. Let us define an $L$-structure $\mathcal{M}$ as follows. We let $M = \{[c] : c \in C\}$. If $d \in L$ is a constant symbol, we choose $c \in C$ such that $(d = c, \mathbf{II}) \in H$ and define $d^{\mathcal{M}} = [c]$. If $f \in L$ and $\#(f) = n$ we let $f^{\mathcal{M}}([c_{i_1}], \ldots, [c_{i_n}]) = [c]$ for some $c$ such that $(fc_{i_1} \ldots c_{i_n} = c, \mathbf{II}) \in H$. For any term $t$ and any $u$ there is a $c \in C$ such that $(t = c, \mathbf{II}) \in H$. It is easy to see that $t^{\mathcal{M}} = [c]$. For a relation symbol $R$ we let $\langle [c_1], \ldots, [c_n] \rangle \in R^{\mathcal{M}}$ if and only if $(Rc_1 \ldots c_n, \mathbf{II})$ is in $H$.

---

[4] To be precise, we can choose once and for all an element $a_0$ of $M$ always to be used at this step.

We prove by induction on $\phi$ that if $\phi$ is an $L$-formula then

$$(\star\star) \qquad \text{If } (\phi, \alpha) \in H, \text{then}(\phi, \emptyset, \alpha) \in \mathcal{T}.$$

(i) $\phi$ is an equation $t = t'$. If $(t = t', \mathbf{II}) \in H$, then there are $c$ and $c'$ in $C$ such that $(t = c, \mathbf{II}), (t' = c) \in H$, and hence $t^{\mathcal{M}} = [c] = [c'] = (t')^{\mathcal{M}}$. Thus $(t = t', \emptyset, \mathbf{II}) \in \mathcal{T}$. Similarly, if $(t = t', \mathbf{I}) \in H$, then $(t = t', \mathbf{II}) \notin H$, whence $[c] \neq [c']$, and $(t = t', \emptyset, \mathbf{I}) \in \mathcal{T}$ follows.

(ii) $\phi$ is an atomic sentence $Rt_1 \ldots t_n$. If $(Rt_1 \ldots t_n, \mathbf{II}) \in H$, then there are $c_1, \ldots, c_n$ in $C$ such that $(t_i = c_i, \mathbf{II}) \in H$ for $i = 1, \ldots, n$, whence $(Rc_1 \ldots c_n, \mathbf{II}) \in H$, and hence by the definition of $\mathcal{M}$, $([c_1], \ldots, [c_n]) \in R^{\mathcal{M}}$. Thus $(Rt_1 \ldots t_n, \emptyset, \mathbf{II}) \in \mathcal{T}$. If $(Rt_1 \ldots t_n, \mathbf{I}) \in H$, then the pair $(Rc_1 \ldots c_n, \mathbf{II})$ is not in $H$, and hence by the definition of $\mathcal{M}$, $([c_1], \ldots, [c_n]) \notin R^{\mathcal{M}}$. Thus $(Rt_1 \ldots t_n, \emptyset, \mathbf{I}) \in \mathcal{T}$.

(iii) Negation: if $(\neg\phi, \alpha) \in H$, then $(\phi, \alpha^*) \in H$, hence by the induction hypothesis $(\phi, \emptyset, \alpha^*) \in \mathcal{T}$, i.e. $(\phi, \emptyset, \alpha) \in \mathcal{T}$.

(iv) Disjunction: if $(\phi \vee \psi, \mathbf{II}) \in H$, then $(\phi, \mathbf{II}) \in H$ or $(\psi, \mathbf{II}) \in H$, hence by the induction hypothesis $(\phi, \emptyset, \mathbf{II}) \in \mathcal{T}$ or $(\psi, \emptyset, \mathbf{II}) \in \mathcal{T}$, i.e. $(\phi \vee \psi, \emptyset, \mathbf{II}) \in \mathcal{T}$. If $(\phi \vee \psi, \mathbf{I}) \in H$, then $(\phi, \mathbf{I}) \in H$ and $(\psi, \mathbf{I}) \in H$, hence by the induction hypothesis $(\phi, \emptyset, \mathbf{I}) \in \mathcal{T}$ and $(\psi, \emptyset, \mathbf{I}) \in \mathcal{T}$, i.e. $(\phi \vee \psi, \emptyset, \mathbf{I}) \in \mathcal{T}$.

(v) Quantifier: if $(\exists x_n \phi(x_n), \mathbf{II}) \in H$, there is a $c \in C$ such that the pair $(\phi(c), \mathbf{II})$ is in $H$. By the induction hypothesis $(\phi(c), \emptyset, \mathbf{II}) \in \mathcal{T}$. Thus the triple $(\phi, \{(n, [c])\}, \mathbf{II})$ is in $\mathcal{T}$, whence the triple $(\exists x_n \phi(x_n), \emptyset, \mathbf{II})$ is in $\mathcal{T}$. If the pair $(\exists x_n \phi(x_n), \mathbf{I})$ is in $H$, and $[c] \in M$ is arbitrary, then $(\phi(c), \mathbf{I}) \in H$. By the induction hypothesis $(\phi(c), \emptyset, \mathbf{I}) \in \mathcal{T}$. Thus $(\phi, \{(n, [c])\}, \mathbf{I}) \in \mathcal{T}$, whence $(\exists x_n \phi(x_n), \emptyset, \mathbf{I}) \in \mathcal{T}$.

In particular, since $H$ is a Hintikka set for $T$, $\mathcal{M} \models T$.                           $\square$

**Corollary 6.34** *Suppose $L$ is a countable vocabulary, $T$ is a set of $L$-sentences, and $\phi$ is any $L$-sentence. Then the following are equivalent:*

(i) $T \models \phi$;
(ii) *player I has a winning strategy in* MEG$(T \cup \{\neg\phi\}, L)$.

*Proof* The game MEG$(T \cup \{\neg\phi\}, L)$ is determined. So, by Theorem 6.33, condition (ii) is equivalent to $T \cup \{\neg\phi\}$ not having a model, which is exactly what condition 1 says (remember that $\phi$ is assumed to be first order).         $\square$

In first order logic, condition (i) of Corollary 6.34 is equivalent to $\phi$ having a *formal proof* from $T$ (see, e.g., ref. [8] for a definition of formal proof). We can think of a winning strategy of player **I** in MEG$(T \cup \{\neg\phi\}, L)$ as a *semantic* *proof* of $\phi$ from $T$. In the literature this concept, or its close relatives, occur under the name of *semantic tree* or *Beth tableaux*.

The model existence theorem has many applications in first order logic. We mention, in particular, the Compactness Theorem, the Omitting Types Theorem and the Interpolation Theorem.

### 6.5.2 Model existence game for Skolem Normal Form

Suppose $\phi$ is a $\Sigma_1^1$-sentence in Skolem normal form,

$$\exists f_1 \ldots \exists f_n \forall x_1 \ldots \forall x_m \psi, \tag{6.11}$$

where $\psi$ is quantifier-free and $f_1, \ldots, f_n$ are function symbols not in $L$. We denote the extension of $L$ by the various new function symbols $f_1, \ldots, f_n$ occurring in the formulas in Eq. (6.11) by $L^*$. The model existence game that we now define for $\Sigma_1^1$ is shockingly simple in that its positions involve quantifier-free sentences only.

**Definition 6.35** *The* model existence game $\mathrm{MEG}_{\Sigma_1^1}(T, L)$ *of the set $T$ of $\Sigma_1^1$-sentences in Skolem normal form of the vocabulary $L$ is defined as follows. Let $C$ be a countably infinite set of new constant symbols. Let $L' = L^* \cup C$. A position of the game is a pair $(\phi, \alpha)$, where $\phi$ is a quantifier-free first order $L'$-formula and $\alpha \in \{\mathbf{I}, \mathbf{II}\}$. At the beginning of the game, the position is $(\top, \mathbf{II})$. The rules of the game are as follows.*

(1) Theory move.       *Player **I** chooses*

$$\exists f_1 \ldots \exists f_n \forall x_{i_1} \ldots \forall x_{i_m} \psi(x_{i_1}, \ldots, x_{i_m}) \in T$$

                              *and $c_1, \ldots, c_n \in C$, and then the game continues from the position $(\psi(c_1, \ldots, c_n), \mathbf{II})$.*

(2) Identity move.       *Player **I** chooses an $L'$-term $t$. The game continues from the position $(t = t, \mathbf{II})$.*

(3) Substitution move. *Player **I** chooses previous positions $(\phi(t), \alpha)$, $\phi(t)$ atomic, and $(t = t', \mathbf{II})$, and then the game continues from the position $(\phi(t'), \alpha)$.*

(4) Negation move.       *Player **I** chooses a previous position $(\neg\phi, \alpha)$. Then the game continues from the position $(\phi, \alpha^*)$.*

(5) Disjunction move. *Player **I** chooses a previous position $(\phi_0 \vee \phi_1, \alpha)$. Then player $\alpha$ decides whether the game continues from the position $(\phi_0, \alpha)$ or from the position $(\phi_1, \alpha)$.*

(6) Constant move.     *Player* **I** *chooses an* $L'$-*term* $t$. *Then player* **II** *chooses* $c \in C$ *and the game continues from the position* $(t = c, \mathbf{II})$.

*Player* **I** *wins if at any point* **II** *has played* $(\phi, \mathbf{II})$ *and* $(\phi, \mathbf{I})$ *for some atomic* $L'$-*sentence* $\phi$. *Otherwise* **II** *wins.*

Note that **I** can play the Theory move several times during the game; indeed, he typically may play it infinitely many times.

The following result follows immediately from Theorem 6.33.

**Theorem 6.36 (Model Existence Theorem for Skolem Normal Form)** *Suppose* $L$ *is a countable vocabulary and* $T$ *is a set of* $\Sigma_1^1$-*sentences of vocabulary* $L$ *in Skolem Normal Form. The following are equivalent.*

 (i) *There is an* $L$-*structure* $\mathcal{M}$ *such that* $\mathcal{M} \models T$.
(ii) *Player II has a winning strategy in* $\mathrm{MEG}_{\Sigma_1^1}(T, L)$.

Since $\Sigma_1^1$ is not closed under negation, we do not get a concept of "semantic proof" for $\Sigma_1^1$ from $\mathrm{MEG}_{\Sigma_1^1}(T, L^*)$. It does not make sense to talk about $\mathrm{MEG}_{\Sigma_1^1}(T \cup \{\neg\phi\}, L^*)$, since we cannot in general write $\neg\phi$ in Skolem Normal Form, unless it is actually first order. Instead we can view Theorem 6.36 as a useful criterion of *consistency* of a set of $\Sigma_1^1$ sentences. So the model existence game provides $\Sigma_1^1$ a criterion for consistency but no criterion of logical consequence and no concept of proof.

We know from Theorem 6.15 that every sentence $\phi$ of dependence logic is logically equivalent to a $\Sigma_1^1$-sentence $\tau_{1,\phi}$. Moreover, also $\neg\phi$ is logically equivalent to a $\Sigma_1^1$-sentence $\tau_{0,\phi}$. If $T$ is a set of sentences of dependence logic, let $\tau_T$ be the set of all $\tau_{1,\phi}$, where $\phi \in T$. Thus we obtain from the above the following.

**Corollary 6.37** *Suppose* $L$ *is a countable vocabulary,* $T$ *is a set of sentences of dependence logic of vocabulary* $L$, *and* $\phi$ *is any sentence of dependence logic in vocabulary* $L$. *Then the following are equivalent.*

 (i) *No model of* $T$ *is a model of* $\neg\phi$.
(ii) *Player* **I** *has a winning strategy in* $\mathrm{MEG}_{\Sigma_1^1}(\tau_T \cup \{\tau_{0,\phi}\}, L)$.

Unfortunately, condition (i) in Corollary 6.37 is not equivalent to $T \models \phi$, as $\phi$ may be non-determined in some models of $T$. Certainly, if $T \models \phi$, condition (i) holds. So, condition (i) is a weak form of provability for $\mathcal{D}$. It would be strong enough to give $T \models \phi$ if $\phi$ was determined, but then $\phi$ would be first order. As it is, condition (ii) gives a complete criterion for deciding whether a first order sentence is a logical consequence of a set of sentences of $\mathcal{D}$.

**Exercise 6.38** *Describe a winning strategy of player* **II** *in* MEG$(T, L)$, *when.*

(i) $T = \{\forall x_0(Px_0 \vee Qx_0), \exists x_0 Px_0, \exists x_0 \neg Qx_0\}$;

(ii) $T = \{\forall x Rxfx, \neg \exists x \forall y \neg Rxy\}$.

**Exercise 6.39** *Describe a winning strategy of player* I *in the game* MEG$(T, L)$, *when*

(i) $T = \{\forall x_0(Px \to Qx), \exists x_0 Px_0, \neg \exists x_0 Qx_0\}$;

(ii) $T = \{\forall x Rxfx, \neg \forall x \exists y Rxy\}$.

**Exercise 6.40** *Let $L$ be let a countable vocabulary, let $C$ be a countable set of new constant symbols and $L' = L \cup C$. Let $\Delta$ be the set of finite sets $S$ of $L'$-sentences such that for each $S$ there is a model $\mathcal{M}_S$ in which every element is an interpretation of some constant $c \in C$, and which satisfies*

$$(\phi, \mathbf{II}) \in S \text{ implies } \mathcal{M}_S \models \phi, \text{ and } (\phi, \mathbf{I}) \in S \text{ implies } \mathcal{M}_S \models \neg\phi.$$

*Show that $\Delta$ is a consistency property.*

**Exercise 6.41** *Let $\mathcal{M}_\omega$ be a model of Peano's axioms. Let $T$ be the set of first order $L_{\{+, \times\}}$-sentences that are true in $\mathcal{M}_\omega$. Let $d$ be a new constant symbol and $L = L_{\{+, \times\}} \cup \{d\}$. Let $T'$ be the union of $T$ and the sentences $\underline{n} < d$ for all $n < \omega$. Show that* **II** *can win the game* MEG$(T', L)$ *with the following strategy: if the played positions are $(\phi_i, \alpha_i)$, $i = 1, \ldots, n$, then $\mathcal{M}_\omega$ has an expansion that is a model of*

$$\{\phi_i : \alpha_i = \mathbf{II}\} \cup \{\neg\phi_i : \alpha_i = \mathbf{I}\}.$$

**Exercise 6.42** *Prove in detail that $(i) \to (ii)$ in Lemma 6.29.*

**Exercise 6.43** *Prove in detail that $(ii) \to (i)$ in Lemma 6.29.*

**Exercise 6.44** *Suppose $H$ is a Hintikka set. Define $c \sim c'$ if $(c = c', \mathbf{II}) \in H$. Prove that $\sim$ is an equivalence relation on $C$.*

**Exercise 6.45** *Suppose $S_1$ and $S_2$ are sets of first order sentences. Let the vocabulary of $S_1$ be $L_1$ and let the vocabulary of $S_2$ be $L_2$. If $L \subseteq L_i$, we denote by $S_i \upharpoonright L$ the set of $\phi \in S_i$, the vocabulary of which is contained in $L$. We say that an $L_1 \cap L_2$-sentence $\theta$ separates $S_1$ and $S_2$ if $S_1 \models \theta$ and $S_2 \cup \{\theta\}$ has no models, and that $S_1$ and $S_2$ are separable if some first order $\theta$ separates them, otherwise they are inseparable.*

*Suppose $L_1$ and $L_2$ are vocabularies consisting (for simplicity) of relation symbols only. Suppose $\phi$ is a first order $L_1$-sentence and $\psi$ is a first order $L_2$-sentence. Let $C$ be a countably infinite set of new constant symbols. Let $\Delta$ be the set of all finite sets $S$ of pairs $(\theta, \alpha)$ satisfying the following conditions.*

(i) *If* $(\theta, \alpha) \in S$, *then* $\theta$ *is an* $L_1 \cup C$*-sentence or an* $L_2 \cup C$*-sentence and* $\alpha \in \{\mathbf{I}, \mathbf{II}\}$.

(ii) *If* $T(S) = \{\theta : (\theta, \mathbf{II}) \in S\} \cup \{\neg\theta : (\theta, \mathbf{I}) \in S\}$, *then* $T(S){\upharpoonright}(L_1 \cup C)$ *and* $T(S){\upharpoonright}(L_2 \cup C)$ *are inseparable.*

*Show that* $\Delta$ *is a consistency property if conditions (i) and (viii) of Definition 6.28 are modified to refer only to terms t that are either* $L_1 \cup C$*-terms or* $L_2 \cup C$*-terms. (This exercise can be used to prove the Separation Theorem (Theorem 6.7).)*

**Exercise 6.46** *Let T be a set of sentences of* $\mathcal{D}$ *of a countable vocabulary L such that every finite subset of T has a model. Describe, without appealing to the Compactness Theorem (Theorem 6.4), a winning strategy of* **II** *in* $\mathrm{MEG}_{\Sigma_1^1}(\tau_T, L)$. *This, combined with the Model Existence Theorem (Theorem 6.36) gives a proof of the Compactness Theorem (Theorem 6.4).*

**Exercise 6.47** *Give sentences* $\phi$ *and* $\psi$ *of* $\mathcal{D}$ *such that* $\phi$ *and* $\psi$ *have no finite models in common, and, if we denote the vocabulary of* $\phi$ *by L and the vocabulary of* $\psi$ *by* $L'$, *then there is no first order sentence* $\theta$ *of vocabulary* $L \cap L'$ *such that every finite model of* $\phi$ *is a model of* $\theta$, *and* $\theta$ *and* $\psi$ *have no finite models in common.*

**Exercise 6.48** *Suppose we have a countable vocabulary L and T is a set of first order L-sentences. A type of T is a countable sequence* $\psi_0(x_0), \psi_1(x_0), \ldots$ *of formulas with just one free variable x such that for all n:* $T \cup \{\exists x(\psi_0(x) \wedge \ldots \wedge \psi_n(x))\}$ *has a model. A model* $\mathcal{M}$ *of T satisfies a type p if some element a of M satisfies* $\mathcal{M} \models_{\{(x_0, a)\}} \psi(x_0)$ *for all* $\psi(x_0) \in p$. *A model* $\mathcal{M}$ *of T omits the type p if no element of M satisfies p. A type p of T is* principal *if there is a formula* $\theta(x_0)$ *such that*

- $T \models \theta(x_0) \to \psi(x_0)$ *for all* $\psi(x_0) \in p$;
- *every finite subset of* $T \cup \{\exists x_0 \theta(x_0)\}$ *has a model.*

*Suppose* $p = \{\psi_n(x_0) : n \in \mathbb{N}\}$ *is a non-principal type of T. By Exercise 6.46, player* **II** *has a winning strategy in the game* $\mathrm{MEG}(T, L)$. *Let* $\Delta$ *be a consistency property for T. Show that there is a function* $f : \mathbb{N} \to \mathbb{N}$ *and an increasing sequence* $S_n$ *in* $\Delta$ *such that* $\bigcup_n S_n$ *is a Hintikka set for* $T \cup \{\neg\phi_{f(n)}(c_n) : n \in \mathbb{N}\}$. *Derive the Omitting Types Theorem: if L is a countable vocabulary, T a set of L-sentences such that T has a model, and p a non-principal type of T, then T has a countable model which omits p.*

**Exercise 6.49** (*Failure of Omitting Types Theorem in* $\mathcal{D}$) *This is a continuation of Exercise 6.48. Give an example of a first order theory T in a countable* vocabulary $L$ and type $p = \{\psi_n(x_0) : n \in \mathbb{N}\}$ of formulas of $\mathcal{D}$ in the vocabulary $L$ such that

- $p$ is a non-principal type of $T$ in the sense of Exercise 6.48;
- no model of $T$ omits $p$.

**Exercise 6.50** *Let us write*

$$T \; \Vdash_{\mathcal{D}} \phi$$

*if player* **I** *has a winning strategy in* $\mathrm{MEG}_{\Sigma^1_1}(\tau_T \cup \{\tau_{0,\phi}\}, L)$, *or equivalently, if no model of* $T$ *is a model of* $\neg\phi$. *Show that*

(i) *if* $T \Vdash_{\mathcal{D}} \phi$ *and* $T \Vdash_{\mathcal{D}} \psi$, *then* $T \Vdash_{\mathcal{D}} (\phi \wedge \psi)$;
(ii) *if* $\phi \Vdash_{\mathcal{D}} \theta$ *and* $\psi \Vdash_{\mathcal{D}} \theta$, *then* $(\phi \vee \psi) \Vdash_{\mathcal{D}} \theta$;
(iii) $\phi \nVdash_{\mathcal{D}} \neg\phi$ *and* $\neg\phi \nVdash_{\mathcal{D}} \phi$.

**Exercise 6.51** *Continuation of Exercise 6.50. Show that*

(i) $T$ *has a model if and only if* $T \nVdash_{\mathcal{D}} \neg\top$;
(ii) $T \Vdash_{\mathcal{D}} (\phi \wedge \neg\phi)$ *if and only if* $\phi$ *is non-determined in models of* $T$.

**Exercise 6.52** *Continuation of Exercise 6.50. Show that* $\phi \Vdash_{\mathcal{D}} \psi$ *and* $\psi \Vdash_{\mathcal{D}} \theta$ *do not necessarily imply* $\phi \Vdash_{\mathcal{D}} \theta$.

**Exercise 6.53** *Continuation of Exercise 6.50. Give* $T$ *and* $\phi$ *such that* $T \Vdash_{\mathcal{D}} \phi$ *but there is no finite* $T_0 \subseteq T$ *such that* $T_0 \Vdash_{\mathcal{D}} \phi$.

## 6.6 Ehrenfeucht–Fraïssé game for dependence logic

We define the concept of elementary equivalence of models and give this concept a game characterization. In this section we assume that vocabularies do not contain function symbols, for simplicity.

**Definition 6.38** *Two models* $\mathcal{M}$ *and* $\mathcal{N}$ *of the same vocabulary are* $\mathcal{D}$-*equivalent in symbols* $\mathcal{M} \equiv_{\mathcal{D}} \mathcal{N}$ *if they satisfy the same sentences of dependence logic.*

It is easy to see that isomorphic structures are $\mathcal{D}$-equivalent and that $\mathcal{D}$-equivalence is an equivalence relation. However, despite the quite strong Löwenheim–Skolem Theorem of dependence logic, we have the following negative result about $\mathcal{D}$-equivalence.

**Proposition 6.39** *There is an uncountable model* $\mathcal{M}$ *in a finite vocabulary such that* $\mathcal{M}$ *is not* $\mathcal{D}$-*equivalent to any countable models.*

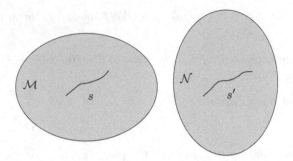

Fig. 6.1. A position in the ordinary Ehrenfeucht–Fraïssé game.

*Proof* Let $\mathcal{M} = (\mathbb{R}, \mathbb{N})$ be a model for the vocabulary $\{P\}$. Suppose $\mathcal{M} \equiv_{\mathcal{D}} \mathcal{N}$, where $\mathcal{N}$ is countable. It is clear that there is a one-to-one function from $N$ into $P^{\mathcal{N}}$. Let $\phi$ be a sentence of $\mathcal{D}$ which says exactly this. Since $\mathcal{N} \models \phi$, we have $\mathcal{M} \models \phi$, a contradiction.                                                □

It turns out that the following, more basic, concept is a far better concept to start with.

**Definition 6.40** *Suppose $\mathcal{M}$ and $\mathcal{N}$ are structures for the same vocabulary. We say that $\mathcal{M}$ is $\mathcal{D}$-semiequivalent to $\mathcal{N}$, in symbols $\mathcal{M} \Rightarrow_{\mathcal{D}} \mathcal{N}$, if $\mathcal{N}$ satisfies every sentence of dependence logic that is true in $\mathcal{M}$.*

Note that equivalence and semiequivalence are equivalent concepts if dependence logic is replaced by first order logic.

**Proposition 6.41** *Every infinite model in a countable vocabulary is $\mathcal{D}$-semiequivalent to models of all infinite cardinalities.*

*Proof* Suppose $\mathcal{M}$ is an infinite model with a countable vocabulary $L$. Let $T$ be the set of sentences of dependence logic in the vocabulary $L$ that are true in $\mathcal{M}$. By Theorem 6.5 the theory $T$ has models of all infinite cardinalities (Theorem 6.5 is formulated for sentences only, but the proof for countable theories is the same).                                                □

We now introduce an Ehrenfeucht–Fraïssé game adequate for dependence logic and use this game to characterize $\Rightarrow_{\mathcal{D}}$. See Figs. 6.1 and 6.2.

**Definition 6.42** *Let $\mathcal{M}$ and $\mathcal{N}$ be two structures of the same vocabulary. The game $\mathrm{EF}_n$ has two players and n moves. The position after move m is a pair $(X, Y)$, where $X \subseteq M^m$ and $Y \subseteq N^m$ for some m. In the beginning, the position*

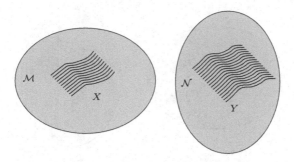

Fig. 6.2. A position in the Ehrenfeucht–Fraïssé game for $\mathcal{D}$.

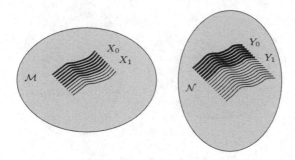

Fig. 6.3. A splitting move.

is $(\{\emptyset\}, \{\emptyset\})$ and $i_0 = 0$. Suppose the position after move number $m$ is $(X, Y)$. There are the following possibilities for the continuation of the game.

Splitting move. Player **I** represents $X$ as a union $X = X_0 \cup X_1$. Then player **II** represents $Y$ as a union $Y = Y_0 \cup Y_1$. Now player **I** chooses whether the game continues from the position $(X_0, Y_0)$ or from the position $(X_1, Y_1)$; see Fig. 6.3.

Duplication move. Player **I** decides that the game should continue from the new position

$$(X(M/x_{i_m}), Y(N/x_{i_m}));$$

see Fig. 6.4.

Supplementing move. Player **I** chooses a function $F : X \to M$. Then player **II** chooses a function $G : Y \to N$. Then the game continues from the position $(X(F/x_{i_m}), Y(G/x_{i_m}))$; see Fig. 6.5.

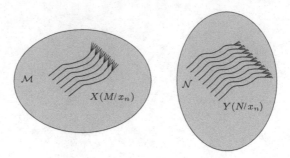

Fig. 6.4. A duplication move.

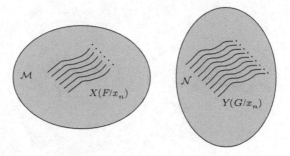

Fig. 6.5. A supplementing move.

*After n moves the position $(X_n, Y_n)$ is reached and the game ends. Player **II** is the winner if*

$$\mathcal{M} \models_{X_n} \phi \Rightarrow \mathcal{N} \models_{Y_n} \phi$$

*holds for all atomic and negated atomic and dependence formulas of the form $\phi(x_0, \ldots, x_{i_n-1})$. Otherwise player **I** wins.*

This is a game of perfect information, and the concept of winning strategy is defined as usual. By the Gale–Stewart Theorem, the game is determined.

### Definition 6.43

(i) $\mathrm{qr}(\phi) = 0$ *if $\phi$ is atomic or a dependence formula.*

(ii) $\mathrm{qr}(\phi \vee \psi) = \max(\mathrm{qr}(\phi), \mathrm{qr}(\psi)) + 1.$

(iii) $\mathrm{qr}(\exists x_n \phi) = \mathrm{qr}(\phi) + 1.$

(iv) $\mathrm{qr}(\neg \phi)$ :

    (a) $\mathrm{qr}(\neg \phi) = 0$ *if $\phi$ is atomic or a dependence formula;*

    (b) $\mathrm{qr}(\neg\neg\phi) = \mathrm{qr}(\phi);$

    (c) $\mathrm{qr}(\neg(\phi \vee \psi)) = \max(\mathrm{qr}(\neg\phi), \mathrm{qr}(\neg\psi));$

    (d) $\mathrm{qr}(\neg\exists x_n \phi) = \mathrm{qr}(\neg\phi) + 1.$

*Let* $\mathrm{Fml}_n^m$ *be the set of formulas* $\phi$ *of* $\mathcal{D}$ *with* $\mathrm{qr}\,\phi \leq m$ *and with free variables among* $x_0, \ldots, x_{n-1}$. *We write* $\mathcal{M} \Rightarrow_{\mathcal{D}}^n \mathcal{N}$ *if* $\mathcal{M} \models \phi$ *implies* $\mathcal{N} \models \phi$ *for all* $\phi$ *in* $\mathrm{Fml}_0^n$, *and* $\mathcal{M} \equiv_{\mathcal{D}}^n \mathcal{N}$ *if* $\mathcal{M} \models \phi$ *is equivalent to* $\mathcal{N} \models \phi$ *for all* $\phi$ *in* $\mathrm{Fml}_0^n$.

Note that there are for each $n$ and $m$, up to logical equivalence, only finitely many formulas in $\mathrm{Fml}_n^m$.

**Theorem 6.44** *Suppose* $\mathcal{M}$ *and* $\mathcal{N}$ *are models of the same vocabulary. Then the following are equivalent:*

(1) *player* **II** *has a winning strategy in the game* $\mathrm{EF}_n(\mathcal{M}, \mathcal{N})$;
(2) $\mathcal{M} \Rightarrow_{\mathcal{D}}^n \mathcal{N}$.

*Proof* We prove the equivalence, for all $n$, of the following two statements.

$(3)_m$ Player **II** has a winning strategy in the game $\mathrm{EF}_m(\mathcal{M}, \mathcal{N})$ in position $(X, Y)$, where $X \subseteq M^n$ and $Y \subseteq N^n$.
$(4)_m$ If $\phi$ is a formula in $\mathrm{Fml}_n^m$, then

$$\mathcal{M} \models_X \phi \Rightarrow \mathcal{N} \models_Y \phi. \tag{6.12}$$

The proof is by induction on $m$. For each $m$ we prove the claim simultaneously for all $n$. The case $m = 0$ is true by construction. Let us then assume $(3)_m \iff (4)_m$ as an induction hypothesis. Assume now $(3)_{m+1}$ and let $\phi$ be a formula in $\mathrm{Fml}_n^{m+1}$ such that $\mathcal{M} \models_X \phi$. As part of the induction hypothesis, we assume that the claim in Eq. (6.12) holds for formulas shorter than $\phi$.

Case (1) $\phi = \psi_0 \vee \psi_1$, where $\psi_0, \psi_1 \in \mathrm{Fml}_n^m$. Since $\mathcal{M} \models_X \phi$, there are $X_0$ and $X_1$ such that $X = X_0 \cup X_1$, $\mathcal{M} \models_{X_0} \psi_0$ and $\mathcal{M} \models_{X_1} \psi_1$. We let **I** play $\{X_0, X_1\}$. Then **II** plays according to her winning strategy $\{Y_0, Y_1\}$. Since the next position in the game can be either one of $(X_0, Y_0)$, $(X_1, Y_1)$, we can apply the induction hypothesis to both. This yields $\mathcal{N} \models_{Y_0} \psi_0$ and $\mathcal{N} \models_{Y_1} \psi_1$. Thus $\mathcal{N} \models_Y \phi$.
Case (2) $\phi = \exists x_n \psi$, where $\psi \in \mathrm{Fml}_{n-1}^m$. Since $\mathcal{M} \models_X \phi$, there is a function $F : X \to M$ such that $\mathcal{M} \models_{X(F/x_n)} \psi$. We let **I** play $F$. Then **II** plays according to her winning strategy a function $G : Y \to N$ and the game continues in position $(X(F/x_n), Y(G/x_n))$. The induction hypothesis gives $\mathcal{N} \models_{Y(G/x_n)} \psi$. Now $\mathcal{N} \models_Y \phi$ follows.
Case (3) $\phi = \neg\neg\psi$, $n = \max(n_0, n_1)$. Since $\mathcal{M} \models_X \phi$, we have $\mathcal{M} \models_X \psi$. By the induction hypothesis, $\mathcal{N} \models_Y \psi$. Thus $\mathcal{N} \models_Y \phi$.
Case (4) $\phi = \neg(\psi_0 \vee \psi_1)$, where $\psi_0, \psi_1 \in \mathrm{Fml}_n^m$. Since $\mathcal{M} \models_X \phi$, we have $\mathcal{M} \models_X \neg\psi_0$ and $\mathcal{M} \models_X \neg\psi_1$. By the induction hypothesis, $\mathcal{N} \models_Y \neg\psi_0$ and $\mathcal{N} \models_Y \neg\psi_1$. Thus $\mathcal{N} \models_Y \phi$.

Case (5) $\phi = \neg\exists x_n \psi$, where $\neg\psi \in \mathrm{Fml}^m_{n+1}$. By assumption, $\mathcal{M} \models_{X(M/x_n)} \neg\psi$. We let now **I** demand that the game continues in the duplicated position $(X(M/x_n), Y(N/x_n))$. The induction hypothesis gives $\mathcal{N} \models_{Y(N/x_n)} \neg\psi$. Now $\mathcal{N} \models_Y \phi$ follows trivially.

To prove the converse implication, assume $(4)_{m+1}$. To prove $(3)_{m+1}$ we consider the possible moves that player **I** can make in the position $(X, Y)$.

Case (i) Player **I** writes $X = X_0 \cup X_1$. Let $\phi_j$, $j < k$, be a complete list (up to logical equivalence) of formulas in $\mathrm{Fml}^m_n$. Since

$$\mathcal{M} \models_{X_0} \neg \bigvee_{\mathcal{M}\models_{X_0}\phi_j} \neg\phi_j$$

and

$$\mathcal{M} \models_{X_1} \neg \bigvee_{\mathcal{M}\models_{X_1}\phi_j} \neg\phi_j,$$

we have

$$\mathcal{M} \models_X \left( \neg \bigvee_{\mathcal{M}\models_{X_0}\phi_j} \neg\phi_j \right) \vee \left( \neg \bigvee_{\mathcal{M}\models_{X_1}\phi_j} \neg\phi_j \right).$$

Note that

$$\mathrm{qr}\left( \neg \bigvee_{\mathcal{M}\models_{X_1}\phi_j} \neg\phi_j \right) = \max_{\mathcal{M}\models_{X_1}\phi_j} (\neg\neg\phi_j) = \max_{\mathcal{M}\models_{X_1}\phi_j} \phi_j \le m.$$

Therefore, by $(4)_{m+1}$,

$$\mathcal{N} \models_Y \left( \neg \bigvee_{\mathcal{M}\models_{X_0}\phi_j} \neg\phi_j \right) \vee \left( \neg \bigvee_{\mathcal{M}\models_{X_1}\phi_j} \neg\phi_j \right).$$

Thus $Y = Y_0 \cup Y_1$ such that

$$\mathcal{N} \models_{Y_0} \neg \bigvee_{\mathcal{M}\models_{X_0}\phi_j} \neg\phi_j$$

and

$$\mathcal{N} \models_{Y_1} \neg \bigvee_{\mathcal{M}\models_{X_1}\phi_j} \neg\phi_j.$$

By this and the induction hypothesis, player **II** has a winning strategy in the positions $(X_0, Y_0)$, $(X_1, Y_1)$. Thus she can play $\{Y_0, Y_1\}$ and maintain her winning strategy.

Case (ii) Player **I** decides that the game should continue from the new position $(X(M/x_n), Y(m/x_n))$. We claim that

$$\mathcal{M} \models_{X(M/x_n)} \phi \Rightarrow \mathcal{N} \models_{Y(N/x_n)} \phi$$

for all $\phi \in \mathrm{Fml}_{n+1}^m$. From this the induction hypothesis would imply that **II** has a winning strategy in the position $(X(M/x_n), Y(N/x_n))$. So let us assume $\mathcal{M} \models_{X(M/x_n)} \phi$, where $\phi \in \mathrm{Fml}_{n+1}^m$. By definition,

$$\mathcal{M} \models_X \neg \exists x_n \neg \phi.$$

Since $\neg \exists x_n \neg \phi \in \mathrm{Fml}_n^{m+1}$, $(4)_{m+1}$ gives $\mathcal{N} \models_Y \neg \exists x_n \neg \phi$ and $\mathcal{N} \models_{Y(N/x_n)} \phi$ follows.

Case (iii) Player **I** chooses a function $F : X \to M$. Let $\phi_i, i < M$ be a complete list (up to logical equivalence) of formulas in $\mathrm{Fml}_{n+1}^m$. Now

$$\mathcal{M} \models_X \exists x_n \neg \bigvee_{\mathcal{M} \models_{X(F/x_n)} \phi_i} \neg \phi_i.$$

Note that

$$\mathrm{qr}\left( \exists x_n \neg \bigvee_{\mathcal{M} \models_{X(F/x_n)} \phi_j} \neg \phi_j \right) = \mathrm{qr}\left( \neg \bigvee_{\mathcal{M} \models_{X(F/x_n)} \phi_j} \neg \phi_j \right) + 1$$

$$= \left( \max_{\mathcal{M} \models_{X(F/x_n)} \phi_j} \mathrm{qr}(\neg \neg \phi_j) \right) + 1$$

$$= \left( \max_{\mathcal{M} \models_{X(F/x_n)} \phi_j} \mathrm{qr}(\phi_j) \right) + 1 \leq m + 1,$$

and hence, by $(4)_{m+1}$,

$$\mathcal{N} \models_Y \exists x_n \neg \bigvee_{\mathcal{M} \models_{X(F/x_n)} \phi_i} \neg \phi_i.$$

Thus there is a function $G : Y \to N$ such that

$$\mathcal{N} \models_{Y(G/x_n)} \neg \bigvee_{\mathcal{M} \models_{X(F/x_n)} \phi_i} \neg \phi_i.$$

The game continues from position $(X(F/x_n), Y(G/x_n))$. Given that now

$$\mathcal{M} \models_{X(F/x_n)} \phi \Rightarrow \mathcal{N} \models_{Y(G/x_n)} \phi$$

for all $\phi \in \mathrm{Fml}_{n+1}^m$, the induction hypothesis implies that **II** has a winning strategy in position $(X(F/x_n), Y(G/x_n))$. $\qquad\square$

**Corollary 6.45** *Suppose $\mathcal{M}$ and $\mathcal{N}$ are models of the same vocabulary. Then the following are equivalent.*

(1) $\mathcal{M} \Rightarrow_{\mathcal{D}} \mathcal{N}$.
(2) *For all natural numbers n, player* **II** *has a winning strategy in the game* $\mathrm{EF}_n(\mathcal{M}, \mathcal{N})$.

**Corollary 6.46** *Suppose $\mathcal{M}$ and $\mathcal{N}$ are models of the same vocabulary. Then the following are equivalent.*

(1) $\mathcal{M} \equiv_{\mathcal{D}} \mathcal{N}$.
(2) *For all natural numbers n, player* **II** *has a winning strategy both in the game* $\mathrm{EF}_n(\mathcal{M}, \mathcal{N})$ *and in the game* $\mathrm{EF}_n(\mathcal{N}, \mathcal{M})$.

The two games $\mathrm{EF}_n(\mathcal{M}, \mathcal{N})$ and $\mathrm{EF}_n(\mathcal{N}, \mathcal{M})$ can be put together into one game simply by making the moves of the former symmetric with respect to $\mathcal{M}$ and $\mathcal{N}$. Then player **II** has a winning strategy in this new game if and only if $\mathcal{M} \equiv_{\mathcal{D}}^n \mathcal{N}$. Instead of a game, we could have used a notion of a back-and-forth sequence.

The Ehrenfeucht–Fraïssé game can be used to prove non-expressibility results for $\mathcal{D}$, but we do not yet have examples where a more direct proof using compactness, interpolation, and Löwenheim–Skolem theorems would not be simpler.

**Proposition 6.47** *There are countable models $\mathcal{M}$ and $\mathcal{N}$ such that $\mathcal{M} \Rightarrow_{\mathcal{D}} \mathcal{N}$, but $\mathcal{N} \not\equiv_{\mathcal{D}} \mathcal{M}$.*

*Proof* Let $\mathcal{M}$ be the standard model of arithmetic. Let $\Phi_n$, $n \in \omega$, be the list of all $\Sigma_1^1$-sentences true in $\mathcal{M}$. Suppose

$$\Phi_n = \exists R_1^n \ldots \exists R_{k_n}^n \phi_n.$$

Let $\mathcal{M}^*$ be an expansion of $\mathcal{M}$ in which each $\phi_n$ is true. Let $\mathcal{N}^*$ be a countable non-standard elementary extension of $\mathcal{M}^*$. Let $\mathcal{N}$ be the reduct of $\mathcal{N}^*$ to the language of arithmetic. By construction, $\mathcal{M} \Rightarrow_{\mathcal{D}} \mathcal{N}$. On the other hand, $\mathcal{N} \not\Rightarrow_{\mathcal{D}} \mathcal{M}$ as non well-foundedness of the integers in $\mathcal{N}$ can be expressed by a sentence of $\mathcal{D}$. $\square$

**Proposition 6.48** *Suppose K is a model class[5] and n is a natural number. Then the following are equivalent:*

---

[5] A model class is a class of models, closed under isomorphism, of the same vocabulary.

Table 6.5.

| X | | |
|---|---|---|
| $x_0$ | $x_1$ | $x_2$ |
| 0 | 2 | 1 |
| 1 | 0 | 0 |
| 2 | 1 | 1 |

(1) *K is definable in dependence logic by a sentence in* $\mathrm{Fml}_0^n$;
(2) *K is closed under the relation* $\Rightarrow_{\mathcal{D}}^n$.

*Proof* Suppose $K$ is the class of models of $\phi \in \mathrm{Fml}_0^n$. If $\mathcal{M} \models \phi$ and $\mathcal{M} \Rightarrow_{\mathcal{D}}^n$ $\mathcal{N}$, then, by definition, $\mathcal{N} \models \phi$. Conversely, suppose $K$ is closed under $\Rightarrow_{\mathcal{D}}^n$. Let

$$\phi_{\mathcal{M}} = \neg \bigvee \{\neg\phi : \phi \in \mathrm{Fml}_0^n, \mathcal{M} \models \phi\},$$

where the conjunction is taken over a finite set which covers all such $\phi$ up to logical equivalence. Let $\theta$ be the disjunction of all $\phi_{\mathcal{M}}$, where $\mathcal{M} \in K$. Again we take the disjunction over a finite set up to logical equivalence. We show that $K$ is the class of models of $\theta$. If $\mathcal{M} \in K$ then $\mathcal{M} \models \phi_{\mathcal{M}}$, whence $\mathcal{M} \models \theta$. On the other hand suppose $\mathcal{M} \models \phi_{\mathcal{N}}$ for some $\mathcal{N} \in K$. Now $\mathcal{N} \Rightarrow_{\mathcal{D}}^n \mathcal{M}$, for if $\mathcal{N} \models \phi$ and $\phi \in \mathrm{Fml}_\emptyset^n$, then $\phi$ is logically equivalent with one of the conjuncts of $\phi_{\mathcal{N}}$, whence $\mathcal{M} \models \phi$. As $K$ is closed under $\Rightarrow_{\mathcal{D}}^n$, we have $\mathcal{M} \in K$.        □

**Corollary 6.49** *Suppose $K$ is a model class. Then the following are equivalent:*

(1) *K is definable in dependence logic.*
(2) *There is a natural number n such that K is closed under the relation* $\Rightarrow_{\mathcal{D}}^n$.

Corollary 6.49 also gives a characterization of $\Sigma_1^1$-definability in second order logic. No assumptions about cardinalities are involved, so if we restrict ourselves to finite models we get a characterization of NP-definability.

**Exercise 6.54** *Suppose $L = \emptyset$ and $\mathcal{M}$ is an L-structure such that $M = \{0, 1, 2\}$. Consider the team in Table 6.5. List the formulas $\phi$ in $\mathrm{Fml}_3^0$ such that X is of type $\phi$ in $\mathcal{M}$.*

Table 6.6.

| | X | |
|---|---|---|
| $x_0$ | $x_1$ | $x_2$ |
| 30 | 30 | 10 |
| 1 | 1 | 0 |
| 2 | 2 | 1 |
| 30 | 30 | 1 |

Table 6.7.

| | X | |
|---|---|---|
| $x_0$ | $x_1$ | $x_2$ |
| 0 | 0 | 0 |
| 1 | 2 | 0 |
| 2 | 4 | 1 |
| 3 | 8 | 1 |
| 3 | 16 | 1 |
| 3 | 32 | 1 |

**Exercise 6.55** *Suppose $L = \emptyset$ and $\mathcal{M}$ is an L-structure such that $M = \mathbb{N}$. Consider the team in Table 6.6. List the formulas $\phi$ in $\mathrm{Fml}_3^0$ such that X is of type $\phi$ in $\mathcal{M}$.*

**Exercise 6.56** *Suppose $L = \{P\}$ and $\mathcal{M}$ is an L-structure such that $M = \mathbb{N}$ and $P^{\mathcal{M}} = \{2n : n \in \mathbb{N}\}$. Consider the team in Table 6.7. List the formulas $\phi$ in $\mathrm{Fml}_3^0$ such that X is of type $\phi$ in $\mathcal{M}$.*

**Exercise 6.57** *Suppose $L = \emptyset$ and $\mathcal{M}$ and $\mathcal{N}$ are L-structures such that $M = \{0, 1, 2\}$ and $N = \{a, b, c, d\}$, where a, b, c, and d are distinct. The game $\mathrm{EF}_3(\mathcal{M}, \mathcal{N})$ ends in the position shown in Table 6.8. Who won the game?*

**Exercise 6.58** *Suppose $L = \emptyset$ and $\mathcal{M}$ and $\mathcal{N}$ are L-structures such that $M = \{0, 1, 2\}$ and $N = \{a, b, c, d\}$, where a, b, c, and d are distinct. The game $\mathrm{EF}_3(\mathcal{M}, \mathcal{N})$ ends in the position shown in Table 6.9. Who won the game?*

Table 6.8.

| | X | | | Y | | |
|---|---|---|---|---|---|---|
| $x_0$ | $x_1$ | $x_2$ | $x_0$ | $x_1$ | $x_2$ | |
| 0 | 2 | 2 | a | c | b | |
| 1 | 0 | 0 | b | b | c | |
| 2 | 1 | 1 | c | a | a | |
| | | | d | d | d | |

Table 6.9.

| X | | | Y | | |
|---|---|---|---|---|---|
| $x_0$ | $x_1$ | $x_2$ | $x_0$ | $x_1$ | $x_2$ |
| 0 | 1 | 1 | a | c | b |
| 1 | 1 | 0 | b | c | b |
| 2 | 1 | 0 | c | c | b |
| | | | d | c | a |

Table 6.10.

| X | | Y | |
|---|---|---|---|
| $x_0$ | $x_1$ | $x_0$ | $x_1$ |
| 0 | 2 | a | b |
| 1 | 3 | b | c |
| 2 | 3 | c | d |
| 3 | 2 | d | a |

**Exercise 6.59** *Suppose* $L = \emptyset$ *and* $\mathcal{M}$ *and* $\mathcal{N}$ *are L-structures such that* $M = \{0, 1, 2, 3\}$ *and* $N = \{a, b, c, d\}$, *where a, b, c, and d are distinct. The game* $EF_3(\mathcal{M}, \mathcal{N})$ *is in the position shown in Table 6.10. Can you spot a good splitting move for player* **I**?

**Exercise 6.60** *Suppose* $L = \{P\}$ *and* $\mathcal{M}$ *and* $\mathcal{N}$ *are L-structures such that* $M = \{0, 1, 2\}$, $P^{\mathcal{M}} = \{0\}$, $N = \{a, b, c\}$, *and* $P^{\mathcal{N}} = \emptyset$, *where a, b, and c are distinct. The game* $EF_3(\mathcal{M}, \mathcal{N})$ *is in the position shown in Table 6.11. Can you spot a good supplementing move for player* **I**?

Table 6.11.

| X | | Y | |
|---|---|---|---|
| $x_0$ | $x_1$ | $x_0$ | $x_1$ |
| 0 | 2 | a | b |
| 1 | 1 | b | b |
| 2 | 1 | c | a |

**Exercise 6.61** *Let $L = \emptyset$. Show that there is no sentence $\phi$ of vocabulary $L$ in* $\text{Fml}_0^2$ *such that $M \models \phi$ if and only if $M$ is infinite.*

**Exercise 6.62** *Let $L = \emptyset$. Show that there is no sentence $\phi$ of vocabulary $L$ in* $\text{Fml}_0^2$ *such that for finite models $M$ we have $M \models \phi$ if and only if $|M|$ is even.*

**Exercise 6.63** *Show that there is a countable model which is not $\mathcal{D}$-equivalent to any uncountable models.*

**Exercise 6.64** *Show that $(\mathbb{R}, \mathbb{N}) \Rrightarrow_{\mathcal{D}} (\mathbb{Q}, \mathbb{N})$.*

**Exercise 6.65** *Show that $(\omega + \omega^* + \omega, <) \not\Rrightarrow_{\mathcal{D}} (\omega, <)$.*

**Exercise 6.66** *Show that if $M \Rrightarrow_{\mathcal{D}} \mathcal{N}$ and $\mathcal{N}$ is a connected graph, then so is $M$.*

**Exercise 6.67** *Give three models $M$, $M'$, and $M''$ such that $M \Rrightarrow_{\mathcal{D}} M'$ and $M \Rrightarrow_{\mathcal{D}} M''$, but $M' \not\Rrightarrow_{\mathcal{D}} M''$ and $M'' \not\Rrightarrow_{\mathcal{D}} M'$.*

**Exercise 6.68** *Give models $M_n$ such that for all $n \in \mathbb{N}$ we have $M_{n+1} \Rrightarrow_{\mathcal{D}} M_n$, but $M_n \not\Rrightarrow_{\mathcal{D}} M_{n+1}$.*

**Exercise 6.69** *Let $\text{Part}(M, \mathcal{N})$ be the set of pairs $(X, Y)$, where, for some $n$, both $X \subseteq M^n$ and $Y \subseteq N^n$, and, moreover, $Y$ is of the type of any atomic, negated atomic formula, or atomic dependence formulas, $\phi(x_0, \ldots, x_{n-1})$ in $\mathcal{N}$ and $X$ is in $M$. Suppose $M$ and $\mathcal{N}$ are $L$-structures. A dependence back-and-forth set from $M$ to $\mathcal{N}$ is any non-empty set $P \subseteq \text{Part}(M, \mathcal{N})$ such that for all $(X, Y) \in P$:*

(i) *for any $X_1$ and $X_2$ such that $X = X_1 \cup X_2$ there are $Y_1$ and $Y_2$ such that $Y = Y_1 \cup Y_2$ and both $(X_1, Y_1) \in P$ and $(X_2, Y_2) \in P$;*
(ii) *$(X(M/x_n), Y(N/x_n)) \in P$;*

(iii) *for any mapping $F : X \to M$ there is a mapping $G : Y \to N$ such that*
$(X(F/x_n), Y(G/x_n)) \in P$.

*The structure $\mathcal{M}$ is said to be* partially dependence isomorphic *to $\mathcal{N}$, in symbols $\mathcal{M} \gtrsim_D \mathcal{N}$, if there is a back-and-forth set from $\mathcal{M}$ to $\mathcal{N}$. Show that $\mathcal{M} \gtrsim_D \mathcal{N}$ implies $\mathcal{M} \Rightarrow_D \mathcal{N}$.*

**Exercise 6.70** *This continues Exercise 6.69. Show that if $\mathcal{M} \cong \mathcal{N}$, then $\mathcal{M} \gtrsim_D \mathcal{N}$.*

**Exercise 6.71** *This continues Exercise 6.69. Show that if $\mathcal{M} \gtrsim_D \mathcal{M}'$ and $\mathcal{M}' \gtrsim_D \mathcal{M}''$, then $\mathcal{M} \gtrsim_D \mathcal{M}''$.*

**Exercise 6.72** *This continues Exercise 6.69. Give dense linear orders without end points $\mathcal{M}$ and $\mathcal{M}'$ such that $\mathcal{M} \not\gtrsim_D \mathcal{M}'$.*

# 7

# Complexity

## 7.1 Decision and other problems

The basic problem this chapter considers is how difficult is it to decide some basic questions concerning the relation $\mathcal{M} \models \phi$, when $\mathcal{M}$ is a structure and $\phi$ is a $\mathcal{D}$-sentence? Particular questions we will study are as follows:

**Decision Problem**. Is $\phi$ valid, i.e. does $\mathcal{M} \models \phi$ hold for *all* $\mathcal{M}$? See Fig. 7.1.

**Non-validity Problem**. Is $\phi$ *nonvalid*, i.e. does $\phi$ *avoid* some model, i.e. does $\mathcal{M} \models \phi$ fail for some model $\mathcal{M}$?

**Consistency Problem**. Is $\phi$ *consistent*, i.e. does $\phi$ have a model, i.e. does $\mathcal{M} \models \phi$ hold for *some* $\mathcal{M}$?

**Inconsistency Problem**. Is $\phi$ *inconsistent*, i.e. does $\phi$ *avoid* all models, i.e. is $\mathcal{M} \models \phi$ true for no model $\mathcal{M}$ at all?

Obviously such questions depend on the vocabulary. In a unary vocabulary it may be easier to answer some of the above questions. On the other hand, if at least one binary predicate is allowed, then the questions are as hard as for any other vocabulary, as there are coding techniques that allow us to code bigger vocabularies into one binary predicate. We can ask the same questions about models of particular theories such as like groups, linear orders, fields, graphs, and so on. Furthermore, we can ask these questions in the framework of finite models.

By definition,

$$\phi \text{ is non-valid if and only if } \phi \text{ is not valid,}$$

$$\phi \text{ is inconsistent if and only if } \phi \text{ is not consistent.}$$

So it suffices to concentrate on the Decision Problem and the Consistency Problem. In first order logic we have the further equivalence

$$\phi \text{ is consistent if and only if } \neg\phi \text{ is not valid.}$$

Table 7.1.

| First order logic | |
| --- | --- |
| Problem | Complexity |
| Decision Problem | $\Sigma_1^0$ |
| Non-validity Problem | $\Pi_1^0$ |
| Consistency Problem | $\Pi_1^0$ |
| Inconsistency Problem | $\Sigma_1^0$ |

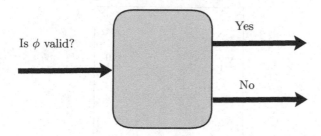

Fig. 7.1. Machine model of complexity.

So, if we crack the Decision Problem for first order logic, everything else follows. Indeed, the Gödel Completeness Theorem tells us that a first order sentence is valid if and only if it is provable. Hence the Decision Problem of first order logic is $\Sigma_1^0$ (i.e. recursively enumerable); see Fig 7.2. We obtain Table 7.1.

The Consistency Problem for dependence logic can be reduced to that of first order logic by the following equivalence:

$\phi$ is consistent if and only if $\tau_{1,\phi}$ is consistent.

So we obtain Table 7.2.

In Section 7.2 we will replace the two question marks with a complexity class.

## 7.2 Some set theory

The complexity of the Decision Problem and of the Non-validity Problem of dependence logic is so great that we have to move from complexity measures

Table 7.2.

| Dependence logic | |
|---|---|
| Problem | Complexity |
| Decision Problem | ? |
| Nonvalidity Problem | ? |
| Consistency Problem | $\Pi_1^0$ |
| Inconsistency Problem | $\Sigma_1^0$ |

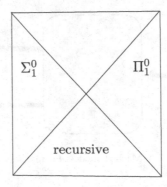

Fig. 7.2. Degrees of decidability.

on the integers to complexity measures in set theory. With this in mind, we recall some elementary concepts from set theory.

A set $a$ is *transitive* if $c \in b$ and $b \in a$ imply $c \in a$ for all $a$ and $b$. The *transitive closure* $TC(a)$ of a set $a$ is the intersection of all transitive supersets of $a$, or, in other words, $a \cup (\cup a) \cup (\cup \cup a) \ldots$ Intuitively, $TC(a)$ consists of elements of $a$, elements of elements of $a$, elements of elements of elements of $a$, etc. We define $H_\kappa = \{a : |TC(a)| < \kappa\}$.

*A priori* it is not evident that $H_\kappa$ is a set. However, this can be easily proved with another useful concept from set theory, namely the concept of rank. The *rank* $\mathrm{rk}(a)$ of a set $a$ is defined recursively as follows:

$$\mathrm{rk}(a) = \sup\{\mathrm{rk}(b) + 1 : b \in a\}.$$

Recall the definition of the cumulative hierarchy in Eq. (4.1); see Fig. 7.3. Now, $V_\alpha = \{x : \mathrm{rk}(x) < \alpha\}$ and we can prove that $H_\kappa$ is indeed a set.

**Lemma 7.1** *For all infinite cardinals $\kappa$ we have $H_\kappa \subseteq V_\kappa$.*

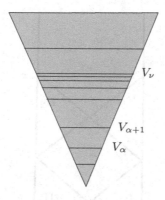

Fig. 7.3. The $V$-hierarchy of sets.

*Proof* Suppose $x \in H_\kappa$, i.e. $|TC(x)| < \kappa$. We claim that $\mathrm{rk}(x) < \kappa$, i.e. $|\mathrm{rk}(x)| < \kappa$. It suffices to show that $|\mathrm{rk}(x)| \leq |TC(x)|$. This follows by the Axiom of Choice, if we show that there is, for all $x$, an onto function from $TC(x)$ onto $\mathrm{rk}(x)$. This function is in fact the function $z \mapsto \mathrm{rk}(z)$. Suppose the claim that $\mathrm{rk} \restriction TC(y)$ maps $TC(y)$ onto $\mathrm{rk}(y)$ holds for all $y \in x$. We show that it holds for $x$. To this end, suppose $\alpha < \mathrm{rk}(x)$. By definition, $\alpha \leq \mathrm{rk}(y)$ for some $y \in x$. If $\alpha < \mathrm{rk}(y)$, then, by the induction hypothesis, there is $z \in TC(y)$ such that $\mathrm{rk}(z) = \alpha$. Now $z \in TC(x)$, so we are done. The other case is that $\alpha = \mathrm{rk}(y)$. Since $y \in TC(x)$, we are done again. $\qquad\square$

The converse of Lemma 7.1 is certainly not true in general. For example, the set $V_{\aleph_1}$ has sets such as $\mathcal{P}(\omega)$ which cannot be in $H_{\aleph_1}$. However, let us define the *Beth numbers* as follows: $\beth_0 = \omega$, $\beth_{\alpha+1} = 2^{\beth_\alpha}$, and $\beth_\nu = \lim_{\alpha<\nu} \beth_\alpha$ for limit $\nu$. For any $\kappa$ there is $\lambda \geq \kappa$ such that $\lambda = \beth_\lambda$, as follows. Let $\kappa_0 = \kappa$, $\kappa_{n+1} = \beth_{\kappa_n}$ and $\lambda = \lim_{n<\omega} \kappa_n$. Then $\lambda = \beth_\lambda$. It is easy to see by induction on $\alpha$ that $|V_{\omega+\alpha}| = \beth_\alpha$.

**Lemma 7.2** *If $\kappa = \beth_\kappa$, then $H_\kappa = V_\kappa$.*

*Proof* The claim $H_\kappa \subseteq V_\kappa$ follows from Lemma 7.1. On the other hand, if $x \in V_\kappa$, say $x \in V_\alpha$, where $\alpha = \omega + \alpha < \kappa$, then $|TC(x)| \leq |V_\alpha| = \beth_\alpha < \beth_\kappa = \kappa$, so $x \in H_\kappa$. $\qquad\square$

We now recall an important hierarchy in set theory.

**Definition 7.3** *(ref. [30]) The* Levy Hierarchy *of formulas of set theory is obtained as follows. The $\Sigma_0$-formulas, which are also called $\Pi_0$-formulas, are all formulas in the vocabulary $\{\in\}$ obtained from atomic formulas by the*

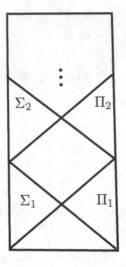

Fig. 7.4. Levy Hierarchy of formulas.

operations ¬, ∨, ∧, and the following bounded quantifiers:

$$\exists x_0 (x_0 \in x_1 \wedge \phi)$$

and

$$\forall x_0 (x_0 \in x_1 \rightarrow \phi).$$

The $\Sigma_{n+1}$-formulas are obtained from $\Pi_n$-formulas by existential quantification (see Fig 7.4). The $\Pi_{n+1}$-formulas are obtained from $\Sigma_n$-formulas by existential quantification.

A basic property of the $\Sigma_1$-formulas is captured by the following lemma.

**Lemma 7.4** For any uncountable cardinal $\kappa$, $a_1, \ldots, a_n \in H_\kappa$ and $\Sigma_1$-formula $\phi(x_1, \ldots, x_n)$: $(H_\kappa, \in) \models \phi(a_1, \ldots, a_n)$ if and only if $\phi(a_1, \ldots, a_n)$.

*Proof* The "only if" part is very easy (see Exercise 7.2). For the more difficult direction suppose $\phi(x_1, \ldots, x_n)$ is of the form $\exists x_0 \psi(x_0, x_1, \ldots, x_n)$, where $\psi(x_0, x_1, \ldots, x_n)$ is $\Sigma_0$ and there is an $a_0$ such that $\phi(a_0, a_1, \ldots, a_n)$. Let $\alpha$ be large enough that $a_0, \ldots a_n \in V_\alpha$. Then $(V_\alpha, \in) \models \psi(a_0, \ldots, a_n)$ by Exercise 7.2. Thus $(V_\alpha, \in) \models \phi(a_1, \ldots, a_n)$. Let $\mathcal{M}$ be an elementary submodel of the model $(V_\alpha, \in)$ such that $TC(\{a_1, \ldots, a_n\}) \subseteq M$ and $|M| < \kappa$. By Mostowski's Collapsing Lemma (see Exercise 7.3) there is a transitive model $(N, \in)$ and an isomorphism $\pi : (N, \in) \cong \mathcal{M}$ such that $a_1, \ldots, a_n \in N$ and $\pi(a_i) = a_i$ for each $i$. Thus $(N, \in) \models \phi(a_1, \ldots, a_n)$. Since $x \in N$ implies

Table 7.3.

| Dependence logic | |
|---|---|
| Problem | Complexity |
| Decision Problem | $\Pi_2$ |
| Nonvalidity Problem | $\Sigma_2$ |
| Consistency Problem | $\Pi_1^0$ |
| Inconsistency Problem | $\Sigma_1^0$ |

$TC(x) \subseteq N$, and $|N| < \kappa$, we have $N \subseteq H_\kappa$, and hence again, by absoluteness, $(H_\kappa, \epsilon) \models \phi(a_1, \ldots, a_n)$. $\qquad\square$

An intuitive picture of a $\Sigma_1$-statement $\exists x_0 \phi(a, x_0)$, where $\phi$ is $\Sigma_0$, is that we search through the universe for an element $b$ such that the relatively simple statement $\phi(a, b)$ becomes true. By Lemma 7.4 we need only look near where $a$ is. This means that satisfying a $\Sigma_1$-sentence is, from a set theoretic point of view, not very complex, although it may still be at least as difficult as checking whether a recursive binary relation on $\mathbb{N}$ is well-founded (which is a $\Pi_1^1$-complete problem).

In contrast, to check whether a $\Sigma_2$-sentence $\exists x_0 \forall x_1 \phi(a, x_0, x_1)$ is true, one has to search through the whole universe for a $b$ such that $\forall x_1 \phi(a, b, x_1)$ is true. Now, we cannot limit ourselves to search close to $a$ as we may have to look close to $b$, too. So, in the end, we have to go through the whole universe in search of $b$. This means that checking the truth of a $\Sigma_2$-sentence is of extremely high complexity. Any of the following statements can be expressed as the truth of a $\Sigma_2$-sentence:

(i) the Continuum Hypothesis, i.e. $2^{\aleph_0} = \aleph_1$;
(ii) the failure of the Continuum Hypothesis, i.e. $2^{\aleph_0} \neq \aleph_1$;
(iii) $V \neq L$;
(iv) there is an inaccessible cardinal;
(v) there is a measurable cardinal.

We now have the notions at hand for filling in the complexity table (Table 7.3) for dependence logic, even if we have not yet proved anything. The proof is given in Section 7.3.

**Exercise 7.1** *Show that the following predicates of set theory can be defined with a $\Sigma_0$-formula:*

(i) $x = y \cup z$;

(ii) $x = \{y, z\}$;

(iii) $x$ *is transitive;*

(iv) $x$ *is an ordinal (i.e. a transitive set of transitive sets);*

(v) $x$ *is a function* $y \to z$.

**Exercise 7.2** *(Absoluteness of $\Sigma_0$) For any transitive $M$, $a_1, \ldots, a_n \in M$ and $\Sigma_0$-formula $\phi(x_1, \ldots, x_n)$: $(M, \epsilon) \models \phi(a_1, \ldots, a_n)$ if and only if $\phi(a_1, \ldots, a_n)$.*

**Exercise 7.3** *(Mostowski's Collapsing Lemma) Suppose $(M, E)$ is a well-founded model of the Axiom of Extensionality: $\forall x_0 \forall x_1 (\forall x_2 (x_2 \in x_0 \leftrightarrow x_2 \in x_1) \to x_0 = x_1)$. Show that the equation $\pi(x) = \{\pi(y) : yEx\}$ defines an isomorphism between $(M, E)$ and $(N, \epsilon)$, where $N$ is a transitive set. Show also that if $E = \epsilon$, then $\pi(x) = x$ for every $x \in M$ which is transitive.*

## 7.3 $\Sigma_2$-completeness in set theory

Let us now go a little deeper into details. We identify problems with sets $X \subseteq \mathbb{N}$. The problem $X$ in such a case is really the problem of deciding whether a given $n$ is in $X$ or not. A problem $P \subseteq \mathbb{N}$ is called $\Sigma_n$-definable if there is a $\Sigma_n$-formula $\phi(x_0)$ of set theory such that

$$n \in P \iff \phi(n).$$

A problem $P \subseteq \mathbb{N}$ is called $\Sigma_n$-complete if the problem itself is $\Sigma_n$-definable and, moreover, for every $\Sigma_n$-definable set $X \subseteq \mathbb{N}$ there is a recursive function $f : \mathbb{N} \to \mathbb{N}$ such that

$$n \in X \iff f(n) \in P.$$

The concepts of a $\Pi_n$-definable set and a $\Pi_n$-complete set are defined analogously.

**Theorem 7.5** *The Decision Problem of dependence logic is $\Pi_2$-complete in set theory.*

*Proof* Without loss of generality, we consider the decision problem in the context of a vocabulary consisting of just one binary predicate. Let us first observe that the predicate $x = \mathcal{P}(y)$ is $\Pi_1$-definable:

$$x = \mathcal{P}(y) \iff \forall z(z \in x \leftrightarrow \forall u \in z(u \in y)).$$

Let $On(x)$ be the $\Sigma_0$-predicate "$x$ is an ordinal," i.e. $x$ is a transitive set of transitive sets. We can $\Pi_1$-define the property $R(x)$ of $x$ of being equal to some $V_\alpha$, where $\alpha$ is a limit ordinal. Let $Str(x)$ be the first order formula in the language of set theory which says that $x$ is a structure of the vocabulary containing just one binary predicate symbol. If $\phi$ is a sentence of $\mathcal{D}$, let $Sat_\phi(x)$ be the first order formula in the language of set theory which says "$Str(x)$ and $\phi$ is true in the structure $x$." Let $Relsat_\phi(x, y)$ be the first order formula in the language of set theory which says "$x \in y$ and if $Str(x)$, then $Sat_\phi(x)$ is true when relativized to the set $y$." Thus $\models \phi$ if and only if $\forall x(Str(x) \to Sat_\phi(x))$. Note that for limit $\alpha$ and $a \in V_\alpha$:

$$Sat_\phi(a) \iff (V_\alpha \models Sat_\phi(a)).$$

Thus a sentence $\phi$ of $\mathcal{D}$ is valid if and only if

$$\forall x(R(x) \to \forall y \in x \, Relsat_\phi(y, x)).$$

We have proved that the Decision Problem of $\mathcal{D}$ is $\Pi_2$-definable. Suppose then that $A$ is an arbitrary $\Pi_2$-definable set of integers. Let $\forall x \exists y \psi(n, x, y)$ be the $\Pi_2$-definition. Let $\phi_n$ be the first order sentence $\forall x \exists y \psi(n, x, y)$, where $n$ is a defined term. We claim

$$n \in A \iff \models \theta \vee \phi_n,$$

where $\theta$ is a sentence of $\mathcal{D}$ which is true in every model except the models $(V_\kappa, \in)$, $\kappa = \beth_\kappa$. Suppose first $n \in A$, i.e. $\forall x \exists y \psi(n, x, y)$. Suppose $(V_\kappa, \in)$ is a given model in which $\theta$ is not true. We prove $V_\kappa \models \forall x \exists y \psi(n, x, y)$. Suppose $a \in V_\kappa$. By the above lemmas, there is $b \in V_\kappa$ such that $\psi(n, a, b)$. We have proved $\models \theta \vee \phi_n$. Conversely, suppose $\models \theta \vee \phi_n$. To prove $\forall x \exists y \psi(n, x, y)$, let $a$ be given. Let $\kappa$ be an infinite cardinal such that $\kappa = \beth_\kappa$ and $a \in H_\kappa$. Then there is $b \in H_\kappa$ with $H_\kappa \models \psi(n, a, b)$. Now $\psi(n, a, b)$ follows. $\qquad \square$

**Corollary 7.6** *The Decision Problem of dependence logic is not $\Sigma_2$-definable in set theory.*

The fact that the Decision Problem of dependence logic is not $\Sigma_n^m$ for any $m, n < \omega$ follows easily from this. Moreover, it follows that we cannot in general express "$\phi$ is valid," for $\phi \in \mathcal{D}$, even by searching through the whole set theoretical universe for a set $x$ such that a universal quantification over the subsets of $x$ would guarantee the validity of $\phi$. In contrast, to check the validity of a first order sentence, one needs only to search through all natural numbers and then perform a finite polynomial calculation on that number. Figures 7.5 and 7.6 illustrate the difference between categories of validity in first order logic and dependence logic.

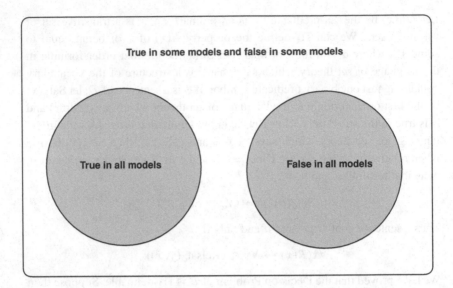

Fig. 7.5. Categories of validity in first order logic.

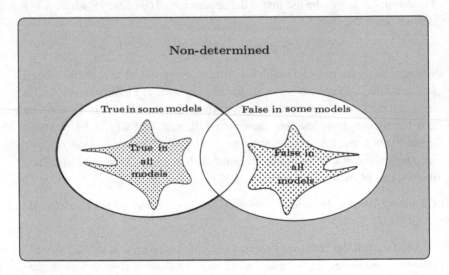

Fig. 7.6. Categories of validity in dependence logic.

**Exercise 7.4** *Give a $\Sigma_2$-definition of the property of $x$ being equal to some $V_\kappa$, where $\kappa = \beth_\kappa$.*

**Exercise 7.5** *Show that for limit $\alpha$ and $a \in V_\alpha$:*

$$\mathrm{Sat}_\phi(a) \iff (V_\alpha \models \mathrm{Sat}_\phi(a)).$$

**Exercise 7.6** *Give a sentence of $\mathcal{D}$ which is true in every model except (mod $\cong$) the models $(V_\kappa, \in)$, $\kappa = \beth_\kappa$.*

**Exercise 7.7** *Use Theorem 7.5 to prove that the Decision Problem of $\mathcal{D}$ is not arithmetical, i.e. not first order definable in $(\mathbb{N}, +, \cdot, <, 0, 1)$. (In fact, the same proof shows that it is not $\Sigma_n^m$-definable for any $m, n < \omega$.)*

**Exercise 7.8** *Show that the problem*

$$\text{"does } \mathcal{M} \models \phi \text{ hold for all countable } \mathcal{M}?"$$

*is not arithmetical and that the problem*

$$\text{"does } \mathcal{M} \models \phi \text{ hold for some countable } \mathcal{M}?"$$

*is a $\Pi_1^0$-property of $\phi \in \mathcal{D}$.*

**Exercise 7.9** *Show that there is a sentence $\phi$ of $\mathcal{D}$ such that $\phi$ avoids some model if and only if there is an inaccessible cardinal.*

# 8

# Team logic

The negation ¬ of dependence logic does not satisfy the Law of Excluded Middle and is therefore not the classical Boolean negation. This is clearly manifested by the existence of non-determined sentences $\phi$ in $\mathcal{D}$. In such cases, the failure of $\mathcal{M} \models \phi$ does not imply $\mathcal{M} \models \neg\phi$. Hintikka [19] introduced *extended independence friendly logic* by taking the Boolean closure of his independence friendly logic. We take the further action of making classical negation $\sim$ one of the logical operations on a par with other propositional operations and quantifiers. This yields an extension TL of the Boolean closure of $\mathcal{D}$. We call the new logic *team logic*.

The basic concept of both team logic TL and dependence logic $\mathcal{D}$ is the concept of dependence $=(t_1, \ldots, t_n)$. In very simple terms, what happens is that, while we can say "$x_1$ *is* a function of $x_0$" with $=(x_0, x_1)$ in $\mathcal{D}$, we will be able to say "$x_1$ is *not* a function of $x_0$" with $\sim =(x_0, x_1)$ in team logic.

While we define team logic we have to restrict ¬. The game theoretic intuition behind ¬$\phi$ is that it says something about "the other player." The introduction of $\sim$ unfortunately ruins the basic game theoretic intuition, and there is no "other player" anymore. If $\phi$ is in $\mathcal{D}$, then $\sim \phi$ has the meaning "**II** does not have a winning strategy," but it is not clear what the meaning of ¬$\sim \phi$ would be. We also change notation by using $\phi \otimes \psi$ ("tensor") instead of $\phi \vee \psi$. The reason for this is that by means of $\sim$ and $\wedge$ we can actually define the classical Boolean disjunction $\phi \vee \psi$, which really says that the team is of type $\phi$ or of type $\psi$. Likewise, we adopt the notation $!x_n\phi$ ("shriek") for the quantifier that in $\mathcal{D}$ was denoted by $\forall x_n\phi$.

## 8.1 Preorder of determination

Let us consider teams of soccer players as an example. In this case, players are the agents. We consider only the colors of their outfits as relevant

Table 8.1. *Team 1*

|   | Shirt | Shorts |
|---|-------|--------|
| 1 | yellow | white |
| 2 | yellow | white |
| 3 | yellow | white |
| 4 | yellow | white |
| 5 | red | black |
| 6 | red | black |
| 7 | red | black |

Table 8.2. *Team 2*

|   | Shirt | Shorts | Socks |
|---|-------|--------|-------|
| 1 | yellow | white | green |
| 2 | yellow | white | green |
| 3 | yellow | white | green |
| 4 | red | white | black |
| 5 | red | white | black |
| 6 | red | black | green |
| 7 | red | black | green |

features. We imagine a spectator trying to figure out how the different colors are determined.

Consider Team 1 of Table 8.1, consisting of seven members. It is clear that in this team both the shirt color depends on the shorts color and vice versa. What about Team 2 of Table 8.2? We can observe that the shirt color depends on shorts color and socks' color in the sense that it can be computed if both shorts color and socks color are known. Interestingly, it cannot be computed from either shorts or socks alone. It seems correct to say that shirt color is *dependent* on shorts color, and that shirt color is dependent on socks color. However, if the team was as in Table 8.3, we could compute shirt color from socks color alone. It would seem correct to say that shirt color is, in this case, *independent* of shorts color, as the only addition that together with the shorts color would determine shirt color, achieves this by itself. Let us finally consider Table 8.4. Now shirt color can be computed alternatively from shorts color alone or from the combination of socks and shoes. We certainly cannot say

Table 8.3. *Team 3*

|   | Shirt | Shorts | Socks |
|---|-------|--------|-------|
| 1 | yellow | white | green |
| 2 | yellow | white | green |
| 3 | yellow | white | green |
| 4 | red | white | black |
| 5 | red | white | black |
| 6 | red | black | red |
| 7 | red | black | red |

Table 8.4. *Team 4*

|   | Shirt | Shorts | Socks | Shoes |
|---|-------|--------|-------|-------|
| 1 | yellow | white | green | yellow |
| 2 | yellow | white | green | yellow |
| 3 | yellow | white | green | yellow |
| 4 | red | blue | green | red |
| 5 | red | blue | green | red |
| 6 | red | black | yellow | yellow |
| 7 | red | black | yellow | yellow |

shirt color is independent of shorts color, even if it can be computed without it. Similarly, shirt color is not independent of socks color, as socks color (together with shoes color) gives away shirt color, but shoes color alone does not. We can say that shirt color depends on shorts color, but not so strongly that it could not be computed without it.

Another example is the human genome "team." Here an agent is the DNA sequence of the set of chromosomes of any one individual human being. Thus there are over 6.5 billion agents from people alive today. A team is any collection of such agents. Features are any of the 25 000 genes that make up the DNA sequence of the individual person. Extra features such as diseases (diabetes, asthma, heart disease, etc.) can be added to genome teams to study how faulty genes are associated with particular diseases. Then a team would be a patient database with fields for certain genes and certain diseases. In anticipation of the discussion in Section 8.2, of the different kinds of dependencies, we may ask the following kinds of questions.

Table 8.5. *A team*

|   | $x_1$ | $x_2$ | ... | $x_n$ |
|---|---|---|---|---|
| 1 | $m_1^1$ | $m_2^1$ | ... | $m_n^1$ |
| 2 | $m_1^2$ | $m_2^2$ | ... | $m_n^2$ |
| 3 | $m_1^3$ | $m_2^3$ | ... | $m_n^3$ |
| 4 | $m_1^4$ | $m_2^4$ | ... | $m_n^4$ |
| 5 | $m_1^5$ | $m_2^5$ | ... | $m_n^5$ |
| ... | ... | ... | ... | ... |
| $k$ | $m_1^k$ | $m_2^k$ | ... | $m_n^k$ |

(i) Does a certain gene or gene combination (significantly) *determine* a given hereditary disease in the sense that a patient with (a fault in) those genes has a high risk of the disease?
(ii) Is a disease *totally dependent* on a gene in the sense that every gene combination that (significantly) determines the disease contains that particular gene?
(iii) Is a gene (merely) *dependent* on a gene in the sense that the disease is (significantly) determined by some gene combination with the gene but not without?
(iv) Is a disease *totally independent* of a gene in the sense that no gene combination that (significantly) determines the disease contains that particular gene?
(v) Is a gene (merely) *independent* of a gene in the sense that some gene combinations (significantly) determine the disease without containing that particular gene?

Let $X$ be a team, i.e. a set of assignments (agents) of some finitely many variables $\{x_1, \ldots, x_n\}$ into a domain $M$, as in Table 8.5. Equivalently, $X$ could be a database with fields $\{x_1, \ldots, x_n\}$ and values of the fields from a fixed domain $M$. There is an obvious partial order in the powerset of $\{x_1, \ldots, x_n\}$, namely the set theoretical subset relation $\subseteq$. We now define a new relation, called the *preorder of determination*:

$V \leq W$    $W$ determines $V$, i.e. features in $V$ can be determined if the values of the features in $W$ are known. In symbols, $\forall s, s' \in$
$X((\forall y \in W(s(y) = s'(y)) \to (\forall x \in V(s(x) = s'(x)))))$.

Equivalently, $\{w_1, \ldots, w_n\}$ determines $V$, if, for all $y \in V$, there is a function $f_y$ such that for all $s$ in $X$: $s(y) = f_y(s(w_1), \ldots, s(w_n))$.

It is evident from the definition that the preorder of determination is weaker than the partial order of inclusion in the sense that a set obviously determines every subset of itself. It is more interesting that sometimes a set determines a set disjoint from itself, and a singleton set may determine another singleton set. Some sets may be determined by the empty set (in that case the feature has to have a constant value). Every set is certainly determined by the whole universe.

The Armstrong Axioms of functional dependence (see ref. [1]) state exactly the following lemma.

**Lemma 8.1**

(i) $V \leq W$ is a preorder, i.e. reflexive and transitive.

(ii) If $V \subseteq W$, then $V \leq W$.

(iii) If $V \leq W$ and $U$ is arbitrary, then $V \cup U \leq W \cup U$.

*Moreover,*

(iv) If $V \leq W$, then there is a minimal $U \subseteq W$ such that $V \leq U$.

## 8.2 Dependence and independence

Now we can define two versions of dependence, by using the preorder of determination. Suppose $W \cap V = \emptyset$. We define the following.

| | |
|---|---|
| $V$ is *dependent* on $W$ | There is some minimal $U \geq V$ such that $U \cap V = \emptyset$ and $W \subseteq U$. |
| $V$ is *totally dependent* on $W$ | For every $U \geq V$ such that $U \cap V = \emptyset$, we have $W \subseteq U$. |
| $V$ is *independent* of $W$ | There is some $U \geq V$ such that $U \cap V = \emptyset$ and $W \cap U = \emptyset$. |
| $V$ is *totally independent* of $W$ | For every minimal $U \geq V$ such that $U \cap V = \emptyset$, we have $W \cap U = \emptyset$. |
| $V$ is *non-determined* | There is no $U \geq V$ such that $U \cap V = \emptyset$. In the opposite case, $V$ is called determined. |

Note that the above concepts are defined with respect to the features that are present in the setup. Indeed, it seems meaningless to define what independence means in a setup where any *new* feature can be introduced. The presence of new features can change independence to dependence, as in Team 2 of Table 8.2,

Table 8.6. *Team for*
*Exercise 8.3*

|       | $x_1$ | $x_2$ | $x_3$ |
|-------|-------|-------|-------|
| $s_1$ | 0     | 1     | 0     |
| $s_2$ | 0     | 0     | 1     |
| $s_3$ | 1     | 1     | 1     |

where shirt color is independent of both shorts and socks if either one of them is missing, but is dependent on either if both are present.

In the following, we present some immediate relationships between the introduced concepts of dependence and independence.

**Lemma 8.2**

  (i) *Every V is totally dependent on and totally independent of Ø.*
 (ii) *If V is (totally) dependent on W, then V is (totally) dependent on every subset of W.*
(iii) *If V is dependent on W, it can still be also independent of W, but not totally, unless W = Ø.*
(iv) *V is independent of $\{x\}$ if and only if V is not totally dependent on $\{x\}$.*
 (v) *V is totally independent of $\{x\}$ if and only if V is not dependent on $\{x\}$.*

**Exercise 8.1** *Prove Lemma 8.1.*

**Exercise 8.2** *Prove Lemma 8.2.*

**Exercise 8.3** *Consider the team given in Table 8.6. Which $W \subseteq \{x_2, x_3\}$ is $\{x_1\}$ (a) dependent on, (b) totally dependent on, (c) independent of, (d) totally independent of?*

**Exercise 8.4** *Consider the team given in Table 8.7. Which subsets W of $\{x_2, x_3, x_4, x_5\}$ is $\{x_1\}$ (a) dependent on, (b) totally dependent on, (c) independent of, (d) totally independent of?*

**Exercise 8.5** *Consider the team given in Table 8.8. Find a value for a in natural numbers such that (a) $\{x_3, x_4\}$ determines $\{x_1\}$, (b) $\{x_3, x_4\}$ does not determine $\{x_1\}$, (c) $\{x_2, x_4\}$ determines $\{x_1\}$, (d) $\{x_2, x_4\}$ does not determine $\{x_1\}$?*

**Exercise 8.6** *Consider the team given in Table 8.9. Fill in values for $x_3$ and $x_4$ from the set $\{0, 1\}$ such that $\{x_4\}$ is (a) dependent but not totally dependent*

Table 8.7. *Team for Exercise 8.4*

|       | $x_1$ | $x_2$ | $x_3$ | $x_4$ | $x_5$ |
|-------|-------|-------|-------|-------|-------|
| $s_1$ | 2     | 3     | 0     | 1     | 0     |
| $s_2$ | 2     | 2     | 4     | 0     | 3     |
| $s_3$ | 8     | 0     | 4     | 0     | 0     |
| $s_4$ | 0     | 3     | 4     | 1     | 1     |
| $s_5$ | 0     | 0     | 0     | 0     | 0     |
| $s_6$ | 2     | 4     | 0     | 1     | 1     |

Table 8.8. *Team for Exercise 8.5*

|       | $x_1$ | $x_2$ | $x_3$ | $x_4$ |
|-------|-------|-------|-------|-------|
| $s_1$ | 15    | 15    | 3     | 6     |
| $s_2$ | 1     | 1     | 1     | 1     |
| $s_3$ | 6     | 0     | 2     | $a$   |
| $s_4$ | 4     | 3     | 2     | 2     |
| $s_5$ | 0     | 0     | 0     | 5     |

Table 8.9. *Team for Exercise 8.6*

|       | $x_1$ | $x_2$ | $x_3$ | $x_4$ |
|-------|-------|-------|-------|-------|
| $s_1$ | 0     | 0     |       |       |
| $s_2$ | 0     | 1     |       |       |
| $s_3$ | 1     | 1     |       |       |

*on, (b) totally dependent on, (c) independent but not totally independent of, (d) totally independent of $\{x_1\}$.*

## 8.3 Formulas of team logic

In this section we give the syntax and semantics of team logic and indicate some basic principles.

Table 8.10.

| Atomic | Name |
|---|---|
| $t_1 = t_n$ | equation |
| $\neg t_1 = t_n$ | dual equation |
| $Rt_1 \ldots t_n$ | relation |
| $\neg Rt_1 \ldots t_n$ | dual relation |
| $=(t_1, \ldots, t_n)$ | dependence |
| $\neg =(t_1, \ldots, t_n)$ | dual dependence |

Table 8.11.

| Compound | Name |
|---|---|
| $\phi \otimes \psi$ | tensor |
| $\phi \wedge \psi$ | conjunction |
| $\sim \phi$ | negation |
| $\exists x_n \phi$ | existential |
| $!x_n \phi$ | shriek |

**Definition 8.3** *Suppose L is a vocabulary. The formulas of team logic TL are of the atomic form, as in Table 8.10, or compound form, as in Table 8.11.*

The semantics of team logic is defined as follows.

**Definition 8.4**

(TL1) $\mathcal{M} \models_X t_1 = t_2$ *if and only if for all* $s \in X$ *we have* $t_1^{\mathcal{M}}\langle s \rangle = t_2^{\mathcal{M}}\langle s \rangle$.
(TL2) $\mathcal{M} \models_X \neg t_1 = t_2$ *if and only if for all* $s \in X$ *we have* $t_1^{\mathcal{M}}\langle s \rangle \neq t_2^{\mathcal{M}}\langle s \rangle$.
(TL3) $\mathcal{M} \models_X =(t_1, \ldots, t_n)$ *if and only if for all* $s, s' \in X$ *such that*

$$t_1^{\mathcal{M}}\langle s \rangle = t_1^{\mathcal{M}}\langle s' \rangle, \ldots, t_{n-1}^{\mathcal{M}}\langle s \rangle = t_{n-1}^{\mathcal{M}}\langle s' \rangle,$$

*we have* $t_n^{\mathcal{M}}\langle s \rangle = t_n^{\mathcal{M}}\langle s' \rangle$.
(TL4) $\mathcal{M} \models_X \neg =(t_1, \ldots, t_n)$ *if and only if* $X = \emptyset$.
(TL5) $\mathcal{M} \models_X Rt_1 \ldots t_n$ *if and only if for all* $s \in X$ *we have* $(t_1^{\mathcal{M}}\langle s \rangle, \ldots, t_n^{\mathcal{M}}\langle s \rangle) \in R^{\mathcal{M}}$.
(TL6) $\mathcal{M} \models_X \neg Rt_1 \ldots t_n$ *if and only if for all* $s \in X$ *we have* $(t_1^{\mathcal{M}}\langle s \rangle, \ldots, t_n^{\mathcal{M}}\langle s \rangle) \notin R^{\mathcal{M}}$.

*Team logic*

(TL7) $\mathcal{M} \models_X \phi \otimes \psi$ *if and only if* $X = Y \cup Z$ *such that* $\mathrm{dom}(Y) = \mathrm{dom}(Z)$,
$\quad\quad \mathcal{M} \models_Y \phi$, *and* $\mathcal{M} \models_Z \psi$.
(TL8) $\mathcal{M} \models_X \phi \wedge \psi$ *if and only if* $\mathcal{M} \models_X \phi$ *and* $\mathcal{M} \models_X \psi$.
(TL9) $\mathcal{M} \models_X \exists x_n \phi$ *if and only if* $\mathcal{M} \models_{X(F/x_n)} \phi$ *for some* $F : X \to M$.
(TL10) $\mathcal{M} \models_X \,!x_n \phi$ *if and only if* $\mathcal{M} \models_{X(M/x_n)} \phi$
(TL11) $\mathcal{M} \models_X \sim \phi$ *if and only if* $\mathcal{M} \not\models_X \phi$.

We can easily define a translation $\phi \mapsto \phi^*$ of dependence logic into team logic, but we have to assume the formula $\phi$ of dependence logic is in negation normal form:

$$
\begin{aligned}
(t = t')^* \quad &= t = t', \\
(\neg t = t')^* \quad &= \neg t = t', \\
(Rt_1 \dots t_n)^* \quad &= Rt_1 \dots t_n, \\
(\neg Rt_1 \dots t_n)^* \quad &= \neg Rt_1 \dots t_n, \\
(=(t_1, \dots, t_n))^* \quad &= \,=(t_1, \dots, t_n), \\
(\neg =(t_1, \dots, t_n))^* \quad &= \neg =(t_1, \dots, t_n), \\
(\phi \vee \psi)^* \quad &= \phi^* \otimes \psi^*, \\
(\phi \wedge \psi)^* \quad &= \phi^* \wedge \psi^*, \\
(\exists x_n \phi)^* \quad &= \exists x_n \phi^*, \\
(\forall x_n \phi)^* \quad &= \,!x_n \phi^*.
\end{aligned}
$$

It is an immediate consequence of the definitions that for all $\mathcal{M}$, all $\phi$, and all $X$ we have

$$\mathcal{M} \models_X \phi \text{ in } \mathcal{D} \text{ if and only if } \mathcal{M} \models_X \phi^* \text{ in TL}.$$

So we may consider $\mathcal{D}$ a fragment of TL, and TL an extension of $\mathcal{D}$ obtained by adding classical negation.

Logical consequence $\phi \Rightarrow \psi$ and logical equivalence $\phi \Leftrightarrow \psi$ are defined similarly as for dependence logic. Lemma 8.5 demonstrates that even though $!x_n \phi$ for $\phi \in \mathcal{D}$ acts like what is denoted by $\forall x_n \phi$ in dependence logic, it is, in the presence of $\sim$, not at all like our familiar universal quantifier, as it commutes with negation.

**Lemma 8.5** $\sim !x_n \phi \Leftrightarrow \,!x_n \sim \phi$.

*Proof*

$$
\begin{aligned}
\mathcal{M} \models_X \sim !x_n \phi \quad &\text{if and only if} \quad \mathcal{M} \not\models_X \,!x_n \phi, \\
&\text{if and only if} \quad \mathcal{M} \not\models_{X(M/x_n)} \phi, \\
&\text{if and only if} \quad \mathcal{M} \models_{X(M/x_n)} \sim \phi, \\
&\text{if and only if} \quad \mathcal{M} \models_X \,!x_n \sim \phi.
\end{aligned}
$$

$\qquad\qquad\qquad\qquad\qquad\qquad\qquad\qquad\qquad\qquad\qquad\qquad\qquad\quad$ $\square$

Table 8.12. *The intuition behind the logical operations*

| | |
|---|---|
| $\phi \vee \psi$ | the team is either of type $\phi$ or of type $\psi$ (or both) |
| $\phi \wedge \psi$ | the team is both of type $\phi$ and of type $\psi$ |
| $\phi \otimes \psi$ | the team divides between type $\phi$ and type $\psi$ |
| $\phi \oplus \psi$ | every division of the team yields type $\phi$ or type $\psi$ |
| $\sim\phi$ | the team is not of type $\phi$ |
| $\top$ | any team |
| $\bot$ | no team |
| $\mathbf{1}$ | any non-empty team |
| $\mathbf{0}$ | only the empty team |

We adopt the following abbreviations (see Table 8.12):[1]

$$\phi \vee \psi \ = \sim(\sim\phi \wedge \sim\psi) \quad \text{disjunction;}$$
$$\phi \oplus \psi \ = \sim(\sim\phi \otimes \sim\psi) \quad \text{sum;}$$
$$\phi \to \psi = \sim\phi \vee \psi \quad \text{implication;}$$
$$\phi \multimap \psi = \sim\phi \oplus \psi \quad \text{linear implication;}$$
$$\forall x_n \phi \ \ = \sim \exists x_n \sim\phi \quad \text{universal quantifier.}$$

Thus,

$$\mathcal{M} \models_X \phi \vee \psi \text{ if and only if } \mathcal{M} \models_X \phi \text{ or } \mathcal{M} \models_X \psi.$$

We have recovered the classical disjunction with the properties

$$\phi \vee \psi \qquad \Leftrightarrow \psi \vee \phi,$$
$$\phi \vee (\psi \vee \theta) \Leftrightarrow (\phi \vee \psi) \vee \theta,$$
$$\phi \vee \phi \qquad \Leftrightarrow \phi.$$

Note that

$$\mathcal{M} \models_X \phi \oplus \psi \text{ if and only if whenever } X = Y \cup Z, \text{ then } \mathcal{M} \models_Y \phi \text{ or } \mathcal{M} \models_Z \psi.$$

Thus a team of type $\phi \oplus \psi$ has in a sense $\phi$ or $\psi$ everywhere. We have

$$\phi \oplus \psi \qquad \Leftrightarrow \psi \oplus \phi,$$
$$\phi \oplus (\psi \oplus \theta) \Leftrightarrow (\phi \oplus \psi) \oplus \theta,$$

but

$$\phi \oplus \phi \nLeftrightarrow \phi.$$

Note also that

$$\mathcal{M} \models_X \forall x_n \phi \text{ if and only if } \mathcal{M} \models_{X(F/x_n)} \phi \text{ holds for every } F : X \to M.$$

---

[1] Note that the symbols $\vee$ and $\forall$ have different meanings in $\mathcal{D}$ and in TL.

Table 8.13.

|       | $x_0$ | $x_1$ |
|-------|-------|-------|
| $s_1$ | 3     | 8     |
| $s_2$ | 2     | 5     |
| $s_3$ | 1     | 4     |
| $s_4$ | 9     | 1     |

Table 8.14. *Dependence values in* $\mathcal{D}$

| Sentence | Teams of that type |
|----------|--------------------|
| $\top$   | $\emptyset, \{\emptyset\}$ |
| $\bot$   | (none)             |
| **1**    | $\{\emptyset\}$    |
| **0**    | $\emptyset$        |

Thus a team of type $\forall x_n \phi$ has type $\phi$ whatever we put as values of $x_n$. A team of type $\forall x_2((x_0 + x_1) + x_2 = x_0 + (x_1 + x_2))$ but not of type $\forall x_2(\sim =(x_0, x_1, x_2))$ in the model $(\mathbb{N}, +)$ is shown in Table 8.13.

We have

$$\forall x_n(\phi \wedge \psi) \Leftrightarrow (\forall x_n \phi \wedge \forall x_n \psi),$$
$$\forall x_n \forall x_m \phi \quad \Leftrightarrow \forall x_m \forall x_n \phi,$$

but in general (see Lemma 8.5)

$$\forall x_n \phi \not\Leftrightarrow \, ! \, x_n \phi.$$

**Definition 8.6** *The* dependence values *are the following special sentences of team logic (see Fig. 8.1 and Table 8.14):*

$$\top = \, =(),$$
$$\bot = \sim =(),$$
$$\mathbf{0} = \neg =(),$$
$$\mathbf{1} = \sim \neg =().$$

What about $\neg \sim =()$? This is not a sentence of team logic at all!

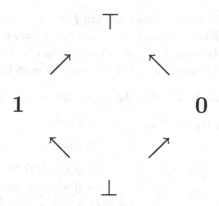

Fig. 8.1. The diamond of dependence values.

**Example 8.7** *Here are some trivial relations between the dependence values:*

(i) $\perp \Rightarrow 0 \Rightarrow \top$;

(ii) $\perp \Rightarrow 1 \Rightarrow \top$;

(iii) $1 = {\sim}0, \perp = {\sim}\top$;

(iv) $0 \Leftrightarrow {\sim}1, \top \Leftrightarrow {\sim}\perp$;

(v) $0 = \neg\top$.

The equation $X = X \cup X$ yields

$$(\phi \wedge \psi) \Rightarrow (\phi \otimes \psi),$$
$$(\phi \oplus \psi) \Rightarrow (\phi \vee \psi).$$

The equation $X = X \cup \emptyset$ yields

$$\phi \Rightarrow (\phi \otimes 0),$$
$$(\phi \oplus \perp) \Rightarrow {\sim}\phi.$$

The logic TL is much stronger than $\mathcal{D}$. Let us immediately note the failure of compactness.

**Proposition 8.8** *The logic TL does not satisfy the Compactness Theorem.*

*Proof* Let $\phi_n$ be the sentence $\forall x_0 \ldots \forall x_n \exists x_{n+1}(\neg x_0 = x_{n+1} \wedge \ldots \wedge \neg x_n = x_{n+1})$. Then, any finite subset of $T = \{\phi_n : n \in \mathbb{N}\} \cup \{{\sim}(\Phi_\infty)^*\}$ has a model, but $T$ itself does not have a model.                                    $\square$

**Proposition 8.9** *The logic* TL *does not satisfy the Löwenheim–Skolem Theorem. There is a sentence $\phi$ of team logic such that $\phi$ has an infinite model, but $\phi$ has no uncountable models. There is also a sentence $\psi$ of team logic such that $\psi$ has an uncountable model, but no countable models.*

*Proof* Recall Lemma 4.1 Let $\phi$ be the conjunction of $P^-$ and

$$\sim \exists x_5 \exists x_4 \,!\, x_0 \exists x_1 \,!\, x_2 \exists x_3 (= (x_2, x_3) \wedge x_4 < x_5$$
$$\wedge \, (((x_0 = x_2 \wedge x_0 < x_5)$$
$$\wedge \, (x_1 = x_3 \wedge x_1 < x_4))$$
$$\wedge \, ((\neg x_0 = x_2 \otimes \neg x_0 < x_5)$$
$$\wedge \, (\neg x_1 = x_3 \otimes \neg x_1 < x_4)))).$$

Then $\phi$ has the infinite model $(\mathbb{N}, +, \cdot, 0, 1, <)$. But since every model of $\phi$ is isomorphic to $(\mathbb{N}, +, \cdot, 0, 1, <)$, it cannot have any uncountable models. For $\psi$, recall the sentence $\Phi_{\mathrm{cmpl}}$ from Section 4.3. Let $\psi$ be the conjunction of the axioms of dense linear order and $\sim \Phi_{\mathrm{cmpl}}$. Now $\psi$ has models, e.g. $(\mathbb{R}, <)$, but every model is a dense complete order and is therefore uncountable.    □

It follows that there cannot be any translation of team logic into $\Sigma_1^1$, as such a translation would yield both the Compactness Theorem and the Löwenheim–Skolem Theorem as a consequence. With a translation to $\Sigma_1^1$ ruled out, it is difficult to imagine what a game theoretical semantics of team logic would look like.

Note that there cannot be a truth definition for TL in TL. Suppose $\tau^*(x_0)$ is in TL and $\mathcal{M} \models \phi$ is equivalent to $\mathcal{M} \models \tau^*(\ulcorner \phi \urcorner)$ in all Peano models $\mathcal{M}$. By using the formula $\sim \tau^*(x_0)$, we can reprove Tarski's Undefinability of Truth Theorem 6.20

Despite apparent similarities, team logic and linear logic [11] have very little to do with each other. In linear logic, resources are split into "consumable" parts. In team logic resources are split into "coherent" parts.

**Exercise 8.7** *Show that every formula of* TL *in which $\sim$ does not occur is logically equivalent with a sentence of* $\mathcal{D}$.

**Exercise 8.8** *Prove Example 8.7.*

**Exercise 8.9** *Prove the following equivalences:*

$$\phi \wedge \top \Leftrightarrow \phi,$$
$$\phi \vee \top \Leftrightarrow \top,$$
$$\top \otimes \top \Leftrightarrow \top,$$
$$\phi \oplus \top \Leftrightarrow \top.$$

**Exercise 8.10** *Prove the following equivalences:*

$$\phi \wedge \bot \Leftrightarrow \bot,$$
$$\phi \vee \bot \Leftrightarrow \phi,$$
$$\phi \otimes \bot \Leftrightarrow \bot,$$
$$\bot \oplus \bot \Leftrightarrow \bot.$$

**Exercise 8.11** *Prove the following equivalences:*

$$\phi^* \wedge 0 \Leftrightarrow \phi^* \qquad if \phi \in \mathcal{D},$$
$$\phi^* \vee 0 \Leftrightarrow \phi^* \qquad if \phi \in \mathcal{D},$$
$$\phi \otimes 0 \Leftrightarrow \phi,$$
$$0 \oplus 0 \Leftrightarrow 0.$$

**Exercise 8.12** *Prove the following equivalences:*

$$1 \otimes 1 \Leftrightarrow 1,$$
$$1 \oplus 1 \Leftrightarrow 1.$$

**Exercise 8.13** *Give an example which demonstrates* $\forall x_n \phi \not\Leftrightarrow \,! \, x_n \phi.$

**Exercise 8.14** *Suppose $\phi$ is a formula. Give a formula $\psi$ with the property that a team $X$ is of type $\psi$ if and only if every subset of $X$ is of type $\phi$.*

**Exercise 8.15** *Suppose $\phi$ is a formula. Give a formula $\psi$ with the property that a team $X$ is of type $\psi$ if and only if every subset of $X$ has a subset which is of type $\phi$.*

**Exercise 8.16** *Give a formula $\phi$ with the property that a team $X$ is of type $\phi$ if and only if every subset of $X$ has a subset which is of type $\phi$, but it is not true that a team $X$ is of type $\phi$ if and only if every subset of $X$ is of type $\phi$.*

**Exercise 8.17** *Show that $\sim$ is not definable from the other symbols in team logic, i.e show that the sentence $\sim P$, where $P$ is a 0-ary predicate symbol, is not logically equivalent to any sentence of team logic of vocabulary $\{P\}$ without $\sim$.*

**Exercise 8.18** *Show that $=(x_1, \ldots, x_n)$ is definable from the other symbols in team logic and formulas of the form $=(t)$. That is, show that there is a formula $\phi(x_1, \ldots, x_n)$ in team logic such that $=(x_1, \ldots, x_n)$ and $\phi(x_1, \ldots, x_n)$ are logically equivalent and $\phi(x_1, \ldots, x_n)$ has no occurrences of atomic formulas of the form $=(t_1, \ldots, t_m)$, where $m \geq 2$.*

Table 8.15.

| Atomic | Name |
| --- | --- |
| $t_1 = t_n$ | *equation* |
| $Rt_1 \ldots t_n$ | *relation* |
| $X_n^m t_1 \ldots t_m$ | *second order atomic* |

Table 8.16.

| Compound | Name |
| --- | --- |
| $\phi \vee \psi$ | *disjunction* |
| $\phi \wedge \psi$ | *conjunction* |
| $\sim \phi$ | *negation* |
| $\exists x_n \phi$ | *existential first order* |
| $\exists X_n^m \phi$ | *existential second order* |

## 8.4 From team logic to $L^2$

One of the main features of dependence logic is its equivalence with the $\Sigma_1^1$ part of second order logic, established in Chapter 6. Since $\Sigma_1^1$ is not closed under classical negation, we cannot do the same for the extension TL of $\mathcal{D}$. Instead we translate team logic into full second order logic $L^2$.

**Definition 8.10** *Suppose $L$ is a vocabulary. Second order logic $L^2$ has, in addition to the (individual) variables $x_n$, another type of variable, namely relation variables $X_n^m$. The variables $X_n^m$ are said to be m-ary, and will range over m-ary relations on the domain. The formulas of $L^2$ are of the atomic form as in Table 8.15 or compound form as in Table 8.16.*

The semantics of second order logic for a model $\mathcal{M}$ is defined by adopting the concept of a second order assignment for $\mathcal{M}$. These are functions $S$ such that, for all $m$ and $n$, $S(X_n^m) \subseteq M^m$. If $A \subseteq M^m$, then the modified assignment $S(A/X_n^m)$ maps $X_j^i$ to $S(X_j^i)$, if $i \neq m$ or $j \neq n$, and $X_n^m$ to $A$.

**Definition 8.11**

- $\mathcal{M} \models_{s,s} t_1 = t_2$ if and only if $t_1^{\mathcal{M}}\langle s \rangle = t_2^{\mathcal{M}}\langle s \rangle$;
- $\mathcal{M} \models_{s,s} Rt_1 \ldots t_n$ if and only if $(t_1^{\mathcal{M}}\langle s \rangle, \ldots, t_n^{\mathcal{M}}\langle s \rangle) \in R^{\mathcal{M}}$;
- $\mathcal{M} \models_{s,s} X_n^m t_1 \ldots t_m$ if and only if $(t_1^{\mathcal{M}}\langle s \rangle, \ldots, t_m^{\mathcal{M}}\langle s \rangle) \in S(m, n)$;

- $\mathcal{M} \models_{s,S} \phi \vee \psi$ *if and only if* $\mathcal{M} \models_{s,S} \phi$ *or* $\mathcal{M} \models_{s,S} \psi$;
- $\mathcal{M} \models_{s,S} \phi \wedge \psi$ *if and only if* $\mathcal{M} \models_{s,S} \phi$ *and* $\mathcal{M} \models_{s,S} \psi$;
- $\mathcal{M} \models_{s,S} {\sim} \phi$ *if and only if* $\mathcal{M} \not\models_{s,S} \phi$;
- $\mathcal{M} \models_{s,S} \exists x_n \phi$ *if and only if* $\mathcal{M} \models_{s(a/x_n),S} \phi$ *for some* $a \in M$;
- $\mathcal{M} \models_{s,S} \exists X_n^m \phi$ *if and only if* $\mathcal{M} \models_{s,S(A/X_n^m)} \phi$ *for some* $A \subseteq M^m$.

The definitions of both the syntax and the semantics of $L^2$ look deceptively similar to those of first order logic. However, $L^2$ is very different from first order logic. For example, $L^2$ does not satisfy the Compactness Theorem, nor the Löwenheim–Skolem Theorem, and its Decision Problem is $\Pi_2$-complete like that of $\mathcal{D}$. We give in this chapter translations of team logic into $L^2$ and, in a weaker sense, of $L^2$ into team logic.

**Theorem 8.12** *We can associate with every formula* $\phi(x_{i_1}, \ldots, x_{i_n})$ *of team logic in vocabulary* $L$ *an* $L^2$-*sentence* $\eta_\phi(U)$, *where* $U$ *is* $n$-*ary, such that for all* $L$-*structures* $\mathcal{M}$ *and teams* $X$ *with* $\mathrm{dom}(X) = \{i_1, \ldots, i_n\}$ *the following are equivalent:*

(i) $\mathcal{M} \models_X \phi$;
(ii) $(\mathcal{M}, \mathrm{rel}(X)) \models \eta_\phi(U)$.

*Proof* Exactly as in the proof of Theorem 6.2.

Case (1) Suppose $\phi(x_{i_1}, \ldots, x_{i_n})$ is $t_1 = t_2$ or $Rt_1 \ldots t_n$. Let $\eta_\phi(U)$ be given by

$$\forall x_{i_1} \ldots \forall x_{i_n} (U x_{i_1} \ldots x_{i_n} \to \phi(x_{i_1}, \ldots, x_{i_n})).$$

Case (2) Suppose $\phi(x_{i_1}, \ldots, x_{i_n})$ is the dependence formula

$$= (t_1(x_{i_1}, \ldots, x_{i_n}), \ldots, t_m(x_{i_1}, \ldots, x_{i_n})),$$

where $i_1 < \ldots < i_n$. We define $\eta_\phi(U)$ as follows.
Subcase (2.1) $m = 0$; we let $\eta_\phi(U) = \top$ and $\eta_{\neg\phi}(U) = \bot$.
Subcase (2.2) $m = 1$; now $\phi(x_{i_1}, \ldots, x_{i_n})$ is the dependence formula $= (t_1(x_{i_1}, \ldots, x_{i_n}))$. We let $\eta_\phi(U)$ be the formula

$$\forall x_{i_1} \ldots \forall x_{i_n} \forall x_{i_n+1} \ldots \forall x_{i_n+n} ((U x_{i_1} \ldots x_{i_n} \wedge U x_{i_n+1} \ldots x_{i_n+n})$$
$$\to t_1(x_{i_1}, \ldots, x_{i_n}) = t_1(x_{i_n+1}, \ldots, x_{i_n+n})),$$

and we further let $\eta_{\neg\phi}(U)$ be the formula $\forall x_{i_1} \ldots \forall x_{i_n} \neg U x_{i_1} \ldots x_{i_n}$.

**Subcase (2.3)** If $m > 1$ we let $\eta_\phi(U)$ be the formula

$$\forall x_{i_1} \ldots \forall x_{i_n} \forall x_{i_n+1} \ldots \forall x_{i_n+n}((U x_{i_1} \ldots x_{i_n} \wedge U x_{i_n+1} \ldots x_{i_n+n}$$
$$\wedge \, t_1(x_{i_1}, \ldots, x_{i_n}) = t_1(x_{i_n+1}, \ldots, x_{i_n+n})$$
$$\wedge \ldots$$
$$t_{m-1}(x_{i_1}, \ldots, x_{i_n}) = t_{m-1}(x_{i_n+1}, \ldots, x_{i_n+n}))$$
$$\rightarrow t_m(x_{i_1}, \ldots, x_{i_n}) = t_m(x_{i_n+1}, \ldots, x_{i_n+n})),$$

and we further let $\eta_{\neg\phi}(U)$ be the formula $\forall x_{i_1} \ldots \forall x_{i_n} \neg U x_{i_1} \ldots x_{i_n}$.

**Case (3)** Suppose $\phi(x_{i_1}, \ldots, x_{i_n})$ is the disjunction

$$(\psi(x_{j_1}, \ldots, x_{j_p}) \otimes \theta(x_{k_1}, \ldots, x_{k_q})),$$

where $\{i_1, \ldots, i_n\} = \{j_1, \ldots, j_p\} \cup \{k_1, \ldots, k_q\}$. We let the sentence $\eta_\phi(U)$ be

$$\exists R \exists T (\eta_\psi(R) \wedge \eta_\theta(T) \wedge \forall x_{i_1} \ldots \forall x_{i_n}(U x_{i_1} \ldots x_{i_n}$$
$$\rightarrow (R x_{j_1} \ldots x_{j_p} \vee T x_{k_1} \ldots x_{k_q}))).$$

**Case (4)** Suppose $\phi(x_{i_1}, \ldots, x_{i_n})$ is the conjunction

$$(\psi(x_{j_1}, \ldots, x_{j_p}) \wedge \theta(x_{k_1}, \ldots, x_{k_q})),$$

where $\{i_1, \ldots, i_n\} = \{j_1, \ldots, j_p\} \cup \{k_1, \ldots, k_q\}$. We let the sentence $\eta_\phi(U)$ be $\eta_\psi(U) \wedge \eta_\theta(U)$.

**Case (5)** $\phi$ is $\sim \psi$; $\eta_\phi(U)$ is the formula $\sim \eta_\psi(U)$.

**Case (6)** Suppose $\phi(x_{i_1}, \ldots, x_{i_n})$ is the formula $\exists x_{i_n+1} \psi(x_{i_1}, \ldots, x_{i_n+1})$. $\tau_{1,\phi}(U)$ is the formula

$$\exists R(\tau_{1,\psi}(R) \wedge \forall x_{i_1} \ldots \forall x_{i_n}(U x_{i_1} \ldots x_{i_n} \rightarrow \exists x_{i_n+1} R x_{i_1} \ldots x_{i_n+1})).$$

**Case (7)** Suppose $\phi(x_{i_1}, \ldots, x_{i_n})$ is the formula $! x_{i_n+1} \psi(x_{i_1}, \ldots, x_{i_n+1})$. $\tau_{1,\phi}(U)$ is the formula

$$\exists R(\eta_\psi(R) \wedge \forall x_{i_1} \ldots \forall x_{i_n+1}(U x_{i_1} \ldots x_{i_n} \leftrightarrow R x_{i_1} \ldots x_{i_n+1})). \qquad \square$$

**Corollary 8.13** *For every sentence $\phi$ of team logic there is an $L^2$-sentence $\eta_\phi$ such that for all models $\mathcal{M}$ we have $\mathcal{M} \models \phi$ if and only if $\mathcal{M} \models \eta_\phi$.*

*Proof* Let $\eta_\phi$ be the result of replacing in $\eta_\phi(U)$ every occurrence of the 0-ary relation symbol $U$ by $\top$. Now the claim follows from Theorem 8.12. $\qquad \square$

With the translation $\phi \mapsto \eta_\phi$ we can consider team logic TL a fragment of second order logic $L^2$, even if the origin of team logic in the dependence relation $=(t_1, \ldots, t_n)$ is totally different from the origin of second order logic, and even if the logical operations of team logic are totally different from those of second order logic.

We can draw many immediate conclusions from Corollary 8.13 and from what is known about second order logic (see, e.g., ref. [40]).

**Corollary 8.14** *The Decision Problem of team logic is $\Pi_2$-complete. The Consistency Problem is $\Sigma_2$-complete.*

**Corollary 8.15** *If $\kappa$ is measurable and $\phi \in$ TL has a model $\mathcal{M}$ of cardinality $\kappa$, then $\phi$ is true in a submodel $\mathcal{N}$ of $\mathcal{M}$ of cardinality $< \kappa$. If $\kappa$ is supercompact and $\phi \in$ TL has a model $\mathcal{M}$ of any cardinality, then $\phi$ is true in a submodel $\mathcal{N}$ of $\mathcal{M}$ of cardinality $< \kappa$.*

For the definition of measurable and supercompact cardinals we refer to ref. [25].

*Proof* This follows from the similar result for $L^2$ [32]. $\qquad\qquad\square$

## 8.5 From $L^2$ to team logic

We have given a translation of team logic into second order logic. Now we give an implicit translation of second and higher order logic into team logic. The translation is implicit in the sense that it uses new predicates and an extension of the universe. However, the new predicates and the new universe are unique up to isomorphism.

**Theorem 8.16** *Suppose $L$ is a vocabulary and $n \in \mathbb{N}$. There is a sentence $\Phi$ of TL in the vocabulary $L' = L \cup \{P, E\}$ such that for all $L$-structures $\mathcal{M}$ there is a unique (mod $\cong$) $\mathcal{N} \models \Phi$ with $(\mathcal{N}{\restriction}L)^{(P^{\mathcal{N}})} = \mathcal{M}$. Moreover, we can associate with every sentence $\phi$ of $L^2$, with second order variables of any arity $\leq n$, in vocabulary $L$ a TL-sentence $\xi_\phi$ in the vocabulary $L'$ such that the following are equivalent for all $L$-structures $\mathcal{M}$:*

(i) $\mathcal{M} \models \phi$;
(ii) $\mathcal{N} \models \xi_\phi$ *for the unique (mod $\cong$) $\mathcal{N}$ such that $\mathcal{N} \models \Phi$ and $(\mathcal{N}{\restriction}L)^{(P^{\mathcal{N}})} = \mathcal{M}$.*

*Proof* To avoid cumbersome notation, we restrict ourselves to second order formulas with only the unary second order variables $X_n^1$. We assume, again for simplicity, that individual variables $x_0, x_2, x_4, \ldots$ have not been used in our second order formulas. Clearly this is harmless as one can change bound variables. The idea is that $x_{2n}$ is a new "first order" symbol for $X_n^1$.

Let $\xi_{t_1=t_2}$ be $t_1 = t_2$ and let $\xi_{Rt_1...t_n}$ be $Rt_1 \ldots t_n$. For second order atomic formula, we let $\xi_{X_n^1 t}$ be $tEx_{2n}$. Proceeding to compound formulas, let $\xi_{\phi \vee \psi} = \xi_\phi \vee \xi_\psi$, $\xi_{\phi \wedge \psi} = \xi_\phi \wedge \xi_\psi$, $\xi_{\neg \phi} = \sim \xi_\phi$, $\xi_{\exists x_n \phi} = \exists x_n (Px_n \wedge \xi_\phi)$, and finally $\xi_{\exists X_n^1 \phi} = \exists z_n \xi_\phi$.

Let $\theta$ be the conjunction of the (first order definable) sentence

$$\forall x_0 \forall x_1 (\neg x_0 E x_1 \vee Px_0)$$

and the *Axiom of Extensionality*

$$\forall x_0 \forall x_1 (\exists x_2 ((x_2 E x_0 \wedge \neg x_2 E x_1) \vee (x_2 E x_1 \wedge \neg x_2 E x_0)) \vee x_0 = x_1).$$

Let $\Phi$ be the following sentence of team logic:

$$\theta \wedge \sim \exists x_6 \ ! \ x_0 \exists x_1 \ ! \ x_4 \exists x_5 \ (= (x_4, x_5)$$
$$\wedge \ (\neg x_4, x_5)$$
$$\wedge \ (\neg x_1 = x_6 \otimes Px_0)$$
$$\wedge \ (\neg x_0 = x_5$$
$$\otimes \ (x_5 E x_4 \wedge \neg x_1 = x_6)$$
$$\otimes \ (\neg x_5 E x_4 \wedge x_1 = x_6))).$$

The sentence $\Phi$ is true in a structure $\mathcal{M}$ if and only if $\mathcal{M} \cong \mathcal{N}$ for some $\mathcal{N}$ such that $E^{\mathcal{N}} = \{(a, b) : a \in P^{\mathcal{N}}, a \in b \in N\}$ and $N = \mathcal{P}(P^{\mathcal{N}})$.

Now one can use induction on $\phi \in L^2$ to prove that if $\mathcal{N}$ is as above then

$$\mathcal{M} \models \phi \text{ if and only if } \mathcal{N} \models \xi_\phi.$$

More precisely, one proves for all formulas $\phi \in L^2$ and all assignments $s$ and $S$ the following:

$$\mathcal{M} \models_{s,S} \phi \text{ if and only if } \mathcal{N} \models_{X_{s,S}} \xi_\phi,$$

where $X_{s,S} = \{s'\}$, $s'(x_{2n+1}) = s(x_{2n+1})$, and $s'(x_{2n}) = S(X_n^1)$.                    □

**Corollary 8.17** *A second order sentence $\phi$ has a model (of cardinality $\kappa$) if and only if the team logic sentence $\Phi \wedge \xi_\phi$ has a model (of cardinality $2^\kappa$).*

With the translation $\phi \mapsto \xi_\phi$ we can consider second order logic $L^2$ as an implicitly defined fragment of team logic TL.

**Corollary 8.18** *The Decision Problems of team logic and second order logic are recursively isomorphic.[2] They have the same Löwenheim and Hanf numbers. They have the same $\Delta$-extension.*

---

[2] Two sets of natural numbers are recursively isomorphic if there is a recursive bijection of $\mathbb{N}$ which maps one to the other.

The method of Theorem 8.16 was used in ref. [29] to prove that second order logic has an implicit translation into the extension of first order logic by the Henkin quantifier (see Exercise 6.20). As is apparent from the proof, we do not need the full power of team logic. It suffices to have the $\sim$-negated dependence logic sentence which forms one conjunct of $\Phi$. This emphasizes the strength of $\sim$ as compared to $\neg$. We note without proof that there is also a direct translation of second order logic into team logic, based on ref. [13].

**Exercise 8.19** *Suppose $\kappa$ is a cardinal number with the following property: if $\phi \in$ TL, then every model $\mathcal{M}$ of TL has a submodel $\mathcal{N}$ such that $\mathcal{N} \models \phi$ and $|N| < \kappa$. Show that $\kappa$ is inaccessible. (In fact, $\kappa$ is supercompact.)*

## 8.6 Ehrenfeucht–Fraïssé game for team logic

We now introduce an Ehrenfeucht–Fraïssé game adequate for team logic and use this game to characterize $\equiv_{\mathrm{TL}}$. This game is simply a "two-directional" version of the Ehrenfeucht–Fraïssé game of $\mathcal{D}$.

**Definition 8.19** *Let $\mathcal{M}$ and $\mathcal{N}$ be two structures of the same vocabulary. The game $\mathrm{EF}_n^{\mathrm{TL}}$ has two players and n moves. The position after move m is a pair $(X, Y)$, where $X \subseteq M^{i_m}$ and $Y \subseteq N^{i_m}$ for some $i_m$. At the beginning, the position is $(\{\emptyset\}, \{\emptyset\})$ and $i_0 = 0$. Suppose the position after move number m is $(X, Y)$. There exist the following possibilities for the continuation of the game.*

**Splitting move.** *Player **I** represents $X$ (or $Y$) as a union $X = X_0 \cup X_1$. Then player **II** represents $Y$ (respectively, $X$) as a union $Y = Y_0 \cup Y_1$. Now player **I** chooses whether the game continues from the position $(X_0, Y_0)$ or from the position $(X_1, Y_1)$.*

**Duplication move.** *Player **I** decides that the game should continue from the new position,*

$$(X(M/x_{i_m}), Y(N/x_{i_m})).$$

**Supplementing move.** *Player **I** chooses a function $F : X \to M$ (or $F : Y \to N$). Then player **II** chooses a function $G : Y \to N$ (respectively, $G : X \to M$). Then the game continues from the position $(X(F/x_{i_m}), Y(G/x_{i_m}))$.*

*After n moves, the position $(X_n, Y_n)$ is reached and the game ends. Player **II** is the winner if*

$$\mathcal{M} \models_{X_n} \phi \Leftrightarrow \mathcal{N} \models_{Y_n} \phi$$

*holds for all atomic, dual atomic, dependence, and dual dependence formulas*
$\phi(x_0, \ldots, x_{i_n-1})$. *Otherwise, player **I** wins.*

This is a game of perfect information and the concept of winning strategy is defined as usual. By the Gale–Stewart Theorem, the game is determined.

**Definition 8.20**

(i) $\mathrm{qr}^{\mathrm{TL}}(\phi) = 0$ *if $\phi$ is atomic, dual atomic, dependence or dual dependence formula.*

(ii) $\mathrm{qr}^{\mathrm{TL}}(\phi \otimes \psi) = \max(\mathrm{qr}^{\mathrm{TL}}(\phi), \mathrm{qr}^{\mathrm{TL}}(\psi))+1.$

(iii) $\mathrm{qr}^{\mathrm{TL}}(\phi \wedge \psi) = \max(\mathrm{qr}^{\mathrm{TL}}(\phi), \mathrm{qr}^{\mathrm{TL}}(\psi)).$

(iv) $\mathrm{qr}^{\mathrm{TL}}(\exists x_n \phi) = \mathrm{qr}^{\mathrm{TL}}(\phi)+1.$

(v) $\mathrm{qr}^{\mathrm{TL}}(!x_n \phi) = \mathrm{qr}^{\mathrm{TL}}(\phi)+1.$

(vi) $\mathrm{qr}^{\mathrm{TL}}(\sim \phi) = \mathrm{qr}^{\mathrm{TL}}(\phi)$

*Let $\mathrm{Fml}_n^{\mathrm{TL},m}$ be the set of formulas $\phi$ of dependence logic with $\mathrm{qr}^{\mathrm{TL}}\phi \leq m$ and with free variables among $x_0, \ldots, x_{n-1}$. We write $\mathcal{M} \equiv_{\mathrm{TL}}^n \mathcal{N}$, if $\mathcal{M} \models \phi$ is equivalent to $\mathcal{N} \models \phi$ for all $\phi$ in $\mathrm{Fml}_0^{\mathrm{TL},n}$.*

Note that there are for each $n$ and $m$, up to logical equivalence, only finitely many formulas in $\mathrm{Fml}_n^{\mathrm{TL},m}$.

**Theorem 8.21** *Suppose $\mathcal{M}$ and $\mathcal{N}$ are models of the same vocabulary. Then the following are equivalent:*

(1) *player **II** has a winning strategy in the game $\mathrm{EF}_n^{\mathrm{TL}}(\mathcal{M}, \mathcal{N})$;*

(2) $\mathcal{M} \equiv_{\mathrm{TL}}^n \mathcal{N}.$

*Proof* As in the proof of Theorem 6.44 we prove the equivalence, for all $n$, of the following two statements.

$(3)_m$ Player **II** has a winning strategy in the game $\mathrm{EF}_m^{\mathrm{TL}}(\mathcal{M}, \mathcal{N})$ in position $(X, Y)$, where $X \subseteq M^n$ and $Y \subseteq N^n$.

$(4)_m$ If $\phi$ is a formula in $\mathrm{Fml}_n^{\mathrm{TL},m}$, then

$$\mathcal{M} \models_X \phi \Leftrightarrow \mathcal{N} \models_Y \phi. \tag{8.1}$$

The proof is by induction on $m$. For each $m$ we prove the claim simultaneously for all $n$. The case $m = 0$ is true by construction. Let us then assume $(3)_m \iff (4)_m$ as an induction hypothesis. Assume now $(3)_{m+1}$ and let $\phi$ be a formula in $\mathrm{Fml}_n^{\mathrm{TL},m+1}$ such that $\mathcal{M} \models_X \phi$. As part of the induction hypothesis we assume that the claim in Eq. (8.1) holds for formulas shorter than $\phi$.

Case (1) $\phi = \psi_0 \otimes \psi_1$, where $\psi_0, \psi_1 \in \mathrm{Fml}_n^{\mathrm{TL},m}$. Since $\mathcal{M} \models_X \phi$, there are $X_0$ and $X_1$ such that $X = X_0 \cup X_1$, $\mathcal{M} \models_{X_0} \psi_0$ and $\mathcal{M} \models_{X_1} \psi_1$. We let

**I** play $\{X_0, X_1\}$. Then **II** plays according to her winning strategy, $\{Y_0, Y_1\}$. Since the next position in the game can be either one of $(X_0, Y_0)$, $(X_1, Y_1)$, we can apply the induction hypothesis to both. This yields $\mathcal{N} \models_{Y_0} \psi_0$ and $\mathcal{N} \models_{Y_1} \psi_1$. Thus $\mathcal{N} \models_Y \phi$.

Case (2)  $\phi = \exists x_n \psi$, where $\psi \in \mathrm{Fml}_{n-1}^{\mathrm{TL},m}$. Since $\mathcal{M} \models_X \phi$, there is a function $F : X \to M$ such that $\mathcal{M} \models_{X(F/x_n)} \psi$. We let **I** play $F$. Then **II** plays according to her winning strategy a function $G : Y \to N$ and the game continues in position $(X(F/x_n), Y(G/x_n))$. The induction hypothesis gives $\mathcal{N} \models_{Y(G/x_n)} \psi$. Now $\mathcal{N} \models_Y \phi$ follows.

Case (3)  $\phi = \sim\psi$. Since $\mathcal{M} \models_X \phi$, we have $\mathcal{M} \not\models_X \psi$. By the induction hypothesis, $\mathcal{N} \not\models_Y \psi$. Thus $\mathcal{N} \models_Y \phi$.

Case (4)  $\phi = \psi_0 \wedge \psi_1$, where $\psi_0, \psi_1 \in \mathrm{Fml}^{\mathrm{TL},m}$. Since $\mathcal{M} \models_X \phi$, we have $\mathcal{M} \models_X \psi_0$ and $\mathcal{M} \models_X \psi_1$. By the induction hypothesis, $\mathcal{N} \models_Y \psi_0$ and $\mathcal{N} \models_Y \psi_1$. Thus $\mathcal{N} \models_Y \phi$.

Case (5)  $\phi = !x_n \psi$, where $\psi \in \mathrm{Fml}_{n+1}^{\mathrm{TL},m}$. By assumption, $\mathcal{M} \models_{X(M/x_n)} \psi$. We let now **I** demand that the game continues in the new position $(X(M/x_n), Y(N/x_n))$. The induction hypothesis gives $\mathcal{N} \models_{Y(N/x_n)} \psi$. Now $\mathcal{N} \models_Y \phi$ follows trivially.

To prove the converse implication, assume $(4)_{m+1}$. To prove $(3)_{m+1}$, we consider the possible moves that player **I** can make in the position $(X, Y)$.

Case (i)  Player **I** writes $X = X_0 \cup X_1$. Let $\phi_j$, $j < k$, be a complete list (up to logical equivalence) of formulas in $\mathrm{Fml}_n^{\mathrm{TL},m}$. Since

$$\mathcal{M} \models_{X_0} \bigwedge_{\mathcal{M} \models_{X_0} \phi_j} \phi_j$$

and

$$\mathcal{M} \models_{X_1} \bigwedge_{\mathcal{M} \models_{X_1} \phi_j} \phi_j,$$

we have

$$\mathcal{M} \models_X \left( \bigwedge_{\mathcal{M} \models_{X_0} \phi_j} \phi_j \right) \vee \left( \bigwedge_{\mathcal{M} \models_{X_1} \phi_j} \phi_j \right).$$

Note that

$$\mathrm{qr}^{\mathrm{TL}} \left( \bigwedge_{\mathcal{M} \models_{X_1} \phi_j} \phi_j \right) = \max_{\mathcal{M} \models_{X_1} \phi_j} (\phi_j) \leq m.$$

Therefore by $(4)_{m+1}$

$$\mathcal{N} \models_Y \left( \bigwedge_{\mathcal{M} \models_{x_0} \phi_j} \phi_j \right) \vee \left( \bigwedge_{\mathcal{M} \models_{x_1} \phi_j} \phi_j \right).$$

Thus $Y = Y_0 \cup Y_1$ such that

$$\mathcal{N} \models_{Y_0} \bigwedge_{\mathcal{M} \models_{x_0} \phi_j} \phi_j$$

and

$$\mathcal{N} \models_{Y_1} \bigwedge_{\mathcal{M} \models_{x_1} \phi_j} \phi_j.$$

By this and the induction hypothesis, player **II** has a winning strategy in the positions $(X_0, Y_0), (X_1, Y_1)$. Thus she can play $\{Y_0, Y_1\}$ and maintain her winning strategy.

Case (ii) Player **I** decides that the game should continue from the new position $(X(M/x_n), Y(m/x_n))$. We claim that

$$\mathcal{M} \models_{X(M/x_n)} \phi \Leftrightarrow \mathcal{N} \models_{Y(N/x_n)} \phi$$

for all $\phi \in \mathrm{Fml}_{n+1}^{\mathrm{TL},m}$. From this the induction hypothesis would imply that **II** has a winning strategy in the position $(X(M/x_n), Y(N/x_n))$. So let us assume $\mathcal{M} \models_{X(M/x_n)} \phi$, where $\phi \in \mathrm{Fml}_{n+1}^{\mathrm{TL},m}$. By definition,

$$\mathcal{M} \models_X \, ! \, x_n \phi.$$

Since $! \, x_n \phi \in \mathrm{Fml}_n^{\mathrm{TL},m+1}$, $(4)_{m+1}$ gives $\mathcal{N} \models_Y \, ! \, x_n \phi$ and $\mathcal{N} \models_{Y(N/x_n)} \phi$ follows.

Case (iii) Player **I** chooses a function $F : X \to M$. Let $\phi_i, i < M$ be a complete list (up to logical equivalence) of formulas in $\mathrm{Fml}_{n+1}^{\mathrm{TL},m}$. Now

$$\mathcal{M} \models_X \exists x_n \bigwedge_{\mathcal{M} \models_{X(F/x_n)} \phi_i} \phi_i.$$

Note that

$$\mathrm{qr}^{\mathrm{TL}} \left( \exists x_n \bigwedge_{\mathcal{M} \models_{X(F/x_n)} \phi_j} \phi_j \right) = \mathrm{qr}^{\mathrm{TL}} \left( \bigwedge_{\mathcal{M} \models_{X(F/x_n)} \phi_j} \phi_j \right) + 1$$

$$= \left( \max_{\mathcal{M} \models_{X(F/x_n)} \phi_j} \mathrm{qr}^{\mathrm{TL}}(\phi_j) \right) + 1 \leq m + 1$$

and hence, by $(4)_{m+1}$,

$$\mathcal{N} \models_Y \exists x_n \bigwedge_{\mathcal{M} \models_{X(F/x_n)} \phi_i} \phi_i.$$

Thus there is a function $G : Y \to N$ such that

$$\mathcal{N} \models_{Y(G/x_n)} \bigwedge_{\mathcal{M} \models_{X(F/x_n)} \phi_i} \phi_i.$$

The game continues from position $(X(F/x_n), Y(G/x_n))$. Given that now

$$\mathcal{M} \models_{X(F/x_n)} \phi \Leftrightarrow \mathcal{N} \models_{Y(G/x_n)} \phi$$

for all $\phi \in \mathrm{Fml}_{n+1}^{\mathrm{TL},m}$, the induction hypothesis implies that **II** has a winning strategy in position $(X(F/x_n), Y(G/x_n))$. □

**Corollary 8.22** *Suppose $\mathcal{M}$ and $\mathcal{N}$ are models of the same vocabulary. Then the following are equivalent:*

(1) $\mathcal{M} \equiv_{\mathrm{TL}} \mathcal{N}$;
(2) *for all natural numbers $n$, player **II** has a winning strategy in the game $\mathrm{EF}_n^{\mathrm{TL}}(\mathcal{M}, \mathcal{N})$.*

**Proposition 8.23** *Suppose $K$ is a model class and $n$ is a natural number. Then the following are equivalent:*

(1) *$K$ is definable in team logic by a sentence in $\mathrm{Fml}_0^{\mathrm{TL},n}$;*
(2) *$K$ is closed under the relation $\equiv_{\mathrm{TL}}^n$.*

*Proof* Suppose $K$ is the class of models of $\phi \in \mathrm{Fml}_0^{\mathrm{TL},n}$. If $\mathcal{M} \models \phi$ and $\mathcal{M} \equiv_{\mathrm{TL}}^n \mathcal{N}$, then, by definition, $\mathcal{N} \models \phi$. Conversely, suppose $K$ is closed under $\equiv_{\mathrm{TL}}^n$. Let

$$\phi_{\mathcal{M}} = \bigwedge \{\phi : \phi \in \mathrm{Fml}_0^{\mathrm{TL},n}, \mathcal{M} \models \phi\},$$

where the conjunction is taken over a finite set which covers all such $\phi$ up to logical equivalence. Let $\theta$ be the disjunction of all $\phi_{\mathcal{M}}$, where $\mathcal{M} \in K$. Again we take the disjunction over a finite set up to logical equivalence. We show that $K$ is the class of models of $\theta$. If $\mathcal{M} \in K$, then $\mathcal{M} \models \phi_{\mathcal{M}}$, whence $\mathcal{M} \models \theta$. On the other hand, suppose $\mathcal{M} \models \phi_{\mathcal{N}}$ for some $\mathcal{N} \in K$. Now $\mathcal{N} \equiv_{\mathrm{TL}}^n \mathcal{M}$, for if $\mathcal{N} \models \phi$ and $\phi \in \mathrm{Fml}_\emptyset^{\mathrm{TL},n}$, then $\phi$ is logically equivalent with one of the conjuncts of $\phi_{\mathcal{N}}$, whence $\mathcal{M} \models \phi$. As $K$ is closed under $\equiv_{\mathrm{TL}}^n$, we have $\mathcal{M} \in K$. □

**Corollary 8.24** *Suppose K is a model class. Then the following are equivalent:*

(1)  *K  is definable in team logic;*
(2)  *there is a natural number n such that K is closed under the relation*

$$\mathcal{M} R \mathcal{N} \iff \text{player } \mathbf{II} \text{ has a winning strategy in } \mathrm{EF}_n^{\pi}(\mathcal{M}, \mathcal{N}).$$

We have obtained, after all, a purely game theoretic definition of team logic.

**Exercise 8.20** *Describe an Ehrenfeucht–Fraïssé  game for the fragment of team logic in which ⊗ does not occur. Give a game theoretic characterization in the spirit of Corollary 8.24 of this fragment.*

# Appendix

## Solutions to selected exercises, by Ville Nurmi

### Chapter 2

#### Exercise 2.1

Let $\mathcal{M}$ be a model and let $M$ be its universe. The task is to prove that for each first order $\phi$ and each assignment $s$ with $\mathrm{Fr}(\phi) \subseteq \mathrm{dom}(s)$ it holds that either $(\phi, s, 1) \in \mathcal{T}$ or $(\phi, s, 0) \in \mathcal{T}$. We prove it by induction on $\phi$.

Assume $\phi$ is of form $t_1 = t_2$. It is a fact that either $t_1^{\mathcal{M}}\langle s \rangle = t_2^{\mathcal{M}}\langle s \rangle$ or $t_1^{\mathcal{M}}\langle s \rangle \neq t_2^{\mathcal{M}}\langle s \rangle$. Thus, by definition of $\mathcal{T}$, either $(t_1 = t_2, s, 1) \in \mathcal{T}$ or $(t_1 = t_2, s, 0) \in \mathcal{T}$.

Assume $\phi$ is of form $Rt_1 \ldots t_n$. Again, either $(t_1^{\mathcal{M}}\langle s \rangle, \ldots, t_n^{\mathcal{M}}\langle s \rangle) \in R^{\mathcal{M}}$ or $(t_1^{\mathcal{M}}\langle s \rangle, \ldots, t_n^{\mathcal{M}}\langle s \rangle) \notin R^{\mathcal{M}}$. Therefore, by the definition of $\mathcal{T}$, either $(Rt_1 \ldots t_n, s, 1) \in \mathcal{T}$ or $(Rt_1 \ldots t_n, s, 0) \in \mathcal{T}$.

For the following inductive steps we assume as our induction hypothesis that the claim holds for subformulas of $\phi$.

Assume $\phi$ is of form $\neg\psi$. By the induction hypothesis, either $(\psi, s, 1) \in \mathcal{T}$ or $(\psi, s, 0) \in \mathcal{T}$. By definition of $\mathcal{T}$, we get in the first case that $(\neg\psi, s, 0) \in \mathcal{T}$, and in the latter case $(\neg\psi, s, 1) \in \mathcal{T}$.

Assume $\phi$ is of form $\psi \vee \theta$. If $(\psi, s, 0) \in \mathcal{T}$ and $(\theta, s, 0) \in \mathcal{T}$, then by definition of $\mathcal{T}$ also $(\psi \vee \theta, s, 0) \in \mathcal{T}$. Otherwise, either $(\psi, s, 0) \notin \mathcal{T}$ or $(\theta, s, 0) \notin \mathcal{T}$, whence, by the induction hypothesis, either $(\psi, s, 1) \in \mathcal{T}$ or $(\theta, s, 1) \in \mathcal{T}$. By definition of $\mathcal{T}$, we get $(\psi \vee \theta, s, 1) \in \mathcal{T}$.

Finally, assume $\phi$ is of form $\exists x_n \psi$. If there is some $a \in M$ satisfying $(\psi, s(a/x_n), 1) \in \mathcal{T}$, then, by definition of $\mathcal{T}$, we get $(\exists x_n \psi, s, 1) \in \mathcal{T}$. Otherwise, for all $a \in M$ we have $(\psi, s(a/x_n), 1) \notin \mathcal{T}$. By the induction hypothesis, for all $a \in M$ we have $(\psi, s(a/x_n), 0) \in \mathcal{T}$, whence, by definition of $\mathcal{T}$, $(\exists x_n \psi, s, 0) \in \mathcal{T}$.

#### Exercise 2.7

The task is to show $\phi \equiv \phi^{\mathrm{p}}$ and $\neg\phi \equiv \phi^{\mathrm{d}}$ for all first order $\phi$. We prove both claims simultaneously by induction on $\phi$. The induction hypothesis is that the claim holds for subformulas of $\phi$. In the following, the places where the induction hypothesis is used are marked by IH.

169

If $\phi$ is atomic, then $\phi^p = \phi$ and $\phi^d = \neg\phi$, by definition.

If $\phi$ is of form $\neg\psi$, then $(\neg\psi)^p = \psi^d \overset{\text{IH}}{\equiv} \neg\psi$. Also $(\neg\psi)^d = \psi^p \overset{\text{IH}}{\equiv} \psi \equiv \neg\neg\psi$.

If $\phi$ is of form $\psi \vee \theta$, then $(\psi \vee \theta)^p = \psi^p \vee \theta^p \overset{\text{IH}}{\equiv} \psi \vee \theta$. Also, $(\psi \vee \theta)^d = \psi^d \wedge \theta^d \overset{\text{IH}}{\equiv} (\neg\psi) \wedge (\neg\theta) \equiv \neg(\psi \vee \theta)$.

Finally, if $\phi$ is of form $\exists x_n \psi$, then $(\exists x_n \psi)^p = \exists x_n \psi^p \overset{\text{IH}}{\equiv} \exists x_n \psi$. Also, $(\exists x_n \psi)^d = \forall x_n \psi^d \overset{\text{IH}}{\equiv} \forall x_n(\neg\psi) \equiv \neg\exists x_n \psi$.

# Chapter 3

## Exercise 3.2

Let $X \in \text{Team}(M, \{x_0, x_1\})$, i.e., let $X$ be some team of assignments $s : \{0, 1\} \to M$. For $X$ to be of type $\phi$ in $\mathcal{M}$ it means that $\mathcal{M} \models_X \phi$.

Now $\mathcal{M} \models_X =(x_0, x_1) \iff (=(x_0, x_1), X, 1) \in T$. By Proposition 3.8, this is equivalent to the condition that for all $s, s' \in X$ it holds that if $s(x_0) = s'(x_0)$ then $s(x_1) = s'(x_1)$. An alternative characterization can be given in terms of a function $f : A \to M$ for some $A \subseteq M$. Then a team $X$ of type $=(x_0, x_1)$ is of the form $\{\{(0, a), (1, f(a))\} : a \in A\}$ for some $A$ and $f$.

Similarly to the previous case, $\mathcal{M} \models_X =(x_1, x_0) \iff (=(x_1, x_0), X, 1) \in T \iff$ (by Proposition 3.8) for all $s, s' \in X$, it holds that if $s(x_1) = s'(x_1)$ then $s(x_0) = s'(x_0)$. In other words, such teams $X$ are of the form $\{\{(0, f(a)), (1, a)\} : a \in A\}$ for some set $A \subseteq M$ and function $f : M \to M$.

$\mathcal{M} \models_X =(x_0, x_0) \iff (=(x_0, x_0), X, 1) \in T \iff$ (by Proposition 3.8) for all $s, s' \in X$ it holds that if $s(x_0) = s'(x_0)$ then $s(x_0) = s'(x_0)$. But this is a trivial condition, satisfied by any team $X$. Thus all $X \in \text{Team}(M, \{x_0, x_1\})$ are of type $=(x_0, x_0)$.

## Exercise 3.3

The language consists of a constant symbol $c$ and a function symbol $f$. We consider teams $X \in \text{Team}(M, \{x_0\})$.

$\mathcal{M} \models_X =(c, c) \iff$ (by Proposition 3.8) for all $s, s' \in X$ it holds that if $c^{\mathcal{M}} = c^{\mathcal{M}}$ then $c^{\mathcal{M}} = c^{\mathcal{M}}$. This condition is trivially true for all $X$. Thus all teams $X \in \text{Team}(M, \{x_0\})$ are of type $=(c, c)$.

$\mathcal{M} \models_X =(x_0, c) \iff$ (by Proposition 3.8) for all $s, s' \in X$, it holds that if $s(x_0) = s'(x_0)$ then $c^{\mathcal{M}} = c^{\mathcal{M}}$. This condition is also trivially true, so all teams $X \in \text{Team}(M, \{x_0\})$ are of type $=(x_0, c)$.

$\mathcal{M} \models_X =(c, x_0) \iff$ (by Proposition 3.8) for all $s, s' \in X$, it holds that if $c^{\mathcal{M}} = c^{\mathcal{M}}$ then $s(x_0) = s'(x_0)$. As $c^{\mathcal{M}} = c^{\mathcal{M}}$ always holds, the condition states that there is some fixed $a \in M$ such that all $s \in X$ map $s(x_0) = a$. Considering also that $\text{dom}(s) = \{0\}$ for all $s \in X$, we get that either $X = \emptyset$ or it is of the form $\{\{(0, a)\}\}$ for some $a \in M$, i.e., there is at most one agent in team $X$.

$\mathcal{M} \models_X =(c, fx_0) \iff$ (by Proposition 3.8) for all $s, s' \in X$, it holds that if $c^{\mathcal{M}} = c^{\mathcal{M}}$ then $f^{\mathcal{M}}(s(x_0)) = f^{\mathcal{M}}(s'(x_0)) \iff$ for all $s, s' \in X : f^{\mathcal{M}}(s(x_0)) = f^{\mathcal{M}}(s'(x_0)) \iff f^M$ is constant in the set $\{s(x_0) : s \in X\}$.

Table A.1.

| $a$ | $b$ | $a+b$ | $a \cdot b$ | $a$ | $b$ | $a+b$ | $a \cdot b$ |
|-----|-----|-------|-------------|-----|-----|-------|-------------|
| 2 | 2 | 4 | 4 | 3 | 4 | 7 | 12 |
| 2 | 3 | 5 | 6 | 3 | 5 | 8 | 15 |
| 2 | 4 | 6 | 8 | 4 | 4 | 8 | 16 |
| 2 | 5 | 7 | 10 | 4 | 5 | 9 | 20 |
| 3 | 3 | 9 | 9 | 5 | 5 | 10 | 25 |

## Exercise 3.5

Here the language has two function symbols, $f$ and $g$, and we consider teams $X \in$ Team$(M, \{x_0, x_1\})$.

$\mathcal{M} \models_X = (f x_0, x_0) \iff$ (by Proposition 3.8) for all $s, s' \in X$, it holds that if $f^{\mathcal{M}}(s(x_0)) = f^{\mathcal{M}}(s'(x_0))$ then $s(x_0) = s'(x_0) \iff$ the function $f^{\mathcal{M}} \upharpoonright \{s(x_0) : s \in X\}$ is one-to-one. One can also characterize this condition by saying, a bit vaguely, that $f^{\mathcal{M}}$ is one-to-one in $\{s(x_0) : s \in X\}$.

$\mathcal{M} \models_X = (f x_1, x_0) \iff$ (by Proposition 3.8) for all $s, s' \in X$, it holds that if $f^{\mathcal{M}}(s(x_1)) = f^{\mathcal{M}}(s'(x_1))$ then $s(x_0) = s'(x_0)$. This is equivalent to being able to define a function $h : \{f^{\mathcal{M}}(s(x_1)) : s \in X\} \to M$ by the equation $h(f^{\mathcal{M}}(s(x_1))) = s(x_0)$.

$\mathcal{M} \models_X = (f x_0, g x_1) \iff$ (by Proposition 3.8) for all $s, s' \in X$, it holds that if $f^{\mathcal{M}}(s(x_0)) = f^{\mathcal{M}}(s'(x_0))$ then $g^{\mathcal{M}}(s(x_1)) = g^{\mathcal{M}}(s'(x_1))$. This is equivalent to being able to define a function $h : \{f^{\mathcal{M}}(s(x_0)) : s \in X\}$ by the equation $h(f^{\mathcal{M}}(s(x_0))) = g^{\mathcal{M}}(s(x_1))$.

## Exercise 3.7

Let $X_5 = \{\{(0, a), (1, b)\} : 1 < a \leq n, 1 < b \leq n, a \leq b\}$. It is possible to prove in general that, for any natural number $n$, $X_n$ is of type $=(x_0 + x_1, x_0 \cdot x_1, x_0)$. When inspecting just the case $n = 5$ we can save ourselves the trouble of the general proof and just check all cases as they are not that many. Thus we list for each element $s = \{\{(0, a), (1, b)\}\} \in X_5$ the values $a, b, a + b$, and $a \cdot b$ (see Table A.1).

We can see that for no two $s, s' \in X_5$ does it hold that $s(x_0) + s(x_1) = s'(x_0) + s'(x_1)$ and $s(x_0) \cdot s(x_1) = s'(x_0) \cdot s'(x_1)$ and $s(x_0) \neq s'(x_0)$. This proves that $X_5$ is of type $=(x_0 + x_1, x_0 \cdot x_1, x_0)$.

Here is an alternative, general proof that $X_n$ is of type $=(x_0 + x_1, x_0 \cdot x_1, x_0)$, for any $n$. Let $1 < x_0 \leq x_1 \leq n$. Denote $a = x_0 + x_1$ and $b = x_0 \cdot x_1$. Then we have $a^2 - 4b = x_0^2 - 2x_0x_1 + x_1^2 = (x_1 - x_0)^2 \geq 0$. To see that knowing the values of $a$ and $b$ gives us a unique value for $x_0$, we can continue as follows. Because $x_1 = a - x_0$, we get $b = -x_0^2 + ax_0$. This gives the quadratic equation $x_0^2 - ax_0 + b = 0$, which yields $x_0 = \frac{1}{2}(a \pm \sqrt{a^2 - 4b})$. Because $x_0 \leq x_1$, the only solution is $x_0 = \frac{1}{2}(a - \sqrt{a^2 - 4b})$.

## Exercise 3.10

In order to show $\models \phi$ for some sentence $\phi$, one must show that, for all models $\mathcal{M}$, $\mathcal{M} \models_{\{\emptyset\}} \phi$. Therefore, let $\mathcal{M}$ be arbitrary:

$$\mathcal{M} \models_{\{\emptyset\}} \forall x_0 \forall x_1 (x_1 = fx_0 \rightarrow \; = (x_0, x_1))$$
$$\Leftrightarrow \mathcal{M} \models_X \neg x_1 = fx_0 \vee =(x_0, x_1), \text{ where } X = \{s : \{0,1\} \rightarrow M\}$$
$$\Leftrightarrow \text{ there are teams } Y, Z \text{ such that } X = Y \cup Z, \; \mathrm{dom}(Y) = \mathrm{dom}(Z),$$
$$\text{and } \mathcal{M} \models_Y \neg x_1 = fx_0 \text{ and } \mathcal{M} \models_Z =(x_0, x_1). \tag{A.1}$$

So, in order to prove the claim, it suffices to find teams $Y$ and $Z$ satisfying the condition given in Eq. (A.1). We can choose $Y = \{s \in X : s(x_1) \neq f^{\mathcal{M}}(s(x_0))\}$ and $Z = X \setminus Y$. Then, clearly, $X = Y \cup Z$ and $\mathrm{dom}(Y) = \mathrm{dom}(Z)$. It is easy to check that $\mathcal{M} \models_Y \neg x_1 = f(x_0)$. Also $\mathcal{M} \models_Z =(x_0, x_1)$ because, in fact, $Z = \{s \in X : s(x_1) = f^{\mathcal{M}}(s(x_0))\}$, so, whenever we have some $s, s' \in Z$ with $s(x_0) = s'(x_0)$, then $s(x_1) = f^{\mathcal{M}}(s(x_0)) = f^{\mathcal{M}}(s'(x_0)) = s'(x_1)$.

## Exercise 3.11

First note that for all formulas $\phi$ and models $\mathcal{M}$ it holds that $(\phi, \emptyset, 1) \in \mathcal{T}$ and $(\neg \phi, \emptyset, 1) \in \mathcal{T}$, i.e., $\mathcal{M} \models_\emptyset \phi$ and $\mathcal{M} \models_\emptyset \neg \phi$. This means that $\mathcal{M} \models_X \neg \phi \Rightarrow \mathcal{M} \not\models_X \phi$ fails for $X = \emptyset$. Therefore, when studying the claim $\mathcal{M} \models_X \neg \phi \Leftrightarrow \mathcal{M} \not\models_X \phi$ for some model $\mathcal{M}$ and team $X$, we limit ourselves to non-empty teams.

Let $\phi$ be the formula $=(x_0, x_1) \wedge \neg x_0 = x_1$. Assume towards contradiction that for all models $\mathcal{M}$ and teams $X \neq \emptyset$ it holds that $\mathcal{M} \models_X \neg \phi \iff \mathcal{M} \not\models_X \phi$. Let us inspect the model $\mathcal{M}$ of the empty language, with universe $M = \{a, b\}$, where $a$ and $b$ are two different elements. (There are also lots of other possible models to consider, but this is probably the simplest one.) Let $X = \{s, s'\}$, where $s = \{(0, a), (1, a)\}$ and $s' = \{(0, a), (1, b)\}$. Then $\mathcal{M} \not\models_X \phi$ because $\mathcal{M} \not\models_X =(x_0, x_1)$. By assumption, $\mathcal{M} \models_X \neg \phi$. By the Closure Test, $\mathcal{M} \models_{\{s'\}} \neg \phi$. Again, by assumption, $\mathcal{M} \not\models_{\{s'\}} \phi$. But this is a contradiction with the facts that $\mathcal{M} \models_{\{s'\}} =(x_0, x_1)$ and $\mathcal{M} \models_{\{s'\}} \neg x_0 = x_1$. Thus the equivalence does not hold for $=(x_0, x_1) \wedge \neg x_0 = x_1$.

Let $\psi$ be the formula $=(x_0, x_1) \rightarrow x_0 = x_1$, i.e. $\neg =(x_0, x_1) \vee x_0 = x_1$. Assume towards contradiction that for all models $\mathcal{M}$ and teams $X \neq \emptyset$ it holds that $\mathcal{M} \models_X \neg \psi \iff \mathcal{M} \not\models_X \psi$. Let $\mathcal{M}$ and $X = \{s, s'\}$ be as earlier. Now $\mathcal{M} \not\models_X \psi$ because for all $X = Y \cup Z$ either $\mathcal{M} \not\models_Y \neg =(x_0, x_1)$ or $\mathcal{M} \not\models_Z x_0 = x_1$. Namely, $\mathcal{M} \not\models_Z x_0 = x_1$ can hold only for $Z = \emptyset$ and for $Z = \{s\}$ (assuming that $Z \subseteq X$), and, by Proposition 3.8, $\mathcal{M} \not\models_Y \neg =(x_0, x_1)$ can hold only when $Y = \emptyset$. We cannot have $X = Y \cup Z$ with any sets $Y$ and $Z$ satisfying these conditions. Now, by the assumption, $\mathcal{M} \models_X \neg \psi$. By the Closure Test, $\mathcal{M} \models_{\{s\}} \neg \psi$. Again by the assumption, $\mathcal{M} \not\models_{\{s\}} \psi$. But this cannot be true because we can split $\{s\} = \emptyset \cup \{s\}$, and it holds that $\mathcal{M} \models_\emptyset \neg =(x_0, x_1)$ and $\mathcal{M} \models_{\{s\}} x_0 = x_1$. We have ended up in a contradiction, so $\mathcal{M} \not\models_X \psi \Leftrightarrow \mathcal{M} \models_X \neg \psi$ does not hold for all models $\mathcal{M}$ and teams $X \neq \emptyset$.

Let $\theta$ be the formula $=(x_0, x_1) \vee \neg x_0 = x_1$. By definition, $\mathcal{M} \models_X \neg \theta \iff (\neg \theta, X, 1) \in \mathcal{T} \iff (\theta, X, 0) \in \mathcal{T} \iff (=(x_0, x_1), X, 0) \in \mathcal{T}$ and $(\neg x_0 = x_1, X, 0) \in \mathcal{T} \iff X = \emptyset$ (by Proposition 3.8, $(=(x_0, x_1), X, 0) \in \mathcal{T}$ iff $X = \emptyset$), and we have excluded this case from our investigations. Therefore $\mathcal{M} \models_X \neg \theta$ fails for all models $\mathcal{M}$ and teams $X \neq \emptyset$. On the other hand, for any model $\mathcal{M}$ and team

$X \neq \emptyset$, letting $Z = \{s \in X : s(x_0) \neq s(x_1)\}$ and $Y = X \setminus Z$, we see that $X = Y \cup Z$, $\mathrm{dom}(Y) = \mathrm{dom}(Z)$, $\mathcal{M} \models_Y =(x_0, x_1)$, and $\mathcal{M} \models_Z \neg x_0 = x_1$. Therefore $\mathcal{M} \not\models_X \theta$ fails for all $\mathcal{M}$ and $X \neq \emptyset$. Hereby we have proven the equivalence $\mathcal{M} \not\models_X \theta \Leftrightarrow \mathcal{M} \models_X \neg\theta$ for all models $\mathcal{M}$ and teams $X \neq \emptyset$.

## Exercise 3.13

Let $\mathcal{M}$ be a model. Assume that each $T_i$ for $i \in I$ satisfies conditions (D1)–(D12) of Definition 3.5. Our goal is to show that in that case $\bigcap_{i \in I} T_i$ satisfies the conditions also. The proof proceeds condition by condition. At each step we consider some arbitrary $X \in \mathrm{Team}(M, V)$, where $V$ contains all variable indices that appear in the formula in question.

(D1) Assume that for all $s \in X$ we have $t_1^{\mathcal{M}}\langle s \rangle = t_2^{\mathcal{M}}\langle s \rangle$. Then, because each $T_i$ satisfies (D1), we have $(t_1 = t_2, X, 1) \in T_i$. Therefore $(t_1 = t_2, X, 1) \in \bigcap_{i \in I} T_i$, so $\bigcap_{i \in I} T_i$ satisfies (D1).

(D2) Assume that for all $s \in X$ we have $t_1^{\mathcal{M}}\langle s \rangle \neq t_2^{\mathcal{M}}\langle s \rangle$. Then, because each $T_i$ satisfies (D2), we have $(t_1 = t_2, X, 0) \in T_i$. Therefore $(t_1 = t_2, X, 0) \in \bigcap_{i \in I} T_i$, so $\bigcap_{i \in I} T_i$ satisfies (D2).

(D3) Assume that for all $s, s' \in X$ it holds that if $t_i^{\mathcal{M}}\langle s \rangle = t_i^{\mathcal{M}}\langle s' \rangle$ for $i < n$, then $t_n^{\mathcal{M}}\langle s \rangle = t_n^{\mathcal{M}}\langle s' \rangle$. Then, because each $T_i$ satisfies (D3), we have $(=(t_1, \ldots, t_n), X, 1) \in T_i$. Therefore $(=(t_1, \ldots, t_n), X, 1) \in \bigcap_{i \in I} T_i$, so $\bigcap_{i \in I} T_i$ satisfies (D3).

(D4) Because each $T_i$ satisfies (D4), we have $(=(t_1, \ldots, t_n), \emptyset, 0) \in T_i$. Therefore $(=(t_1, \ldots, t_n), \emptyset, 0) \in \bigcap_{i \in I} T_i$, so $\bigcap_{i \in I} T_i$ satisfies (D4).

(D5) Assume that for all $s \in X$ we have $(t_1^{\mathcal{M}}\langle s \rangle, \ldots, t_n^{\mathcal{M}}\langle s \rangle) \in R^{\mathcal{M}}$. Then, because each $T_i$ satisfies (D5), we have $(Rt_1 \ldots t_n, X, 1) \in T_i$. Therefore the triple $(Rt_1 \ldots t_n, X, 1)$ is in $\bigcap_{i \in I} T_i$, so $\bigcap_{i \in I} T_i$ satisfies (D5).

(D6) Assume that for all $s \in X$ we have $(t_1^{\mathcal{M}}\langle s \rangle, \ldots, t_n^{\mathcal{M}}\langle s \rangle) \notin R^{\mathcal{M}}$. Then, because each $T_i$ satisfies (D6), we have $(Rt_1 \ldots t_n, X, 0) \in T_i$. Therefore the triple $(Rt_1 \ldots t_n, X, 0)$ is in $\bigcap_{i \in I} T_i$, so $\bigcap_{i \in I} T_i$ satisfies (D6).

(D7) Assume that $(\phi, X, 1) \in \bigcap_{i \in I} T_i$ and $(\psi, Y, 1) \in \bigcap_{i \in I} T_i$ and $\mathrm{dom}(X) = \mathrm{dom}(Y)$. Then $(\phi, X, 1) \in T_i$ and $(\psi, Y, 1) \in T_i$ for each $i \in I$. Because $T_i$ satisfies (D7), $(\phi \vee \psi, X \cup Y, 1) \in T_i$. Therefore $(\phi \vee \psi, X \cup Y, 1) \in \bigcap_{i \in I} T_i$, so $\bigcap_{i \in I} T_i$ satisfies (D7).

(D8) Assume that $(\phi, X, 0) \in \bigcap_{i \in I} T_i$ and $(\psi, X, 0) \in \bigcap_{i \in I} T_i$. Then $(\phi, X, 0) \in T_i$ and $(\psi, X, 0) \in T_i$ for each $i \in I$. As $T_i$ satisfies (D8), $(\phi \vee \psi, X, 0) \in T_i$ for each $i \in I$. Therefore $(\phi \vee \psi, X, 0) \in \bigcap_{i \in I} T_i$, so $\bigcap_{i \in I} T_i$ satisfies (D8).

(D9) Assume that $(\phi, X, 1) \in \bigcap_{i \in I} T_i$. Then $(\phi, X, 1) \in T_i$ for each $i \in I$. Because each $T_i$ satisfies (D9), $(\neg\phi, X, 0) \in T_i$. Therefore $(\neg\phi, X, 0) \in \bigcap_{i \in I} T_i$, so $\bigcap_{i \in I} T_i$ satisfies (D9).

(D10) Assume that $(\phi, X, 0) \in \bigcap_{i \in I} T_i$. Then $(\phi, X, 0) \in T_i$ for each $i \in I$. Because each $T_i$ satisfies (D10), $(\neg\phi, X, 1) \in T_i$. Therefore $(\neg\phi, X, 1) \in \bigcap_{i \in I} T_i$, so $\bigcap_{i \in I} T_i$ satisfies (D10).

(D11) Let $F : X \to M$, and assume that $(\phi, X(F/x_n), 1) \in \bigcap_{i \in I} T_i$. Then $(\phi, X(F/x_n), 1) \in T_i$ for each $i \in I$. Because each $T_i$ satisfies (D11), $(\exists x_n \phi, X, 1) \in T_i$. Therefore $(\exists x_n \phi, X, 1) \in \bigcap_{i \in I} T_i$, so $\bigcap_{i \in I} T_i$ satisfies (D11).

Table A.2. *The teams $X$ and $X(F/x_0)$ for Exercise 3.15*

| | $X$ | | | $X(F/x_0)$ | | |
|---|---|---|---|---|---|---|
| | $x_0$ | $x_1$ | $x_2$ | $x_0$ | $x_1$ | $x_2$ |
| $s$ | | $a$ | $a$ | $c$ | $a$ | $a$ |
| $s'$ | | $b$ | $a$ | $c$ | $b$ | $a$ |

(D12)   Assume that $(\phi, X(M/x_n), 0) \in \bigcap_{i \in I} T_i$. Then $(\phi, X(M/x_n), 0) \in T_i$ for each $i \in I$. Because each $T_i$ satisfies (D12), $(\exists x_n \phi, X, 0) \in T_i$. Therefore $(\exists x_n \phi, X, 0) \in \bigcap_{i \in I} T_i$, so $\bigcap_{i \in I} T_i$ satisfies (D12).

## Exercise 3.15

Denote by $\psi$ the formula $\exists x_0 \big( =(x_2, x_0) \wedge \neg(x_0 = x_1) \big)$ and by $\phi$ the formula $\exists x_0 \big( =(x_2, x_0) \wedge x_0 = x_1 \big)$. To show that $M \models_X \phi \iff M \models_X \psi$ does not hold, let $M = \{a, b, c\}$, where $a$, $b$, and $c$ are three different elements, and let $X = \{s, s'\} \in \mathrm{Team}(M, \{x_1, x_2\})$, where $s = \{(x_1, a), (x_2, a)\}$ and $s' = \{(x_1, b), (x_2, a)\}$ (see Table A.2). Then $M \models_X \psi$ and $M \not\models_X \phi$. We can see this by noting that $x_2$ is constant in $X$. Because, on the other hand, both $\phi$ and $\psi$ contain $=(x_2, x_0)$, any function $F : X \to M$ that unravels $\exists x_0$ in $\phi$ or $\psi$ must be a constant function. We can choose a constant that avoids both $a$ and $b$, but we cannot choose a constant that equals both $a$ and $b$.

More precisely, let $F : X \to M$ map $F(s) = c$ for all $s \in X$. Then $M \models_{X(F/x_0)} \neg(x_0 = x_1)$, and $M \models_{X(F/x_0)} =(x_2, x_0)$. Thus $M \models_{X(F/x_0)} \neg(x_0 = x_1) \wedge =(x_2, x_0)$, so $M \models_X \psi$. On the other hand, if $M \models_X \phi$ was to hold, there would be some $G : X \to M$ such that $M \models_{X(G/x_0)} =(x_2, x_0)$ and $M \models_{X(G/x_0)} x_0 = x_1$. From the former we get that when $s, s' \in X(G/x_0)$ and $s(x_2) = s'(x_2)$, then $s(x_0) = s'(x_0)$. But $s(x_2) = s'(x_2)$ holds for all $s, s' \in X(G/x_0)$. Therefore $G$ must be a constant function; there is some $x \in M$ such that for all $s \in X : G(s) = x$. But then by $M \models_{X(G/x_0)} x_0 = x_1$ it must be that $x = a$, and $x = b$, which is a contradiction. Thus $M \not\models_X \phi$.

## Exercise 3.17

Denote by $\phi$ the formula $\exists x_1 \big( =(f x_0, x_1) \wedge f x_1 = f x_0 \big)$. We want to falsify the following claim:

$$M \models_X \phi \iff f^M \text{ is one-to-one in } \{s(x_0) : s \in X\}.$$

Let $M = \{a, b\}$, where $a$ and $b$ are distinct elements, interpret $f^M(x) = a$ for all $x \in M$, and let $X = \{s, s'\} \in \mathrm{Team}(M, \{x_0\})$, where $s = \{(0, a)\}$ and $s' = \{(0, b)\}$ (see Table A.3). Clearly, $f^M$ is not one-to-one in $\{a, b\}$ because it is a constant function, but $M \models_X \phi$; namely, let $F : X \to M$ map $F(s) = a$ for all $s \in X$. Then for all $s \in X(F/x_1)$ it holds that $f^M(s(x_1)) = f^M(s(x_0))$ simply because $f^M$ is a constant function, and thus $M \models_{X(F/x_1)} f x_1 = f x_0$. Also, when $s, s' \in X(F/x_1)$, it holds that $s(x_1) = s'(x_1)$, so $M \models_{X(F/x_1)} =(f x_0, x_1)$. Then $M \models_{X(F/x_1)} =(f x_0, x_1) \wedge f x_1 = f x_0$, so $M \models_X \phi$.

Table A.3. *The teams $X$ and $X(F/x_1)$ for*
*Exercise 3.17*

| X | | | X(F/x_1) | | |
|---|---|---|---|---|---|
| | $x_0$ | $x_1$ | | $x_0$ | $x_1$ |
| $s$ | $a$ | | $s$ | $a$ | $a$ |
| $s'$ | $b$ | | $s'$ | $b$ | $a$ |

Table A.4. *The teams $X$ and $X(F/x_1)$*
*for Exercise 3.18*

| X | | | X(F/x_1) | | |
|---|---|---|---|---|---|
| | $x_0$ | $x_1$ | | $x_0$ | $x_1$ |
| $s$ | $a$ | | $s$ | $a$ | $a$ |

In fact, it is possible to prove that any team is of type $\phi$ in any model. Then by simply defining a model where $f^M$ is not one-to-one we can falsify the claim.

## Exercise 3.18

Denote by $\phi$ the formula $\exists x_1 (=(x_0, x_1) \wedge Rx_0x_1)$. We want to falsify the following claim:

$$M \models_X \phi \iff R^M \text{ is a function and } \mathrm{dom}(R^M) = \{s(x_0) : s \in X\}.$$

Let $M = \{a, b\}$ have two distinct elements, let $R^M = \{(a, a), (a, b)\}$, and let $X = \{s\} \in$ Team$(M, \{x_0\})$, where $s = \{(0, a)\}$ (see Table A.4). Clearly, $R^M$ is not a function with domain $\{a\}$, but $M \models_X \phi$; namely, let $F : X \to M$ map $F(s) = a$. Then $M \models_{X(F/x_1)} =(x_0, x_1)$ simply because $X(F/x_1)$ is a singleton. Also $M \models_{X(F/x_1)} Rx_0x_1$ because $X(F/x_1) = \{\{(0, a), (1, a)\}\}$ and $(a, a) \in R^M$. Thus $M \models_{X(F/x_1)} =(x_0, x_1) \wedge Rx_0x_1$, so $M \models_X \phi$.

## Exercise 3.21

We prove only items (i) and (ii).

(i) Our first goal is to find some $\mathcal{D}$-formula $\phi$ such that $\phi \vee \neg\phi \equiv^* \top$ fails. Let $M = \{a, b\}$ be a two-element universe and $X = \{s, s'\} \in$ Team$(M, \{x_0\})$, where $s : 0 \mapsto a$, and $s' : 0 \mapsto b$. Let us investigate the formula $=(x_0)$. If a team is of type $=(x_0)$, then $x_0$ is constant over the team.

Now, by Proposition 3.8, for a team $Y \in$ Team$(M, \{x_0\})$ it holds that $M \models_Y =(x_0)$ iff for all $s, s' \in Y$: $s(x_0) = s'(x_0)$ iff $Y$ has at most one agent. Also, for a team $Z \in$

Team$(M, \{x_0\})$ it holds that $M \models_Z \neg =(x_0)$ iff $Z = \emptyset$. Therefore we cannot split $X$ into some $Y \cup Z$ so that $M \models_Y =(x_0)$ and $M \models_Z \neg =(x_0)$. Thus $M \not\models_X =(x_0) \vee \neg =(x_0)$. However, it is the case that $M \models_X \top$. Hereby we see that the equivalence

$$M \models_X \phi \vee \neg\phi \iff M \models_X \top$$

fails, when we choose $\phi$ to be $=(x_0)$. Thus $\phi \vee \neg\phi \equiv \top$ fails, and furthermore $\phi \vee \neg\phi \equiv^* \top$ fails.

We have, by definition, $\phi \wedge \neg\phi = \neg(\phi \vee \neg\phi)$ and $\bot = \neg\top$. Therefore the failure of $\phi \vee \neg\phi \equiv^* \top$ gives us the failure of $\phi \wedge \neg\phi \equiv^* \bot$.

Let us then prove $\phi \wedge \neg\phi \equiv \bot$. It is shown simply by letting $M$ be an arbitrary model, letting $X$ be an arbitrary team, and letting $\phi$ be an arbitrary $\mathcal{D}$-formula. Then we apply Proposition 3.8:

$$(\phi \wedge \neg\phi, X; 1) \in \mathcal{T} \iff (\phi, X, 1) \in \mathcal{T} \text{ and } (\neg\phi, X, 1) \in \mathcal{T}$$
$$\iff (\phi, X, 1) \in \mathcal{T} \text{ and } (\phi, X, 0) \in \mathcal{T}$$
$$\iff (\bot, X, 1) \in \mathcal{T}.$$

The final equality holds because both sides are true only when $X = \emptyset$.

[Note: When we speak about teams $X \in$ Team$(M, \{x_0\})$, we can use a shorthand notation and write $X \subseteq M$. That is, we can think that a team that interprets only one variable is a subset of the universe. In reality, this is never the case, but the shorthand notation $X \subseteq M$ would instead mean the team $X' = \{s_a : a \in X\}$, where $s_a = \{(x_0, a)\}$ for each $a \in X$. The shorthand notation is okay if we are certain which variable the team is meant to interpret. When we talk about a formula with one free variable, we can be certain of the variable.]

(ii) Let us then prove $\phi \wedge \phi \equiv \phi$. Let $M$ be an arbitrary model, let $X$ be an arbitrary team, and let $\phi$ be an arbitrary $\mathcal{D}$-formula. We apply Proposition 3.8: $(\phi \wedge \phi, X, 1) \in \mathcal{T} \iff (\phi, X, 1) \in \mathcal{T}$ and $(\phi, X, 1) \in \mathcal{T} \iff (\phi, X, 1) \in \mathcal{T}$.

To show that $\phi \wedge \phi \not\equiv^* \phi$ it suffices to find some model $M$, team $X$, and $\mathcal{D}$-formula $\phi$ for which $(\phi \wedge \phi, X, 0) \in \mathcal{T}$ and $(\phi, X, 0) \notin \mathcal{T}$. To this end, let $M = \{a, b\}$ be a two-element universe, $X = \{s, s'\} \in$ Team$(M, \{x_0\})$, where $s = \{(0, a)\}$ and $s' = \{(0, b)\}$ (or with shorthand notation, $X = \{a, b\}$), and let $\phi$ be the formula $\neg =(x_0)$. Now, because there are $s, s' \in X$ with $s(x_0) \neq s'(x_0)$, by Proposition 3.8 $(=(x_0), X, 1) \notin \mathcal{T}$, and so $(\phi, X, 0) \notin \mathcal{T}$. But, on the other hand, any team containing at most one agent is of the type of any determination formula $=(t_1, \ldots, t_n)$. This can be seen from Proposition 3.8. In particular, $(=(x_0), \{s\}, 1) \in \mathcal{T}$, so $(\phi, \{s\}, 0) \in \mathcal{T}$, and likewise also $(\phi, \{s'\}, 0) \in \mathcal{T}$. Thus, by the truth definition, $(\phi \wedge \phi, \{s\} \cup \{s'\}, 0) \in \mathcal{T}$, so we get $(\phi \wedge \phi, X, 0) \in \mathcal{T}$, which completes the proof.

## Exercise 3.23

We prove only item (iv). First we will show $(\phi \wedge \psi) \wedge \theta \equiv^* \phi \wedge (\psi \wedge \theta)$. Let $\phi$, $\psi$, and $\theta$ be $\mathcal{D}$-formulas, let $M$ be a model, and let $X \in$ Team$(M, V)$, where $V = \text{Fr}(\phi) \cup$

$Fr(\psi) \cup Fr(\theta)$. Then

$$((\phi \wedge \psi) \wedge \theta, X, 1) \in T$$
$$\iff (\phi \wedge \psi, X, 1) \in T \text{ and } (\theta, X, 1) \in T$$
$$\iff (\phi, X, 1) \in T \text{ and } (\psi, X, 1) \in T \text{ and } (\theta, X, 1) \in T$$
$$\iff (\phi, X, 1) \in T \text{ and } (\psi \wedge \theta, X, 1) \in T$$
$$\iff (\phi \wedge (\psi \wedge \theta), X, 1) \in T.$$

Also,

$$((\phi \wedge \psi) \wedge \theta, X, 0) \in T$$
$$\iff (\phi \wedge \psi, Y, 0) \in T \text{ and } (\theta, Z, 0) \in T \text{ for some } Y \cup Z = X$$
$$\iff (\phi, A, 0) \in T \text{ and } (\psi, B, 0) \in T \text{ and } (\theta, Z, 0) \in T \text{ for some } A \cup B = Y$$
$$\iff (\phi, A, 0) \in T \text{ and } (\psi \wedge \theta, B \cup Z, 0) \in T$$
$$\iff (\phi \wedge (\psi \wedge \theta), A \cup B \cup Z, 0) \in T.$$

Note that $A \cup B \cup Z = X$, so this completes the proof.

Then we will show $(\phi \vee \psi) \vee \theta \equiv^* \phi \vee (\psi \vee \theta)$. We can prove this with a proof similar to the one above. We can also do it using the Preservation of Equivalence Under Substitution. The identities below are by definition of $\vee$ in terms of $\neg$ and $\wedge$. The strong equivalences, $\equiv^*$, used in the following proof are based on Lemma 3.22, which we proved above:

$$(\phi \vee \psi) \vee \theta = \neg(\neg\phi \wedge \neg\psi) \vee \theta = \neg(\neg\neg(\neg\phi \wedge \neg\psi) \wedge \neg\theta)$$
$$\equiv^* \neg((\neg\phi \wedge \neg\psi) \wedge \neg\theta) \equiv^* \neg(\neg\phi \wedge (\neg\psi \wedge \neg\theta))$$
$$\equiv^* \neg(\neg\phi \wedge \neg\neg(\neg\psi \wedge \neg\theta))$$
$$= \phi \vee \neg(\neg\psi \wedge \neg\theta) = \phi \vee (\psi \vee \theta).$$

## Exercise 3.24

We prove only item (iv), starting from (iv)(a). Let $\phi$ be a $\mathcal{D}$-formula. We have to show $\phi \Rightarrow^* \exists x_n \phi$. In the following solution, it is essential to distinguish between two cases, whether $x_n$ appears free in $\phi$ or not. As we remember from the definition of strong consequence, we have to consider all teams $X$ with $Fr(\phi) \cup Fr(\exists x_n \phi) \subseteq dom(X)$. Because $Fr(\exists x_n \phi) \subseteq Fr(\phi)$, the condition reduces to $Fr(\phi) \subseteq dom(X)$. The distinction if $x_n$ appears free in $\phi$ shows up in the possible domains of the teams.

Let $\mathcal{M}$ be a model. Assume that $x_n$ appears free in $\phi$. We first show $\phi \Rightarrow \exists x_n \phi$, so let $X$ be a team with $Fr(\phi) \subseteq dom(X)$, and assume $(\phi, X, 1) \in T$. Because we assumed that $n \in dom(X)$, we can write $X = X(F/x_n)$, where $F : X \to M$ is defined by $F(s) = s(x_n)$. Thus $(\phi, X(F/x_n), 1) \in T$, so $(\exists x_n \phi, X, 1) \in T$.

Then we show $\neg\exists x_n \phi \Rightarrow \neg\phi$, so let $X$ have $Fr(\phi) \subseteq dom(X)$, and assume $(\exists x_n \phi, X, 0) \in T$. Then $(\phi, X(M/x_n), 0) \in T$, or, in other words, $(\neg\phi, X(M/x_n), 1) \in T$. Because we assumed that $n \in dom(X)$, we have $X \subseteq X(M/x_n)$, namely we can write any $s \in X$ as $s = s(s(x_n)/x_n)$. Then, by the Closure Test we get $(\neg\phi, X, 1) \in T$, so $(\phi, X, 0) \in T$. This completes the proof that $\phi \Rightarrow^* \exists x_n \phi$ in the case that $x_n$ is free in $\phi$.

Assume now that $x_n$ does not appear free in $\phi$. In addition to the proof that we presented for the previous case, we also have to take into account teams $X$ with $x_n \notin \mathrm{dom}(X)$. With these teams the preceding proof does not work because we cannot write things like $s = s(s(x_n)/x_n)$ for all $s \in X$. We need different tricks. First we show $\phi \Rightarrow \exists x_n \phi$, so let $X$ have $\mathrm{Fr}(\phi) \subseteq \mathrm{dom}(X)$, $n \notin \mathrm{dom}(X)$, and assume $(\phi, X, 1) \in \mathcal{T}$. Let $a \in M$, and define $F : X \to M$ by $F(s) = a$ for all $s \in X$. Then we have $(\phi, X(F/x_n), 1) \in \mathcal{T}$ because of Lemma 3.27 and we can write $X = X(F/x_n) \upharpoonright \mathrm{dom}(X)$. Now we get $(\exists x_n \phi, X, 1) \in \mathcal{T}$.

Then we show $\neg \exists x_n \phi \Rightarrow \neg \phi$, so let $X$ have $\mathrm{Fr}(\phi) \subseteq \mathrm{dom}(X)$, $n \notin \mathrm{dom}(X)$, and assume $(\exists x_n \phi, X, 0) \in \mathcal{T}$. Then $(\phi, X(M/x_n), 0) \in \mathcal{T}$, or, in other words, $(\neg \phi, X(M/x_n), 1) \in \mathcal{T}$. Now, because of $X = X(M/x_n) \upharpoonright \mathrm{dom}(X)$ and Lemma 3.27, we get $(\neg \phi, X, 1) \in \mathcal{T}$, whence $(\phi, X, 0) \in \mathcal{T}$. This completes the proof for $\phi \Rightarrow^* \exists x_n \phi$.

Now we prove (iv)(b). Let $\phi$ be a $\mathcal{D}$-formula. We need to show $\forall x_n \phi \Rightarrow^* \phi$. We could write a proof similar to that for (iv)(a), but instead we present a slightly more interesting (and shorter) proof that is based on the duality between $\exists$ and $\forall$. Namely, by (iv)(a), proved above,

$$\forall x_n \phi = \neg \exists x_n \neg \phi \stackrel{\text{(iv)(a)}}{\Rightarrow} \neg \neg \phi \equiv^* \phi$$

and

$$\neg \phi \stackrel{\text{(iv)(a)}}{\Rightarrow} \exists x_n \neg \phi \equiv^* \neg \neg \exists x_n \neg \phi = \neg \forall x_n \phi.$$

Therefore $\forall x_n \phi \Rightarrow \phi$ and $\neg \phi \Rightarrow \neg \forall x_n \phi$, so we have $\forall x_n \phi \Rightarrow^* \phi$.

## Exercise 3.28

We are completing the proof of Proposition 3.31. $\mathcal{M}$ is a model and we are in the middle of an induction on the complexity of a $\mathcal{D}$-formula $\phi$ that is first order. Our induction hypothesis is that for all subformulas $\psi$ of $\phi$, it holds that

(1) if $(\psi, X, 1) \in \mathcal{T}$, then $\mathcal{M} \models_s \psi$ for all $s \in X$;
(2) if $(\psi, X, 0) \in \mathcal{T}$, then $\mathcal{M} \models_s \neg \psi$ for all $s \in X$.

We only have to complete the proof for the cases where $\phi$ is of the form $\neg \psi$ and of the form $\exists x_n \psi$.

If $\phi$ is of form $\neg \psi$ for some first order $\mathcal{D}$-formula $\psi$, then we get from $(\neg \psi, X, 1) \in \mathcal{T}$ that $(\psi, X, 0) \in \mathcal{T}$. Here we can use our induction hypothesis (2) and get $\mathcal{M} \models_s \neg \psi$ for all $s \in X$, which is exactly what we wanted.

If $\phi$ is of form $\neg \psi$, then from $(\neg \psi, X, 0) \in \mathcal{T}$ we get $(\psi, X, 1) \in \mathcal{T}$. Here we can use our induction hypothesis (1) and get that for all $s \in X$, $\mathcal{M} \models_s \psi$, which is by first order logic the same as $\mathcal{M} \models_s \neg \neg \psi$.

If $\phi$ is of form $\exists x_n \psi$ for some first order $\mathcal{D}$-formula $\psi$, then from $(\exists x_n \psi, X, 1) \in \mathcal{T}$ we get $(\psi, X(F/x_n), 1) \in \mathcal{T}$ for some $F : X \to M$. The elements of $X(F/x_n)$ are of form $s(F(s)/x_n)$ for $s \in X$, so by induction hypothesis 1 we get for each $s \in X$ that $\mathcal{M} \models_{s(F(s)/x_n)} \psi$, whence $\mathcal{M} \models_s \exists x_n \psi$.

If $\phi$ is of form $\exists x_n \psi$, then $(\exists x_n \psi, X, 0) \in \mathcal{T}$ gives $(\psi, X(M/x_n), 0) \in \mathcal{T}$. Elements of $X(M/x_n)$ are of form $s(a/x_n)$ for $s \in X$ and $a \in M$, so by induction hypothesis (2) we get for each $s \in X$ that $\mathcal{M} \models_{s(a/x_n)} \neg \psi$ holds for each $a \in M$, whence $\mathcal{M} \models_s \forall x_n \neg \psi$. By first order logic, $\forall x_n \neg \psi \equiv \neg \neg \forall x_n \neg \psi \equiv \neg \exists x_n \psi$, which completes the proof.

## Exercise 3.29

Before getting into the proper exercise, one might ask why in this exercise we ask for a first order formula that is only *logically* *equivalent* to the given $\mathcal{D}$-formula and not strongly logically equivalent. The reason is simply that with first order formulas we always get strong logical equivalence for free if we have logical equivalence. Namely, assuming $\phi \equiv \psi$ and $\mathcal{M} \models_X \neg\phi$, if $\mathcal{M} \models_X \psi$, then $\mathcal{M} \models_X \phi$, which is a contradiction (when $X \neq \emptyset$). Now, because $\psi$ is first order, $\mathcal{M} \models_X \psi \vee \neg\psi$, from which we get $\mathcal{M} \models_X \neg\psi$. Thus we have showed $\neg\phi \Rightarrow \neg\psi$. With a similar proof we see $\neg\psi \Rightarrow \neg\phi$, so we get $\phi \equiv^* \psi$.

(i) We claim that $\exists x_0(=(x_1, x_0) \wedge Px_0) \equiv \exists x_0 Px_0$. The following chain of equivalences is based on definitions, except for the equivalence marked with $(*)$:

$$(\exists x_0(=(x_1, x_0) \wedge Px_0), X, 1) \in \mathcal{T}$$
$$\iff (=(x_1, x_0), X(F/x_0), 1) \in \mathcal{T} \text{ and } (Px_0, X(F/x_0), 1) \in \mathcal{T}$$
$$\text{for some } F : X \to M$$
$$\overset{(*)}{\iff} (Px_0, X(G/x_0), 1) \in \mathcal{T} \text{ for some } G : X \to M$$
$$\iff (\exists x_0 Px_0, X, 1) \in \mathcal{T}.$$

The implication to the right at $(*)$ is clear because we can choose $G$ to be $F$. To show the implication to the left, first note that if $X = \emptyset$ then the claim is clear as the empty team is of any type. Otherwise take some $s_0 \in X$ and assume $(Px_0, X(G/x_0), 1) \in \mathcal{T}$ for some $G : X \to M$. This means in particular that $G(s_0) \in P^{\mathcal{M}}$. Now construct $F : X \to M$ by letting $F(s) = G(s_0)$ for all $s \in X$. In other words, $F$ resembles $G$ but has the additional good property of being constant (it is good concerning determination and dependence). Now $(Px_0, X(F/x_0), 1) \in \mathcal{T}$ because if $s \in X(F/x_0)$ then $s(x_0) = G(s_0) \in P^{\mathcal{M}}$. Also $(=(x_1, x_0), X(F/x_0), 1) \in \mathcal{T}$ because when $s, s' \in X(F/x_0)$, we have $s(x_0) = G(s_0) = s'(x_0)$.

(ii) We claim that $\exists x_0(=(x_1, x_0) \wedge Px_1) \equiv \exists x_0 Px_1$. The following chain of equivalences is based on definitions, except for the equivalence marked with $(*)$:

$$(\exists x_0(=(x_1, x_0) \wedge Px_1), X, 1) \in \mathcal{T}$$
$$\iff (=(x_1, x_0), X(F/x_0), 1) \in \mathcal{T} \text{ and } (Px_1, X(F/x_0), 1) \in \mathcal{T}$$
$$\text{for some } F : X \to M$$
$$\overset{(*)}{\iff} (Px_1, X(G/x_0), 1) \in \mathcal{T} \text{ for some } G : X \to M$$
$$\iff (\exists x_0 Px_1, X, 1) \in \mathcal{T}.$$

The implication to the right at $(*)$ is clear because we can choose $G$ to be $F$. To show the implication to the left, assume $(Px_1, X(G/x_0), 1) \in \mathcal{T}$ for some $G : X \to M$. Fix $a \subset M$ and construct $F : X \to M$ by letting $F(s) = a$ for all $s \in X$. Then $(Px_1, X(F/x_0), 1) \in \mathcal{T}$ because modifying values $s(x_0)$ for $s \in X$ does not affect values $s(x_1)$. Also $(=(x_1, x_0), X(F/x_0), 1) \in \mathcal{T}$ because when $s, s' \in X(F/x_0)$, we have $s(x_0) = a = s'(x_0)$.

Note that we can further refine the result by noticing that in first order logic we have $\exists x_0 Px_1 \equiv Px_1$.

(iii) We claim that $\exists x_0\big((=(x_0, x_1) \wedge Px_0) \to Px_1\big) \equiv \exists x_0(Px_0 \to Px_1)$. First note that, by Lemma 3.25, Preservation of Equivalence under Substitution,

$$\exists x_0\big((\phi \wedge \psi) \to \theta\big) = \exists x_0\big(\neg(\phi \wedge \psi) \vee \theta\big) \equiv^*$$
$$\exists x_0\big(\neg(\neg\neg\phi \wedge \neg\neg\psi) \vee \theta\big) = \exists x_0(\neg\phi \vee \neg\psi \vee \theta).$$

Now we can deduce as follows (the explanation for (∗) is given in the following text):

$$(\exists x_0((=(x_1, x_0) \wedge Px_0) \to Px_1), X, 1) \in \mathcal{T}$$
$$\Longleftrightarrow (\exists x_0(\neg =(x_1, x_0) \vee \neg Px_0 \vee Px_1), X, 1) \in \mathcal{T}$$
$$\Longleftrightarrow (\neg =(x_1, x_0) \vee \neg Px_0 \vee Px_1, X(F/x_0), 1) \in \mathcal{T} \text{ for some } F: X \to M$$
$$\Longleftrightarrow (\neg =(x_1, x_0), A, 1) \in \mathcal{T} \text{ and } (\neg Px_0, B, 1) \in \mathcal{T} \text{ and } (Px_1, C, 1) \in \mathcal{T}$$
$$\quad \text{for some } A \cup B \cup C = X(F/x_0) \text{ and } F: X \to M$$
$$\overset{(*)}{\Longleftrightarrow} (\neg Px_0, B, 1) \in \mathcal{T} \text{ and } (Px_1, C, 1) \in \mathcal{T}$$
$$\quad \text{for some } B \cup C = X(F/x_0) \text{ and } F: X \to M$$
$$\Longleftrightarrow (\neg Px_0 \vee Px_1, X(F/x_0), 1) \in \mathcal{T} \text{ for some } F: X \to M$$
$$\Longleftrightarrow (\exists x_0(Px_0 \to Px_1), X, 1) \in \mathcal{T}.$$

The equivalence (∗) is based on the fact that $(\neg =(x_1, x_0), A, 1) \in \mathcal{T}$ iff $A = \emptyset$. By first order logic we can further refine the formula by noting that $\exists x_0(Px_0 \to Px_1) \equiv \exists x_0(\neg Px_0) \vee Px_1$.

(iv) We claim that $\exists x_0\big(=(x_1, x_0) \wedge Rx_0x_1\big) \equiv \exists x_0 Rx_0x_1$. We can deduce (the explanation for (∗) can be found in the following text):

$$(\exists x_0(=(x_1, x_0) \wedge Rx_0x_1), X, 1) \in \mathcal{T}$$
$$\Longleftrightarrow (=(x_1, x_0), X(F/x_0), 1) \in \mathcal{T} \text{ and } (Rx_0x_1, X(F/x_0), 1) \in \mathcal{T}$$
$$\quad \text{for some } F: X \to M$$
$$\overset{(*)}{\Longleftrightarrow} (Rx_0x_1, X(F/x_0), 1) \in \mathcal{T} \text{ for some } F: X \to M$$
$$\Longleftrightarrow (\exists x_0 Rx_0x_1, X, 1) \in \mathcal{T}.$$

The implication to the right at (∗) is clear. To justify the implication to the left at (∗), assume $(Rx_0x_1, X(F/x_0), 1) \in \mathcal{T}$ for some $F: X \to M$. We can assume that $\mathrm{dom}(X) = \{1\}$ because Lemma 3.27 shows that values of variables that do not appear in the formula do not matter. Now, if we have $s, s' \in X(F/x_0)$ with $s(x_1) = s'(x_1)$, they are of form $s = z(F(z)/x_0)$ and $s' = z'(F(z')/x_0)$ for some $z, z' \in X$. Because $\mathrm{dom}(z) = \{1\}$ and $\mathrm{dom}(z') = \{1\}$, and $s(x_1) = z(1)$ and $s'(x_1) = z'(1)$, we have $z = z'$. Thus $F(z) = F(z')$, so $s(x_0) = F(z) = F(z') = s'(x_0)$. Therefore $(=(x_1, x_0), X(F/x_0), 1) \in \mathcal{T}$.

(v) We claim that $\exists x_0\big(=(x_0) \wedge (Rx_1x_0 \vee Rx_0x_0)\big) \equiv \exists x_0(Rx_1x_2 \vee Rx_0x_0)$. We can deduce as follows (explanation for $(*)$ can be found in the following text):

$$(\exists x_0(=(x_0) \wedge (Rx_1x_2 \vee Rx_0x_0)), X, 1) \in T$$
$$\Longleftrightarrow (=(x_0), X(F/x_0), 1) \in T \text{ and } (Rx_1x_2 \vee Rx_0x_0, X(F/x_0), 1) \in T$$
$$\text{for some } F : X \to M$$
$$\stackrel{(*)}{\Longleftrightarrow} (Rx_1x_2 \vee Rx_0x_0, X(G/x_0), 1) \in T \text{ for some } G : X \to M$$
$$\Longleftrightarrow (\exists x_0(Rx_1x_2 \vee Rx_0x_0), X, 1) \in T.$$

The implication to the right at $(*)$ is clear. To justify the implication to the left, assume $(Rx_1x_2 \vee Rx_0x_0, X(G/x_0), 1) \in T$ for some $G : X \to M$. Then $(Rx_1x_2, A, 1) \in T$ and $(Rx_0x_0, B, 1) \in T$ for some $A \cup B = X(G/x_0)$. In fact we have $A = Y(G/x_0)$ and $B = Z(G/x_0)$ for some $Y \cup Z = X$.

Now we examine two cases. The first case is when $Z \neq \emptyset$. Note that we have

$$\text{for all } s \in Z: (G(s), G(s)) \in R^{\mathcal{M}}. \tag{A.2}$$

Now pick some $s_0 \in Z$ and define $F : X \to M$ by $F(s) = G(s_0)$ for all $s \in X$. Now, if $s, s' \in X(F/x_0)$, it holds that $s(x_0) = s'(x_0)$ because $F$ is a constant function. Therefore $(=(x_0), X(F/x_0), 1) \in T$. We also have $(Rx_1x_2, \emptyset, 1) \in T$ because the empty team is of any type. Finally, we have $(Rx_0x_0, X(F/x_0), 1) \in T$ because if $s \in X(F/x_0)$, then $s(x_0) = G(s_0)$, and by Eq. (A.2) and the fact that $s_0 \in Z$, we get $(s(x_0), s(x_0)) \in R^{\mathcal{M}}$. From these we gather

$$(Rx_1x_2 \vee Rx_0x_0, \underbrace{\emptyset \cup X(F/x_0)}_{X(F/x_0)}, 1) \in T.$$

The second case is when $Z = \emptyset$. Then we have, for all $s \in X$, $(s(x_1), s(x_2)) \in R^{\mathcal{M}}$. Let $a \in M$ and define $F : X \to M$ by $F(s) = a$ for all $s \in X$. Still we have, for all $s \in X$, $(s(x_1), s(x_2)) \in R^{\mathcal{M}}$. Therefore

$$(Rx_1x_2 \vee Rx_0x_0, \underbrace{X(F/x_0) \cup \emptyset}_{X(F/x_0)}, 1) \in T.$$

As in the first case, we have $(=(x_0), X(F/x_0), 1) \in T$ because $F$ is a constant function, and this completes the proof.

## Exercise 3.31

(i) The $\mathcal{D}$-formula $=() \vee \neg =()$ is in itself first order, by definition, because $=()$ is the veritas symbol, $\top$.

(ii) The $\mathcal{D}$-formula $-(x_0)$ is not logically equivalent to a first order formula because it fails the Flatness Test. Let $M = \{a, b\}$ have two different elements, $s = \{(0, a)\}$ and $s' = \{(0, b)\}$. Then $\mathcal{M} \models_{\{s\}} =(x_0)$ and $\mathcal{M} \models_{\{s'\}} =(x_0)$ but $\mathcal{M} \not\models_{\{s,s'\}} =(x_0)$.

(iii) We claim that $=(x_0, x_0) \equiv \top$. We can prove it by noting that $\mathcal{M} \models_X =(x_0, x_0) \Longleftrightarrow$ for all $s, s' \in X$; it holds that if $s(x_0) = s'(x_0)$ then $s(x_0) = s'(x_0)$. This is of course always true, as is $\mathcal{M} \models_X \top$.

## Exercise 3.38

For each of the three given formulas we must find a model and a team that demonstrate that the formula fails the Flatness Test.

(i) Let $\phi$ be $\exists x_0\big(=(x_2, x_0) \wedge Rx_0x_1\big)$. Let $\mathcal{M} = (\mathbb{Z}, <)$, $X = \{s_k : k \in \mathbb{Z}\} \in$ Team$(\mathbb{Z}, \{x_1, x_2\})$, where $s_k(x_1) = k$ and $s_k(x_2) = 0$ for each $k \in \mathbb{Z}$. We have for each $s_k \in X$ that $(\phi, \{s_k\}, 1) \in \mathcal{T}$ because $\big(=(x_2, x_0) \wedge Rx_0x_1, \{s_k\}(F/x_0), 1\big) \in \mathcal{T}$, where $F : s_k \mapsto k - 1$. But $(\phi, X, 1) \notin \mathcal{T}$ because if $G : X \to \mathcal{M}$ satisfies $\big(=(x_2, x_0), X(G/x_0), 1\big) \in \mathcal{T}$ then $G$ must be a constant function because $X$ has a constant value for $x_2$, namely 0. That is, there is some $n \in \mathbb{Z}$ with $G(s_k) = n$ for all $s_k \in X$. But now $G(s_n) = n \not< n = s_n(x_1)$, so $\big(Rx_0x_1, X(G/x_0), 1\big) \notin \mathcal{T}$.

(ii) Let $\psi$ be $\exists x_0\big(=(x_2, x_0) \wedge Rx_1x_0 \wedge Rx_2x_0\big)$. Let $\mathcal{M} = (\mathbb{N}, <)$, $X = \{s_k : k \in \mathbb{N}\}$, where $s_k$ are as above. Now we have, for all $s_k \in X$, $(\psi, \{s_k\}, 1) \in \mathcal{T}$ because $\big(=(x_2, x_0) \wedge Rx_1x_0 \wedge Rx_2x_0, \{s_k\}(F/x_0), 1\big) \in \mathcal{T}$, where $F : s_k \mapsto k + 1$. However, $(\psi, X, 1) \notin \mathcal{T}$ because, as before, if $G : X \to \mathcal{M}$ satisfies $\big(=(x_2, x_0), X(G/x_0), 1\big) \in \mathcal{T}$, then $G$ must be a constant function, mapping $G(s_k) = n$ for all $s_k \in X$ and a fixed $n \in \mathbb{N}$. Then $s_n(x_1) = n \not< n = G(s_n)$, so $\big(Rx_1x_0, X(G/x_0), 1\big) \notin \mathcal{T}$.

(iii) Let $\theta$ be $\exists x_0\big(=(x_2, x_0) \wedge (Rx_1x_0 \leftrightarrow Rx_2x_0)\big)$. Let $\mathcal{M} = (\mathbb{Z}, <)$, $X = \{s_k : k \in \mathbb{Z}\}$, where $s_k$ are as above. Now, for each $s_k \in X$, we have $(\theta, \{s_k\}, 1) \in \mathcal{T}$ because $\big(=(x_2, x_0) \wedge (Rx_1x_0 \leftrightarrow Rx_2x_0), \{s_k\}(F/x_0), 1\big) \in \mathcal{T}$, where $F : s_k \mapsto \max\{k + 1, 1\}$. However, $(\theta, X, 1) \notin \mathcal{T}$ because if $G : X \to \mathcal{M}$ satisfies $\big(=(x_2, x_0), X(G/x_0), 1\big) \in \mathcal{T}$, $G$ has to be a constant function, as before, mapping $G(s_k) = n$ for all $s_k \in X$ and a fixed $n \in \mathbb{Z}$. If $n \leq 0$ then $s_{n-1}(x_1) = n - 1 < n = G(s_{n-1})$ but $s_{n-1}(x_2) = 0 \not< n = G(s_{n-1})$. If $n > 0$ then $s_n(x_1) = n \not< n = G(s_n)$ but $s_n(x_2) = 0 < n = G(s_n)$. In either case we get $\big(Rx_1x_0 \leftrightarrow Rx_2x_0, X(G/x_0), 1\big) \notin \mathcal{T}$.

## Exercise 3.40

In this exercise, we call a set $P = \{s, s'\}$ for $s, s' \in X$ a *pair*. Note that a pair $P$ may have one or two elements.

Let $\psi$ be $\exists x_0\big(=(x_1, x_0) \wedge \phi\big)$, where $\phi$ is first order, and $\text{Fr}(\phi) = \{x_1\}$. The goal is to show that $\psi$ is coherent, i.e.

$$\mathcal{M} \models_X \psi \iff \text{for all pairs } P : \mathcal{M} \models_P \psi. \tag{A.3}$$

The implication to the right in Eq. (A.3) is clear by the Closure Test. Then we prove Eq. (A.3) from right to left. By Lemma 3.27 and the assumption that $\text{Fr}(\phi) = \{x_1\}$, it is sufficient to examine only the case when $\text{dom}(X) = \{1\}$. So, assume that for each pair $P$ we have $(\psi, P, 1) \in \mathcal{T}$. That is, for each pair $P$ we have $F_P : P \to \mathcal{M}$ with

$$\big(=(x_1, x_0) \wedge \phi, P(F_P/x_0), 1\big) \in \mathcal{T}. \tag{A.4}$$

Define $F : X \to \mathcal{M}$ by letting $F(s) = F_{\{s\}}(s)$ for each $s \in X$. We show $\big(=(x_1, x_0), X(F/x_0), 1\big) \in \mathcal{T}$. Let $s, s' \in X(F/x_0)$ and assume $s(x_1) = s'(x_1)$. We can write $s = z(F(z)/x_0)$ and $s' = z'(F(z')/x_0)$ for some $z, z' \in X$. Then $z(1) = z'(1)$, so because $\text{dom}(X) = \{1\}$, $z = z'$ and $s(x_0) = F(z) = F(z') = s'(x_0)$.

Then we show $\big(\phi, X(F/x_0), 1\big) \in \mathcal{T}$. Let $s \in X(F/x_0)$. Then $s = z(F(z)/x_0) = z(F_{\{z\}}(z)/x_0)$ for some $z \in X$, so $\{s\} = \{z\}(F_{\{z\}}/x_0)$. Now, by Eq. (A.4), we get

Table A.5. *The teams X and X(F/x$_0$) for the Exercise 3.40 counterexample, x $\in$ {a, b, c}.*

|  | X | | | X(F/X$_0$) | | |
|---|---|---|---|---|---|---|
|  | $x_0$ | $x_1$ | $x_2$ | $x_0$ | $x_1$ | $x_2$ |
| $s_1$ |  | $a$ | $a$ | $x$ | $a$ | $a$ |
| $s_2$ |  | $a$ | $b$ | $x$ | $a$ | $b$ |
| $s_3$ |  | $a$ | $c$ | $x$ | $a$ | $c$ |

$(\phi, \{s\}, 1) \in \mathcal{T}$. Because $\phi$ is first order, it passes the Flatness Test. Therefore $(\phi, X(F/x_0), 1) \in \mathcal{T}$.

By combining these achievements, we get $(=(x_1, x_0) \wedge \phi, X(F/x_0), 1) \in \mathcal{T}$, whence $(\psi, X, 1) \in \mathcal{T}$. This completes the proof that $\psi$ is coherent.

Note that if we drop the requirement that $Fr(\phi) = \{x_1\}$, the claim does not hold. We can then choose a counterexample by letting $\phi$ be $\neg(x_0 = x_2)$, letting $M = \{a, b, c\}$ have three different elements, and letting $X = \{s_z : z \in M\}$, where $s_z = \{(x_1, a), (x_2, z)\}$ for each $z \in M$ (see Table A.5). We have, for each $P$, $\mathcal{M} \models_P \exists x_0 (=(x_1, x_0) \wedge \phi)$ because for each pair $P$ there is some $z \in M$ with $s_z \notin P$, so we can choose $F_P : P \to M$ to map $F_P(s) = z$. Then $(=(x_1, x_0), P(F_P/x_0), 1) \in \mathcal{T}$ because $F_P$ is constant, and $(\phi, P(F_P/x_0), 1) \in \mathcal{T}$. But we do not have $\mathcal{M} \models_X \exists x_0 (=(x_1, x_0) \wedge \phi)$ because any $F : X \to M$ satisfying $(=(x_1, x_0) \wedge \phi, X(F/x_0), 1) \in \mathcal{T}$ would have to be constant; i.e. for some $z \in M$ we had $F(s) = z$ for all $s \in X$. But then $s_z(2) = F(s_z)$, whence $(\phi, X(F/x_0), 1) \notin \mathcal{T}$.

## Exercise 3.41

Before we get to a straight solution of the exercise, here is an analogy that may help the student to solve the problem alone. Compare the notions "$\phi$ passes the Flatness Test" and "$\phi$ is coherent." Flatness means $\mathcal{M} \models_X \phi \iff \forall s \in X : \mathcal{M} \models_{\{s\}} \phi$, whereas coherence means $\mathcal{M} \models_X \phi \iff \forall s, s' \in X : \mathcal{M} \models_{\{s,s'\}} \phi$. The similarity is clear. The only difference between the two concepts is that one considers singleton teams and the other considers pair teams. Now we can find an answer to this exercise by recalling a typical non-flat formula, $=(x_0)$, and the fact that disjunction splits teams into smaller ones, possibly into singletons or pairs.

To solve the exercise, let $\phi$ be the formula $=(x_0) \vee =(x_0)$. We claim that $\phi$ is not coherent. Let $M = \{a, b, c\}$ have three different elements, and let $X = M$ (shorthand notation meaning $X = \{\{(0, x)\} : x \in M\}$). When $s, s' \in X$, we have $(\phi, \{s, s'\}, 1) \in \mathcal{T}$ because we can split $\{s, s'\} = \{s\} \cup \{s'\}$ and have $(=(x_0), \{s\}, 1) \in \mathcal{T}$ and $(=(x_0), \{s'\}, 1) \in \mathcal{T}$. But we do not have $(\phi, X, 1) \in \mathcal{T}$ because whenever we split $X = Y \cup Z$, either $Y$ or $Z$ has two different elements that do not agree on $x_0$.

As an afterthought, note that $=(x_0)$, while being non-flat, is however coherent. Namely, if $\mathcal{M} \models_{\{s,s'\}} =(x_0)$ for all $s, s' \in X$, then $X$ is constant in $x_0$. Then $\mathcal{M} \models_X =(x_0)$.

## Exercise 3.43

Assume $\mathcal{M} \models_X \phi \to \psi$ and $\mathcal{M} \models_X \neg\psi$. From the first assumption we get $\mathcal{M} \models_X \neg\phi \lor \psi$, whence $\mathcal{M} \models_Y \neg\phi$ and $\mathcal{M} \models_Z \psi$ for some $Y \cup Z = X$. Now $Z \subseteq X$, so by the Closure Test and the second assumption we also have $\mathcal{M} \models_Z \neg\psi$. We can apply Corollary 3.41 to the fact that now $\mathcal{M} \models_Z \psi \land \neg\psi$, and we get $Z = \emptyset$. Hence $Y = X$, which means that $\mathcal{M} \models_X \neg\phi$, as we wanted.

## Exercise 3.44

A straightforward way to show that no non-empty team is of a given type $\phi$ is to apply the truth definition to the formula and see what conditions it poses on the team. There is also another way to check this, namely to flatten the formula and then, remembering that $\phi \Rightarrow \phi^f$, check for the satisfiability of the resulting first order formula $\phi^f$. The flattening technique is sometimes easier, especially if the formula involves determination atoms $=(t_1, \ldots, t_n)$.

(i) $\left( \neg =(x_0, x_1) \right)^f = \neg\top$, which is clearly not satisfiable in first order logic. Therefore no non-empty team is of type $\neg =(x_0, x_1)$.

(ii) Let $\phi$ be $\neg\big( =(x_0, x_1) \to =(x_2, x_1) \big)$. Then $\phi^f = \neg(\top \to \top) \equiv \neg\top$. As $\bot$ is not satisfiable in first order logic, no non-empty team can be of type $\phi$.

(iii) Let $\psi$ be $\neg =(fx_0, x_0) \lor \neg =(x_0, fx_0)$. Then $\phi^f = \neg\top \lor \neg\top = \bot \lor \bot \equiv \bot$. As before, this is not satisfiable, so no non-empty team can be of type $\psi$.

(iv) Let $\theta$ be $\forall x_0 \exists x_1 \forall x_2 \exists x_3 \neg\big(\phi \to =(x_0, x_1)\big)$, where $\phi$ is an arbitrary $\mathcal{D}$-formula. Then $\theta^f = \forall x_0 \exists x_1 \forall x_2 \exists x_3 \neg(\phi^f \to \top) \equiv \forall x_0 \exists x_1 \forall x_2 \exists x_3 \neg\top \equiv \forall x_0 \exists x_1 \forall x_2 \exists x_3 \bot \equiv \bot$. Therefore no non-empty team can be of type $\theta$.

## Exercise 3.45

First, $\mathcal{M} \models_X =(x_0, x_2) \land =(x_1, x_2)$ iff for all $s, s' \in X$ it holds that if $s(x_0) = s'(x_0)$ then $s(x_2) = s'(x_2)$, and if $s(x_1) = s'(x_1)$ then $s(x_2) = s'(x_2)$. In other words, for all $s, s' \in X$ it holds that if $s(x_0) = s'(x_0)$ or $s(x_1) = s'(x_1)$, then $s(x_2) = s'(x_2)$.

On the other hand, $\mathcal{M} \models_X =(x_0, x_1, x_2)$ iff for all $s, s' \in X$ it holds that if $s(x_0) = s'(x_0)$ and $s(x_1) = s'(x_1)$, then $s(x_2) = s'(x_2)$.

Looking at these two observations we can see that the latter is a weaker claim, i.e. $=(x_0, x_1, x_2) \Rightarrow =(x_0, x_2) \land =(x_1, x_2)$, but the converse consequence does not hold.

The team on the left in Table A.6 is of type $=(x_0, x_2) \land =(x_1, x_2)$ because we have a function $f$ mapping the values of the feature $x_0$ to the values of the feature $x_2$, and we also have a function $g$ mapping the values of the feature $x_1$ to the values of the feature $x_2$. These two mappings are $f : \{1, 2\} \to \{1, 2\}$, $f(2) = 1$, $f(1) = 2$, and $g : \{1, 2\} \to \{1, 2, 3, 4\}$, $g(1) = 1$, $g(2) = 1$, $g(3) = 2$, $g(4) = 2$.

The team on the right in Table A.6 is of type $=(x_0, x_1, x_2)$ because we have a function $h$ mapping the value pairs of the features $x_0$ and $x_1$ to the values of the feature $x_2$. This mapping is $h : \{(3, 1), (4, 2), (4, 1), (3, 2)\} \to \{1, 2\}$, $h(3, 1) = 1$, $h(4, 2) = 1$, $h(4, 1) = 2$, $h(3, 2) = 2$.

The team on the right in Table A.6 is not of type $=(x_0, x_2) \land =(x_1, x_2)$ because, for example, there cannot be a mapping $f' : \{3, 4\} \to \{1, 2\}$ that would map $f'(3) = 1$, $f'(4) = 1$, $f'(4) = 2$, $f'(3) = 2$.

Table A.6. *Two example teams for Exercise 3.45,*
*one of type* $=(x_0, x_2) \wedge =(x_1, x_2)$ *and another of*
*type* $=(x_0, x_1, x_2)$

| $=(x_0, x_2) \wedge =(x_1, x_2)$ | | | $=(x_0, x_1, x_2)$ | | |
|---|---|---|---|---|---|
| $x_0$ | $x_1$ | $x_2$ | $x_0$ | $x_1$ | $x_2$ |
| 2 | 1 | 1 | 3 | 1 | 1 |
| 2 | 2 | 1 | 4 | 2 | 1 |
| 1 | 3 | 2 | 4 | 1 | 2 |
| 1 | 4 | 2 | 3 | 2 | 2 |

## Exercise 3.48

Assume $\phi$ is a first order $\mathcal{D}$-formula. Denote by $\psi$ the formula $\big(=(x_0, x_1) \wedge =(x_2, x_3) \wedge \phi\big)$. We want to show that $\forall x_0 \exists x_1 \forall x_2 \exists x_3 \psi \equiv \forall x_2 \exists x_3 \forall x_0 \exists x_1 \psi$. We aim to establish the following chain of equivalences:

$$\forall x_0 \exists x_1 \forall x_2 \exists x_3 \psi$$

$$\equiv \forall x_0 \forall x_2 \exists x_1 \exists x_3 \psi \tag{A.5}$$

$$\equiv \forall x_2 \forall x_0 \exists x_3 \exists x_1 \psi \tag{A.6}$$

$$\equiv \forall x_2 \exists x_3 \forall x_0 \exists x_1 \psi. \tag{A.7}$$

The equivalences in Eqs. (A.5) and (A.7) require a long explanation. The equivalence in Eq. (A.6) is obtained by utilizing Lemma 3.23 twice along with the Preservation of Equivalence under Substitution. Let us now immerse ourselves into Eqs. (A.5) and (A.7).

To deal with some ample notation in solving this exercise, we improve the shorthand notation used in the book. Let $\mathcal{M}$ be a model, let $X$ be a team, and let $s \in X$. We denote $s(a/x_m)(b/x_n)$ by $s(ab/x_m x_n)$ for $a, b \in M$.

To obtain the equivalence in Eq. (A.5), we need to show, for any model $\mathcal{M}$ and team $X$, that $(\forall x_0 \exists x_1 \forall x_2 \exists x_3 \psi, X, 1) \in \mathcal{T} \iff (\forall x_0 \forall x_2 \exists x_1 \exists x_3 \psi, X, 1) \in \mathcal{T}$. By applying the truth definition we get that $(\forall x_0 \exists x_1 \forall x_2 \exists x_3 \psi, X, 1) \in \mathcal{T}$ if and only if $(\psi, Y, 1) \in \mathcal{T}$, where $Y = X(M/x_0)(F/x_1)(M/x_2)(G/x_3)$ for some functions $F : X(M/x_0) \to M$ and $G : X(M/x_0)(F/x_1)(M/x_2) \to M$. This is equivalent to the following condition:

$$(=(x_0, x_1), Y, 1) \in \mathcal{T} \text{ and } (=(x_2, x_3), Y, 1) \in \mathcal{T} \text{ and } (\phi, Y, 1) \in \mathcal{T}. \tag{A.8}$$

We are heading towards $(\forall x_0 \forall x_2 \exists x_1 \exists x_3 \psi, X, 1) \in \mathcal{T}$. Let us unravel it slightly by the truth definition. It is equivalent to $(\psi, Y', 1) \in \mathcal{T}$, where $Y' = X(M/x_0)(M/x_2)(F'/x_1)(G'/x_3)$ for some functions $F' : X(M/x_0)(M/x_2) \to M$ and $G' : X(M/x_0)(M/x_2)(F'/x_1) \to M$. This is equivalent to the following condition:

$$(=(x_0, x_1), Y', 1) \in \mathcal{T} \text{ and } (=(x_2, x_3), Y', 1) \in \mathcal{T} \text{ and } (\phi, Y', 1) \in \mathcal{T}. \tag{A.9}$$

In order to prove Eq. (A.5), we need to show that Eqs. (A.8) and (A.9) are equivalent.

Let us first prove that Eq. (A.8) implies Eq. (A.9). We first need to define $F'$ and $G'$. Let $z(ac/x_0x_2) \in X(M/x_0)(M/x_2)$, where $z \in X$ and $a, c \in M$. We of course want to use the given function $F$ when we define $F'$. A natural way to do this is to set $F'(z(ac/x_0x_2)) = F(z(a/x_0))$. We can do so because $z(a/x_0)$ is in the domain of $F$. Let $z(acb/x_0x_2x_1) \in X(M/x_0)(M/x_2)(F'/x_1)$, where $z \in X$, $a, c \in M$, and $b = F'(z(ac/x_0x_2))$. Again we do the natural thing and set $G'(z(acb/x_0x_2x_1)) = G(z(abc/x_0x_1x_2))$. We can do this because $b = F'(z(ac/x_0x_2)) = F(z(a/x_0))$, so $z(abc/x_0x_1x_2)$ is in the domain of $G$. In fact, $z(acb/x_0x_2x_1)$ and $z(abc/x_0x_1x_2)$ are the same assignment, so $G'$ and $G$ are the same function, but we are not going to pay attention to this.

Now we check that $F'$ and $G'$ satisfy Eq. (A.9). Consider first $=(x_0, x_1)$. Let $s_1 = z_1(a_1c_1b_1d_1/x_0x_2x_1x_3) \in Y'$ and $s_2 = z_2(a_2c_2b_2d_2/x_0x_2x_1x_3) \in Y'$, where, for $i = 1, 2$, $z_i \in X$, $a_i, c_i \in M$, $b_i = F'(z_i(a_ic_i/x_0x_2))$, and $d_i = G'(z_i(a_ic_ib_i/x_0x_2x_1))$. Recall that $b_i = F'(z_i(a_ic_i/x_0x_2)) = F(z_i(a_i/x_0))$ and $d_i = G'(z_i(a_ic_ib_i/x_0x_1x_2)) = G(z_i(a_ib_ic_i/x_0x_1x_2))$. Assume $s_1(x_0) = s_2(x_0)$, i.e. $a_1 = a_2$. Then

$$[z_1(a_1b_1/x_0x_1)](x_0) = a_1 = a_2 = [z_2(a_2b_2/x_0x_1)](x_0).$$

Now, because from Eq. (A.8) we get $(=(x_0, x_1), Y, 1) \in \mathcal{T}$, and because Lemma 3.27 says that satisfaction is unaffected by assignments to variables that do not occur, and because we can extend the assignments $z_i(a_ib_i/x_0x_1)$ to some such assignments that are elements of $Y$, we find

$$b_1 = [z_1(a_1b_1/x_0x_1)](1) = [z_2(a_2b_2/x_0x_1)](1) = b_2.$$

Therefore $s_1(x_1) = b_1 = b_2 = s_2(x_1)$, which proves that $(=(x_0, x_1), Y', 1) \in \mathcal{T}$.

Consider $=(x_2, x_3)$. Let $s_1, s_2 \in Y'$ as above. Assume $s_1(x_2) = s_2(x_2)$, i.e. $c_1 = c_2$. Then

$$[z_1(a_1b_1c_1d_1/x_0x_1x_2x_3)](x_2) = c_1 = c_2 = [z_2(a_2b_2c_2d_2/x_0x_1x_2x_3)](x_2).$$

As above, because from Eq. (A.8) we get $(=(x_2, x_3), Y, 1) \in \mathcal{T}$, and because we have Lemma 3.27, and because we can extend the assignments $z_i(a_ib_ic_id_i/x_0x_1x_2x_3)$ to some such assignments that are elements of $Y$, we find

$$d_1 = [z_1(a_1b_1c_1d_1/x_0x_1x_2x_3)](x_3) = [z_2(a_2b_2c_2d_2/x_0x_1x_2x_3)](x_3) = d_2.$$

Therefore $s_1(x_3) = d_1 = d_2 = s_2(x_3)$, which proves that $(=(x_2, x_3), Y', 1) \in \mathcal{T}$.

Finally, consider $\phi$. Because $\phi$ is first order, we can use the Flatness Test. To show $(\phi, Y', 1) \in \mathcal{T}$ it suffices to show that $(\phi, \{s\}, 1) \in \mathcal{T}$ for each $s \in Y'$. Let $s = z(acbd/x_0x_2x_1x_3) \in Y'$ as above. Because $b = F'(z(ac/x_0x_2)) = F(z(a/x_0))$ and $d = G'(z(acb/x_0x_2x_1)) = G(z(abc/x_0x_1x_2))$, we have $s \in Y$. Therefore $(\phi, \{s\}, 1) \in \mathcal{T}$, by Eq. (A.8) and the Closure Test. This concludes the proof that Eq. (A.8) implies Eq. (A.9). The other direction is similar. Also the proof of Eq. (A.7) is similar to the proof of Eq. (A.5).

# Chapter 4

## Exercise 4.1

In order to come up with a $\mathcal{D}$-sentence that would characterize the oddness of the universe, we can start from the mathematical expression and then modify it into equivalent forms using an informal second order language. We know that in $\mathcal{D}$ we can describe the existence of a function by writing something like $\forall x \exists y (=(x, y) \wedge \phi)$. Here $x$ is the argument for the function, $y$ is the value the function gives to the argument, and $\phi$ describes any conditions that we want to pose on the function.

So we begin. We proceed from the mathematical expression Eq. (A.10) via the equivalent intermediate conditions Eqs. (A.11) and (A.12) to the $\mathcal{D}$-sentence Eq. (A.13). The basic idea for characterizing oddness is similar to that for evenness: a finite set $M$ is even iff there is a one-to-one function $f : M \to M$ that maps no element to itself and that is its own inverse. Here are the three equivalent conditions:

$$|M| \text{ is odd;} \tag{A.10}$$

there is an element $x \in M$ and a function $f : M \to M$ such that $f \circ f = id$

$$\text{and for all } y \in M \text{ we have } f(y) = y \text{ iff } y = x; \tag{A.11}$$

$$\exists x \exists f \Big( \forall y \big( f(f(y)) = y \big) \wedge \forall y \big( f(y) = y \leftrightarrow y = x \big) \Big). \tag{A.12}$$

Now we bring in the $\mathcal{D}$-way of talking about functions. In Eq. (A.13) (see below), the variable $x_1$ acts as an argument to the function and the variable $x_2$ carries the value that the function gives to its argument. Thus we may informally write $y$ for $x_1$ and $f(y)$ for $x_2$.

Actually we need to specify the function $f$ twice because we need to be able to speak about mapping two different elements with it at the same time. Thus we think of $x_3$ as an argument to $f$ and $x_4$ as its image. We can informally write $z$ for $x_3$ and $f(z)$ for $x_4$. Now, in order to express $f(f(y))$ we can and must express it as "$f(z)$ when $z = f(y)$."

Using this intuition we can note how the subformula $x_3 = x_2 \to x_4 = x_1$ in Eq. (A.13) expresses that $x_3 = f(x_1) \to f(x_3) = x_1$, i.e. $f(f(x_1)) = x_1$. Also note how $x_1 = x_3 \to x_2 = x_4$ expresses that $x = y \to f(x) = g(y)$, i.e. $f(x) = g(x)$, i.e. $f = g$. So here is our final $\mathcal{D}$-sentence:

$$\exists x_0 \overbrace{\forall x_1 \exists x_2}^{\exists f} \overbrace{\forall x_3 \exists x_4}^{\exists g} \Big( \overbrace{=(x_1, x_2)}^{x_2 = f(x_1)} \wedge \overbrace{=(x_3, x_4)}^{x_4 = g(x_3)} \wedge \overbrace{(x_1 = x_3 \to x_2 = x_4)}^{f = g} \wedge \underbrace{(x_3 = x_2 \to x_4 = x_1)}_{f(f(y)) = y} \wedge \underbrace{(x_2 = x_1 \leftrightarrow x_1 = x_0)}_{f(x_1) = x_1 \leftrightarrow x_1 = x_0} \Big). \tag{A.13}$$

## Exercise 4.15

Let $G = (V, E)$ be a graph, let $V$ be its vertices, and let $E$ be its edges (a binary relation). We proceed as in Exercise 4.1, starting from the mathematical expression, proceeding to equivalent expressions, and ending up with a $\mathcal{D}$-sentence that describes the mathematical expression. Here are the two equivalent conditions:

$$G \text{ has an infinite clique;} \tag{A.14}$$

$$\text{there is a subset } A \subseteq V \text{such that } A \text{ is a clique and } A \text{ is infinite.} \tag{A.15}$$

We can describe infinity of $A$ by saying that there is a one-to-one function $f : A \to A$ that is not onto

$$\exists A \Big( \forall x \in A \forall y \in A(xEy) \wedge \exists f : A \to A \Big( \forall x \in A \forall y \in A$$

$$\big( f(x) = f(y) \to x = y \big) \wedge \exists z \in A \forall x \in A \big( f(x) \neq z \big) \Big) \Big). \tag{A.16}$$

We need to speak about subsets of the universe. A way to specify a subset, using functions as the only tool, is to speak of the subset as the pre-image of a fixed element in a fixed function. Therefore we convert $\exists A$ into $\exists g \exists w$ and expressions such as $x \in A$ into $f(x) = w$. We also allow $f$ to map $M \to M$, and we require it to be one-to-one only for elements $x, y \in A$:

$$\exists g \exists w \Big( \forall x \forall y \big( (g(x) = w \wedge g(y) = w) \to xEy \big)$$

$$\wedge \exists f \Big( \forall x \forall y \big( (g(x) = w \wedge g(y) = w \wedge f(x) = f(y)) \to x = y \big)$$

$$\wedge \exists z \big( g(z) = w \wedge \forall x (g(x) = w \to f(x) \neq z) \big) \Big) \Big). \tag{A.17}$$

Now we convert our informal and vaguely defined second order sentence into a $\mathcal{D}$-sentence. Note that in order to ensure that $w$ corresponds to a single value, we must express it by $=(w)$, which says that $w$ is a constant. If we leave it out, $w$ could have different values for different arguments of function $g$, i.e. $w$ would be another function. As in Exercise 4.1, we can think of $x_0$ as $x$ and $x_1$ as its value, $g(x)$. Likewise, think of $x_2$ as $y$ and $x_3$ as $g(y)$. So, finally, the $\mathcal{D}$-sentence is given by

$$\overbrace{\forall x_0 \exists x_1}^{\exists g} \overbrace{\forall x_2 \exists x_3}^{\exists g'} \overbrace{\exists x_4}^{\exists w} \Big( \overbrace{=(x_0, x_1)}^{x_1 = g(x_0)} \wedge \overbrace{=(x_2, x_3)}^{x_3 = g'(x_2)} \wedge \overbrace{=(x_4)}^{x_4 \text{ constant}}$$

$$\wedge \overbrace{\big( x_0 = x_2 \to x_1 = x_3 \big)}^{g = g'} \wedge \Big( \overbrace{(x_1 = x_4 \wedge x_3 = x_4)}^{g(x) = w \wedge g(y) = w} \to \overbrace{x_0 E x_2}^{xEy} \Big)$$

$$\wedge \overbrace{\forall x_5 \exists x_6}^{\exists f} \overbrace{\forall x_7 \exists x_8}^{\exists f'} \Big( \overbrace{=(x_5, x_6)}^{x_6 = f(x_5)} \wedge \overbrace{=(x_7, x_8)}^{x_8 = f'(x_7)}$$

$$\wedge \Big( \big( \overbrace{x_1 = x_4 \wedge x_3 = x_4}^{g(x) = w \wedge g(y) = w} \wedge \overbrace{x_0 = x_5 \wedge x_2 = x_7}^{g \text{ and } f \text{map same values}} \wedge \overbrace{x_6 = x_8}^{f(x) = f(y)} \big) \to \overbrace{x_0 = x_2}^{x = y} \Big)$$

$$\wedge \overbrace{\exists x_9}^{\exists z} \Big( =(x_9) \wedge x_2 = x_9 \wedge x_3 = x_4 \wedge$$

$$\big( (x_1 = x_4 \wedge x_5 = x_0) \to (x_6 \neq x_9) \big) \Big) \Big) \Big) \Big). \tag{A.18}$$

## Exercise 4.18

Let $G = (V, E)$ be a graph. We proceed from Eq. (A.19) via equivalent, informal, second order expressions to the $\mathcal{D}$-formula Eq. (A.23). We see that the following are equivalent:

$$G \text{ is 3-colorable.} \qquad (\text{A.19})$$

$G$ can be partitioned into three subsets, $A_0, A_1, A_2,$

such that no edge goes from one subset to the same subset; (A.20)

$$\exists A_0 \subseteq V \, \exists A_1 \subseteq V \, \exists A_2 \subseteq V \Big( \bigwedge_{i<j<3} A_i \cap A_j = \emptyset \, \wedge \, A_0 \cup A_1 \cup A_2 = V$$

$$\wedge \, \forall x \forall y \big( x E y \to \bigwedge_{i<3} \neg (x \in A_i \wedge y \in A_i) \big) \Big). \quad (\text{A.21})$$

As in Exercise 4.15, we express each subset $A_i$ as the pre-image of a fixed element, $f_i^{-1}\{z\}$. As an aside, we could save some quantifiers and take just one function and three elements and write $f^{-1}\{z_i\}$ for $A_i$. We do not do this, however, because then the resulting formula would work only in models with three different elements. That would not be a real problem, because we can then write a separate formula that would describe 3-colorability for the cases when the universe has less than three elements. But as it might look unnecessarily complicated and it would just save us a few quantifiers, we do not do it now:

$$\exists f_0 \exists f_1 \exists x_2 \exists z \Big( \bigwedge_{i<j<3} \neg \exists x \big( f_i(x) = z \wedge f_j(x) = z \big) \, \wedge \, \forall x \bigvee_{i<3} f_i(x) = z$$

$$\wedge \, \forall x \forall y \big( x E y \to \bigwedge_{i<3} \neg (f_i(x) = z \wedge f_i(y) = z) \big) \Big). \quad (\text{A.22})$$

The corresponding $\mathcal{D}$-sentence is given by

$$\forall x_0 \exists x_1 \forall x_2 \exists x_3 \forall x_4 \exists x_5 \exists x_6 \Big( =(x_0, x_1) \, \wedge \, =(x_2, x_3) \, \wedge \, =(x_4, x_5) \, \wedge \, =(x_6)$$

$$\wedge \bigwedge_{i<j<3} (x_{2i} = x_{2j} \to \neg (x_{2i+1} = x_6 \wedge x_{2j+1} = x_6))$$

$$\wedge \big( (x_0 = x_2 \wedge x_2 = x_4) \to (x_1 = x_6 \vee x_3 = x_6 \vee x_5 = x_6) \big) \wedge \forall x_7 \forall x_8$$

$$\Big( x_7 E x_8 \to \bigwedge_{i<3} \neg \big( (x_0 = x_7 \to x_1 = x_6) \wedge (x_0 = x_8 \to x_1 = x_6) \big) \Big) \Big). \quad (\text{A.23})$$

# Chapter 6

## Exercise 6.1

Because $\exists x_1 =(x_1)$ is a sentence, we can omit the relation symbol $S$ from $\iota_{d, \exists x_1 =(x_1)}$ for both $d = 0, 1$:

$$\tau_{1, \exists x_1 =(x_1)}$$
$$= \exists R \big( \tau_{1, =(x_1)}(R) \wedge \exists x_1 R x_1 \big)$$
$$= \exists R \big( \forall x_1 \forall x_2 ((R x_1 \wedge R x_2)$$
$$\to x_1 = x_2) \wedge \exists x_1 R x_1 \big);$$

$$\tau_{0, \exists x_1 =(x_1)}$$
$$= \exists R \big( \tau_{0, =(x_1)}(R) \wedge \forall x_1 R x_1 \big)$$
$$= \exists R \big( \forall x_1 \neg R x_1 \wedge \forall x_1 R x_1 \big).$$

## Exercise 6.2

$\tau_{1,\exists x_1(=(x_1)\vee x_1=x_0)}(S)$

$= \exists R\big(\tau_{1,=(x_1)\vee x_1=x_0}(R) \wedge \forall x_0(Sx_0 \to \exists x_1 Rx_0x_1)\big)$

$= \exists R\exists T_1\exists T_2\big(\tau_{1,=(x_1)}(T_1) \wedge \tau_{1,\vee x_1=x_0}(T_2) \wedge$

$\quad \forall x_0\forall x_1\big(Rx_0x_1 \to (T_1x_1 \vee T_2x_0x_1)\big) \wedge \forall x_0(Sx_0 \to \exists x_1 Rx_0x_1)\big)$

$= \exists R\exists T_1\exists T_2\big(\forall x_1\forall x_2(T_1x_1 \wedge T_1x_2 \to x_1 = x_2) \wedge \forall x_0\forall x_1(T_2x_0x_1 \to x_1 = x_0) \wedge$

$\quad \forall x_0\forall x_1\big(Rx_0x_1 \to (T_1x_1 \vee T_2x_0x_1)\big) \wedge \forall x_0(Sx_0 \to \exists x_1 Rx_0x_1)\big).$

## Exercise 6.5

Assume $\phi(x_{i_1},\ldots,x_{i_n})$ is of form $\exists x_{i_{n+1}}\psi(x_{i_1},\ldots,x_{i_{n+1}})$. Our induction hypothesis is that the claim holds for the subformula, namely that $(\psi, X, d) \in T$ iff $(\mathcal{M}, X) \models \tau_{d,\psi}(S)$. Recall the following definitions from Theorem 6.2:

$$\tau_{1,\phi}(S) = \exists R\big(\tau_{1,\psi}(R) \wedge \forall x_{i_1}\ldots\forall x_{i_n}(Sx_{i_1}\ldots x_{i_n} \to \exists x_{i_{n+1}} Rx_{i_1}\ldots x_{i_{n+1}})\big);$$
$$\tau_{0,\phi}(S) = \exists R\big(\tau_{0,\psi}(R) \wedge \forall x_{i_1}\ldots\forall x_{i_n}(Sx_{i_1}\ldots x_{i_n} \to \forall x_{i_{n+1}} Rx_{i_1}\ldots x_{i_{n+1}})\big).$$

By the truth definition of $\mathcal{D}$, $(\phi, X, 1) \in T$ is equivalent with the pair $(\psi, X(F/x_{i_{n+1}}), 1)$ being in $T$ for some $F : X \to M$. By our induction hypothesis this is equivalent with

$$(\mathcal{M}, X(F/x_{i_{n+1}})) \models \tau_{1,\psi}(S). \tag{A.24}$$

When we choose $X(F/x_{i_{n+1}})$ to witness $R$, we see that Eq. (A.24) implies $(\mathcal{M}, X) \models \tau_{1,\phi}(S)$. Namely, letting $\mathcal{N}$ be the model with universe $N = M$ and interpretations $S^{\mathcal{N}} = X$, $R^{\mathcal{N}} = X(F/x_{i_{n+1}})$, we get $\mathcal{N} \models \tau_{1,\psi}(R)$ and $\mathcal{N} \models \forall x_{i_1}\ldots\forall x_{i_n}(Sx_{i_1}\ldots x_{i_n} \to \exists x_{i_{n+1}} Rx_{i_1}\ldots x_{i_{n+1}})$. The latter comes from the fact that $F$ gives us a suitable value at $\exists x_{i_{n+1}}$. On the other hand, assuming $(\mathcal{M}, X) \models \tau_{1,\phi}(S)$ we get some model $\mathcal{N}$ with universe $N = M$ and relation symbols $R, S$ such that $S^{\mathcal{N}} = X$, $\mathcal{N} \models \tau_{1,\phi}(R)$ and $\mathcal{N} \models \forall x_{i_1}\ldots\forall x_{i_n}(Sx_{i_1}\ldots x_{i_n} \to \exists x_{i_{n+1}} Rx_{i_1}\ldots x_{i_{n+1}})$. The latter gives us a function $F : X \to M$ such that $R^{\mathcal{N}} = X(F/x_{i_{n+1}})$, and (A.24) holds.

Similarly for $d = 0$.

## Exercise 6.6

We solve the exercise by proving a more general claim. The general claim is that when $\phi$ is a $\Sigma_1^1$-formula, $\mathcal{M} \models_s \phi$ for some assignment $s$, and $\pi$ is an isomorphism $\mathcal{M} \cong \mathcal{N}$, then $\mathcal{N} \models_{\pi\circ s} \phi$. We prove this by induction on $\phi$. Our induction hypothesis is that the claim holds for all subformulas of $\phi$. As the first step of the proof, we can point out that if $\phi$ is first order, then it is well known that the claim holds.

Assume then that $\phi$ is of form $\exists R\psi$. Now $\mathcal{M} \models_s \phi$ implies that there is a relation $S$ such that $(\mathcal{M}, S) \models_s \psi$. By $\pi : \mathcal{M} \cong \mathcal{N}$ we get $(\mathcal{N}, S') \models_{\pi\circ s} \psi$, where $S'$ is $\{(\pi(x_1),\ldots,\pi(x_n)) : (x_1,\ldots,x_n) \in S\}$, the image of $S$ under the isomorphism $\pi$. By the truth definition of $\Sigma_1^1$, we get $\mathcal{N} \models_{\pi\circ s} \phi$.

Assume finally that $\phi$ is of form $\exists g \, \psi$. Now $\mathcal{M} \models_s \phi$ implies that there is a function $h : M^n \to M$ such that $(\mathcal{M}, h) \models_s \psi$. By $\pi : \mathcal{M} \cong \mathcal{N}$ we get $(\mathcal{N}, h') \models_{\pi \circ s} \psi$, where $h'$ is $h' : N^n \to N, h'(\pi(x_1), \ldots, \pi(x_n)) = \pi(h(x_1, \ldots, x_n))$ for all $x_1, \ldots, x_n \in M$, the image of $h$ under $\pi$. By the truth definition of $\Sigma_1^1$, we get $\mathcal{N} \models_{\pi \circ s} \phi$.

## Exercise 6.12

Let $\mathcal{M}_n \models \phi_n$ for each $n < \omega$. From

$$\phi_{n+1} \Rightarrow \phi_n \text{ for all } n < \omega$$

we can see with a small inductive proof that

$$\phi_n \Rightarrow \phi_k \text{ for all } k \leq n < \omega. \tag{A.25}$$

We use the Compactness Theorem. Let $T$ be some finite set of sentences $\phi_n$. Then there is a greatest $k$ such that $\phi_k \in T$. We have $\mathcal{M}_k \models \phi_k$, and by Eq. (A.25) this implies $\mathcal{M}_k \models \phi_n$ for all $\phi_n \in T$. Now we have a model for an arbitrary finite set of sentences $\phi_n$. By the Compactness Theorem of $\mathcal{D}$ there is a model $\mathcal{M}$ such that $\mathcal{M} \models \phi_n$ for all $n < \omega$.

## Exercise 6.13

Let $\mathcal{M}$ be some infinite model of $\phi$. Assume towards contradiction that $\mathcal{M} \not\models \psi$. Then $\mathcal{M} \models \neg\psi$ (remember that $\psi$ is first order so negation works in this complementing way). As $\mathcal{M} \models \phi \wedge \neg\psi$, by the Löwenheim–Skolem Theorem of $\mathcal{D}$ there is a countable model $\mathcal{N}$ of $\phi \wedge \neg\psi$. Now we have $\mathcal{N} \models \phi$, and by our assumption we get also $\mathcal{N} \models \psi$. But this is in contradiction with $\mathcal{N} \models \neg\psi$. Therefore we must have $\mathcal{M} \models \psi$. Hereby all models of $\phi$ are models of $\psi$.

## Exercise 6.14

We use the vocabulary $\{<\}$, where $<$ is a binary relation symbol. Let $\theta_{\mathrm{LO}}$ be a first order sentence that lists the axioms of linear order. Let $\psi$ be the first order sentence

$$\theta_{\mathrm{LO}} \wedge \forall x \forall y \big(x < y \to \exists z(x < z \wedge z < y)\big) \wedge \exists x \exists y(x < y).$$

What $\psi$ says is that the model is a dense linear order with at least two elements. Let $\phi$ be the $\mathcal{D}$-sentence

$$\psi \wedge \Phi_{\mathrm{cmpl}}.$$

Recall from Section 4.3 that $\Phi_{\mathrm{cmpl}}$ is true in a linear order iff the linear order is incomplete. Clearly we have $\phi \Rightarrow \psi$. To show that $\mathcal{M} \models \psi$ implies $\mathcal{M} \models \phi$ for countable models $\mathcal{M}$, note first that $\mathcal{M} \models \psi$ implies that $\mathcal{M}$ is an infinite model. We know that there are, up to isomorphism, only four countable dense linear orders, namely the rationals with or without endpoints: $\mathbb{Q}, \mathbb{Q} + 1, 1 + \mathbb{Q}, 1 + \mathbb{Q} + 1$. All of these are incomplete, so we get $\mathcal{M} \models \phi$. However, there are uncountable models that satisfy $\psi$ but not $\phi$, for example the reals, $(\mathbb{R}, <)$.

## Exercise 6.15

Here are Skolem normal forms for the three given first order sentences. There are also other solutions. Here $f_0$ is a 0-ary function symbol, so it works as a constant:

$$\exists f_1 \forall x_0 (x_0 = f_1 x_0),$$
$$\exists f_0 \forall x_1 \neg (f_0 = x_1),$$
$$\exists f_0 \forall x_1 (P f_0 \vee \neg P x_1).$$

## Exercise 6.17

These are the Skolem Normal Forms for the two given $\Sigma_1^1$-sentences, obtained by the process described in the proof of the Skolem Normal Form Theorem:

(i) $\exists f_0 \exists f \forall x_1 \forall x_2 (\neg f x_1 = f_0 \wedge (f f_0 = f x_1 \rightarrow f_0 = x_1))$;

(ii) $\exists f \exists f_0 \exists f_1 \forall x_0 \forall x_1 \forall x_2 \forall x_3 \forall x_4 \forall x_5 \forall x_6$
$$\Big( ((f_0 x_0 x_1 = f_1 \wedge f_0 x_1 x_2 = f_1) \rightarrow f_0 x_0 x_2 = f_1)$$
$$\wedge (f_0 x_3 x_4 = f_1 \vee f_0 x_4 x_3 = f_1 \vee x_3 = x_4) \wedge \neg (f_0 x_5 x_5 = f_1) \wedge f_0 x_6 f x_6 = f_1 \Big).$$

## Exercise 6.18

Since the three given $\Sigma_1^1$-sentences are in Skolem Normal Form and each function symbol appears in them with only one kind of list of arguments, we may straightforwardly transform them into $\mathcal{D}$, following the proof of Theorem 6.15:

(i) $\forall x_0 \forall x_1 \exists x_2 \exists x_3 (=(x_0, x_1, x_2) \wedge =(x_0, x_1, x_3) \wedge \phi(x_0, x_1, x_2, x_3))$;

(ii) $\forall x_0 \forall x_1 \exists x_2 \exists x_3 (=(x_0, x_1, x_2) \wedge =(x_1, x_3) \wedge \phi(x_0, x_1, x_2, x_3))$;

(iii) $\forall x_0 \exists x_1 \exists x_2 (=(x_0, x_1) \wedge =(x_0, x_2) \wedge \phi(x_0, x_1, x_2))$.

## Exercise 6.19

The function symbol $f$ appears in the given $\Sigma_1^1$-sentence with two different lists of arguments. We must therefore first transform the sentence into the following (or similar) equivalent form:

$$\exists f \exists f' \forall x_0 \forall x_1 (\phi(x_0, x_1, f x_0 x_1, f' x_1 x_0) \wedge (x_0 = x_1 \rightarrow f x_0 x_1 = f' x_1 x_0)).$$

Now we can apply the process described in the proof of Theorem 6.15. We obtain the following $\mathcal{D}$-sentence:

$$\forall x_0 \forall x_1 \exists x_2 \exists x_3 (=(x_0, x_1, x_2) \wedge =(x_1, x_0, x_3) \wedge$$
$$\phi(x_0, x_1, x_2, x_3) \wedge (x_0 = x_1 \rightarrow x_2 = x_3)).$$

## Exercise 6.21

Let $\mathcal{M}$ be a model, let $\phi$ define $P \subseteq M^n$, and let $\psi$ define $Q \subseteq M^n$. We claim that $\phi \wedge \psi$ defines $P \cap Q$ and that $\phi \vee \psi$ defines $P \cup Q$. The proofs go as follows:

$$(a_1, \dots, a_n) \in P \cap Q$$
$$\Longleftrightarrow (a_1, \dots, a_n) \in P \text{ and } (a_1, \dots, a_n) \in Q$$
$$\Longleftrightarrow (\mathcal{M}, a_1, \dots, a_n) \models \phi \text{ and } (\mathcal{M}, a_1, \dots, a_n) \models \psi$$
$$\Longleftrightarrow (\mathcal{M}, a_1, \dots, a_n) \models \phi \wedge \psi;$$

$$(a_1, \dots, a_n) \in P \cup Q$$
$$\Longleftrightarrow (a_1, \dots, a_n) \in P \text{ or } (a_1, \dots, a_n) \in Q$$
$$\Longleftrightarrow (\mathcal{M}, a_1, \dots, a_n) \models \phi \text{ or } (\mathcal{M}, a_1, \dots, a_n) \models \psi$$
$$\Longleftrightarrow (\mathcal{M}, a_1, \dots, a_n) \models \phi \vee \psi.$$

The final equivalence follows from the truth definition of $\mathcal{D}$-logic, remembering that $\mathcal{M} \models \phi$ means $\mathcal{M} \models_{\{\emptyset\}} \phi$ and that $\{\emptyset\} = \emptyset \cup \{\emptyset\}$.

To see why $P \setminus Q$ is in general not $\mathcal{D}$-definable even when $P$ and $Q$ are, first assume the contrary. Let $\mathcal{M}_\omega$ be the standard model of arithmetic. By Theorem 6.24 there is a sentence $\tau(c)$ of $\mathcal{D}$ such that for all sentences $\phi$ of $\mathcal{D}$ we have

$$\mathcal{M}_\omega \models \phi \text{ if and only if } \mathcal{M}_\omega \models \tau(\ulcorner \phi \urcorner).$$

Also, let $\theta(c)$ be a sentence of $\mathcal{D}$ for which

$$\mathcal{M}_\omega \models \theta(\underline{k}) \text{ if and only if } k \text{ is the Gödel number of some sentence of } \mathcal{D}.$$

The sentence $\theta(c)$ defines the set $P \subset \omega$ of Gödel numbers of sentences of $\mathcal{D}$, and the sentence $\tau(c)$ defines a set $Q \subset \omega$ that contains the Gödel numbers of true sentences of $\mathcal{D}$. By our assumption, the set $P \setminus Q$ of Gödel numbers of sentences of $\mathcal{D}$ that are not true is definable by a sentence $\pi(c)$ of $\mathcal{D}$. Therefore for all sentences $\phi$ of $\mathcal{D}$ we have the following:

$$\mathcal{M}_\omega \not\models \phi \text{ if and only if } \mathcal{M}_\omega \models \pi(\ulcorner \phi \urcorner).$$

By Theorem 6.19, $\pi$ has a fixed point which is some sentence $\lambda$ of $\mathcal{D}$ such that

$$\mathcal{M}_\omega \models \lambda \text{ if and only if } \mathcal{M}_\omega \models \pi(\ulcorner \lambda \urcorner).$$

Now we have that

$$\mathcal{M}_\omega \models \lambda \text{ if and only if } \mathcal{M}_\omega \models \pi(\ulcorner \lambda \urcorner) \text{ if and only if } \mathcal{M}_\omega \not\models \lambda.$$

This is a contradiction, completing the proof.

## Exercise 6.22

Let $\psi$ be a $\mathcal{D}$-sentence in language $L \cup \{R\}$. Assume we have models $\mathcal{M}$ and $\mathcal{N}$ with $\mathcal{M} \models \psi$, $\mathcal{N} \models \psi$, $\mathcal{M} \upharpoonright L = \mathcal{N} \upharpoonright L$, and $R^\mathcal{M} \neq R^\mathcal{N}$. Also assume that we have a $\mathcal{D}$-sentence $\phi$ in language $L \cup \{c_1, \dots, c_n\}$ that defines $R$ in every model of $\psi$. We can assume w.l.o.g. that there is some $(a_1, \dots, a_n) \in R^\mathcal{M} \setminus R^\mathcal{N}$. Then we have $(a_1, \ldots, a_n) \in R^{\mathcal{M}} \Rightarrow (\mathcal{M}, a_1, \ldots, a_n) \models \phi \Rightarrow (\mathcal{M} \restriction L, a_1, \ldots, a_n) \models \phi \Rightarrow (\mathcal{N} \restriction L, a_1, \ldots, a_n) \models \phi \Rightarrow (\mathcal{N}, a_1, \ldots, a_n) \models \phi \Rightarrow (a_1, \ldots, a_n) \in R^{\mathcal{N}}$, which is a contradiction, proving the claim.

## Exercise 6.23

Let $\psi$ be a $\mathcal{D}$-sentence in the language $L \cup \{R\}$. Assume that for all models $\mathcal{M}$ and $\mathcal{N}$, if $\mathcal{M} \models \psi$, $\mathcal{N} \models \psi$ and $\mathcal{M} \restriction L = \mathcal{N} \restriction L$, then $R^{\mathcal{M}} = R^{\mathcal{N}}$. Let $\psi'$ be obtained from $\psi$ by replacing occurrences of $R$ by $R'$, where $R'$ is a new relation symbol. From the assumption we get that when a model $(\mathcal{M}, A, B)$ satisfies the sentence $\psi \wedge \psi'$ of the language $L \cup \{R, R'\}$, then $A = B$. Therefore the $\mathcal{D}$-sentences $\psi \wedge Rc_1 \ldots c_n$ and $\psi' \wedge \neg R'c_1 \ldots c_n$ have no models in common. We can then use the Separation Theorem and obtain a sentence $\phi$ in the language $L \cup \{c_1, \ldots, c_n\}$ such that every model of $\psi \wedge Rc_1 \ldots c_n$ is a model of $\phi$, and that $\phi$ and $\psi' \wedge \neg R'c_1 \ldots c_n$ have no models in common. Now we have, for an arbitrary model $\mathcal{M} \models \psi$ in the language $L \cup \{R\}$, the following implication:

$$(a_1, \ldots, a_n) \in R^{\mathcal{M}} \Rightarrow (\mathcal{M}, a_1, \ldots, a_n) \models \psi \wedge Rc_1 \ldots c_n$$
$$\Rightarrow (\mathcal{M}, a_1, \ldots, a_n) \models \phi.$$

Also note that $(\mathcal{M}, A) \models \psi \iff (\mathcal{M}, A, A) \models \psi \wedge \psi'$. We obtain the following implication for all models $\mathcal{M} \models \psi \wedge \psi'$:

$$(\mathcal{M}, a_1, \ldots, a_n) \models \phi \Rightarrow (\mathcal{M}, a_1, \ldots, a_n) \not\models \neg R'c_1 \ldots c_n$$
$$\Rightarrow (a_1, \ldots, a_n) \in R'^{\mathcal{M}} \Rightarrow (a_1, \ldots, a_n) \in R^{\mathcal{M}}.$$

## Exercise 6.26

Denote by $S$ the sentence "It is not true that $S$ is true." Assuming that truth is two-valued (which is the case for example in first order logic), $S$ is either true or false. If $S$ is true, then the contents of $S$ state that $S$ is false, which is a contradiction. On the other hand, if $S$ is false, then the contents of $S$ say that $S$ is true, another contradiction. There are no other possibilities, so we have derived the required contradiction.

## Exercise 6.27

If it is raining in both Warsaw and Vienna, then sentence (3) is not paradoxical because it can be false (or true) and still be consistent with sentences (1) and (2). Similarly, when it is not raining in Warsaw nor in Vienna, we can make (3) consistent by making it false.

However, if it is raining in only one of Warsaw and Vienna, we cannot have (3) true or false and still have it consistent. Namely, if (3) is false, then exactly one of (1)–(3) is true, so (3) is actually true, a contradiction. Similarly, if (3) is true, then two of (1)–(3) are true, so (3) does not state a true condition, another contradiction.

## Exercise 6.28

As in the proof of Theorem 6.19, let $\theta(x_0)$ be $\exists x_1\big(\phi(x_1) \wedge \sigma(x_0, x_1, x_0)\big)$, let $k = \ulcorner\theta(x_0)\urcorner$, and let $\psi$ be $\theta(\underline{k})$. Then we have the following equivalence:

$$\mathcal{M}_\omega \models \phi(\ulcorner\psi\urcorner)$$
$$\Longleftrightarrow \mathcal{M}_\omega \models \phi(\ulcorner\theta(\underline{k})\urcorner)$$
$$\Longleftrightarrow \text{there is } a \in M \text{ with } \mathcal{M}_\omega \models \phi(a) \text{ and } (k, a, k) \in \text{Sub}$$
$$\Longleftrightarrow \mathcal{M}_\omega \models \exists x_1\big(\phi(x_1) \wedge \sigma(\underline{k}, x_1, \underline{k})\big)$$
$$\Longleftrightarrow \mathcal{M}_\omega \models \theta(\underline{k})$$
$$\Longleftrightarrow \mathcal{M}_\omega \models \psi.$$

## Exercise 6.31

Let $T$ be a theory in the language $L_{\{+,\times\}} \cup \{c_0, c_1, c_2, c_3\}$, where all the $c_i$ are new constant symbols, so that $T$ consists of Peano's axioms, an axiomatization of our auxiliary tools such as the relations SAT, POS-ID, NEG-ID, TRUE-ID, etc., and the sentences $\text{SAT}c_0c_1 \wedge \text{POS-ID}c_1c_2c_3 \wedge \text{TRUE-ID}c_0c_2c_3$, and $\neg(\underline{n} = c_k)$ for all natural numbers $n$ and all $k = 0, 1, 2, 3$. Then $T$ expresses that the defined truth predicate will also say something about the truth of some non-standard Gödel numbers.

We can satisfy all finite subsets of $T$ by a standard model $(\mathcal{M}_\omega, \text{Sat}_N)$. By the Compactness Theorem of $\mathcal{D}$ we get that $T$ has a model $(\mathcal{M}, S)$. Because of how we built $T$ it now holds that even though $S$ agrees with the standard truth predicate $\text{Sat}_N$ about standard Gödel numbers, there is also a pair of non-standard elements, namely $(c_0^{\mathcal{M}}, c_1^{\mathcal{M}}) \in S$. That is why $S \neq \text{Sat}_N$, even though $(\mathcal{M}, S) \models \theta_L$ and $(\mathcal{M}_\omega, \text{Sat}_N) \models \theta_L$.

## Exercise 6.33

We have the following inference for an arbitrary $\mathcal{D}$-sentence $\phi$ to prove the first part of the exercise. Note that $\phi \leftrightarrow \neg\phi$ is shorthand for $(\neg\phi \vee \neg\phi) \wedge (\neg\neg\phi \vee \phi)$:

$$\big((\neg\phi \vee \neg\phi) \wedge (\neg\neg\phi \vee \phi), \{\emptyset\}, 1\big) \in T$$
$$\Longleftrightarrow (\neg\phi \vee \neg\phi, \{\emptyset\}, 1) \in T \text{ and } (\neg\neg\phi \vee \phi, \{\emptyset\}, 1) \in T$$
$$\Longleftrightarrow \big((\neg\phi, \{\emptyset\}, 1) \in T \text{ or } (\neg\phi, \{\emptyset\}, 1) \in T\big) \text{ and}$$
$$\big((\neg\neg\phi, \{\emptyset\}, 1) \in T \text{ or } (\phi, \{\emptyset\}, 1) \in T\big)$$
$$\Longleftrightarrow (\neg\phi, \{\emptyset\}, 1) \in T \text{ and } (\phi, \{\emptyset\}, 1) \in T$$
$$\Longleftrightarrow (\phi, \{\emptyset\}, 0) \in T \text{ and } (\phi, \{\emptyset\}, 1) \in T$$
$$\Longleftrightarrow \{\emptyset\} = \emptyset.$$

The inference ends in a contradiction.

The reason why we can have a $\mathcal{D}$-sentence $\lambda$ that states "$\lambda$ is not true" is that $\mathcal{D}$ is a three-valued logic, and that gives us more freedom. We avoid contradiction because $\lambda$ is neither true nor false. Note that in $\mathcal{D}$-logic, saying "$\phi$ is false" is different from saying

Table A.7.

| I | II | Rule |
|---|---|---|
| | $\exists x_0 P x_0$ | (1) |
| | $P c_0$ | (6) |
| | $\exists x_0 \neg Q x_0$ | (1) |
| | $\neg Q c_1$ | (6) |
| $Q c_1$ | | (4) |
| | $\forall x_0 (P x_0 \vee Q x_0)$ | (1) |
| | $P c_0 \vee Q c_0$ | (4), (6), (4) |
| | $P c_0$ | (5) |
| | $P c_1 \vee Q c_1$ | (4), (6), (4) |
| | $P c_1$ | (5) |
| | $\vdots$ | |

"$\phi$ is not true." The latter expression is weaker than the former. In other words, when $\phi$ is false it is also not true, but there are cases when $\phi$ is not true but still not false.

## Exercise 6.36

Assume there was some $\mathcal{D}$-sentence $\tau'(c)$ such that for all $\mathcal{D}$-sentences $\phi$ we had $\mathcal{M}_\omega \not\models \phi$ iff $\mathcal{M}_\omega \models \tau'(\ulcorner \phi \urcorner)$. Then by applying the $\mathcal{D}$-alternative of Gödel's Fixed Point Theorem we get a $\mathcal{D}$-sentence $\psi$ such that $\mathcal{M}_\omega \models \psi$ iff $\mathcal{M}_\omega \models \tau'(\ulcorner \underline{\psi} \urcorner)$. This is a contradiction with the first assumption.

## Exercise 6.38

(i) Player **II** can always win by playing as shown in Table A.7. Note that even though we do not have an explicit rule for $\forall x_n \phi$, we can play it by rules (4) (for negation), (6) (for existential quantification), and (4) (for negation again) from Definition 6.25 because $\forall x_n \phi$ is shorthand for $\neg \exists x_n \neg \phi$. Player **II** can always choose $P c_n$ from a disjunction $P c_n \vee Q c_n$.

(ii) Player **II** can always win by playing as shown in Table A.8. All atomic sentences $R t t'$ in the game appear on the side of player **II**, so no contradiction can occur.

## Exercise 6.39

(i) Player **I** can always win by playing as shown in Table A.9. Note that $\phi \to \psi$ is shorthand for $\neg \phi \vee \psi$. At position $(\neg P c_0 \vee Q c_0, \mathbf{II})$, player **II** could have chosen $\neg P c_0$ instead of $Q c_0$, but this would have led to an even easier victory for player **I**.

(ii) Player **I** can always win by playing as shown in Table A.10.

Table A.8.

| I | II | Rule |
|---|---|---|
| | $\neg\exists x\forall y\neg Rxy$ | (1) |
| $\exists x\forall y\neg Rxy$ | | (4) |
| $\forall y\neg Rc_0 y$ | | (6) |
| $\neg Rc_0 c_1$ | | (4), (6), (4) |
| | $Rc_0 c_1$ | (4) |
| | $\forall x\, Rxfx$ | (1) |
| | $Rc_0 fc_0$ | (4), (6), (4) |
| | $fc_0 = c_1$ | (7) |
| | $Rc_0 c_1$ | (3) |

$$\vdots$$

Table A.9.

| I | II | Rule |
|---|---|---|
| | $\exists x_0 Px_0$ | (1) |
| | $Pc_0$ | (6) |
| | $\forall x_0(\neg Px_0 \vee Qx_0)$ | (1) |
| | $\neg Pc_0 \vee Qc_0$ | (4), (6), (4) |
| | $Qc_0$ | (5) |
| | $\neg\exists x_0 Qx_0$ | (1) |
| $\exists x_0 Qx_0$ | | (4) |
| $Qc_0$ | | (6) |

Table A.10.

| I | II | Rule |
|---|---|---|
| | $\neg\forall x\exists y\, Rxy$ | (1) |
| $\forall x\exists y\, Rxy$ | | (4) |
| $\exists y\, Rc_0 y$ | | (4), (6), (4) |
| | $\forall x\, Rxfx$ | (1) |
| | $Rc_0 fc_0$ | (4), (6), (4) |
| | $fc_0 = c_1$ | (7) |
| | $Rc_0 c_1$ | (3) |
| $Rc_0 c_1$ | | (6) |

## Exercise 6.40

Let $\Delta$ be the finite set of sets $S$ of pairs $(\phi, \alpha)$ such that $\alpha \in \{\mathbf{I}, \mathbf{II}\}$, $\phi$ is a sentence in vocabulary $L' = L \cup C$, where $C$ are countably many new constants, and $L$ is a countable vocabulary, and for each $S \in \Delta$ there is a model $\mathcal{M}_S$ such that

(a) the universe consists of constant interpretations, $\mathcal{M}_S = \{c^{\mathcal{M}_S} : c \in C\}$;
(b) if $(\phi, \mathbf{I}) \in S$ then $\mathcal{M}_S \models \neg\phi$;
(c) if $(\phi, \mathbf{II}) \in S$ then $\mathcal{M}_S \models \phi$.

We show that $\Delta$ is a consistency property by going through conditions (i)–(ix) in the definition of a consistency property.

(i) If $S \in \Delta$ and $t$ is a constant $L'$-term, then $\mathcal{M}_S \models t = t$ (simply because any model would satisfy this sentence). Denote $S' = S \cup \{(t = t, \mathbf{II})\}$. Now let us check if $S' \in \Delta$. All we need is to find a model $\mathcal{M}_{S'}$ such that (a), (b), and (c) hold. Choose $\mathcal{M}_S$ to be the candidate for $\mathcal{M}_{S'}$. Clearly (a) holds. To see (b), if $(\phi, \mathbf{I}) \in S'$, then $(\phi, \mathbf{I}) \in S$, so $\mathcal{M}_S \models \neg\phi$. OK. To see (c), if $(\phi, \mathbf{II}) \in S'$, then either $(\phi, \mathbf{II}) \in S$ or $\phi$ is $t = t$. In the first case we get $\mathcal{M}_S \models \phi$. In the second case we again get $\mathcal{M}_S \models \phi$, as we have already seen. This proves condition (i).

(ii) Let $S \in \Delta$ such that $(t = t', \mathbf{II}) \in S$, and let $\phi(t)$ be atomic. If $(\phi(t), \mathbf{II}) \in S \in \Delta$, then $\mathcal{M}_S \models \phi(t)$ and $\mathcal{M}_S \models t = t'$, so $\mathcal{M}_S \models \phi(t')$. Thus $S \cup \{(\phi(t'), \mathbf{II})\} \in \Delta$. On the other hand, if $(\phi(t), \mathbf{I}) \in S \in \Delta$, then $\mathcal{M}_S \models \neg\phi(t)$ and $\mathcal{M}_S \models t = t'$, so $\mathcal{M}_S \models \neg\phi(t')$. Thus $S \cup \{(\phi(t'), \mathbf{I})\} \in \Delta$. This proves condition (ii).

(iii) If $(\neg\phi, \mathbf{II}) \in S \in \Delta$, then $\mathcal{M}_S \models \neg\phi$, so $S \cup \{(\phi, \mathbf{I})\} \in \Delta$. On the other hand, if $(\neg\phi, \mathbf{I}) \in S \in \Delta$, then $\mathcal{M}_S \models \neg\neg\phi$, so $\mathcal{M}_S \models \phi$, so $S \cup \{(\phi, \mathbf{II})\} \in \Delta$. This proves condition (iii).

(iv) If $(\phi \vee \psi, \mathbf{II}) \in S \in \Delta$, then $\mathcal{M}_S \models \phi \vee \psi$, so either $\mathcal{M}_S \models \phi$ or $\mathcal{M}_S \models \psi$. In the first case we get $S \cup \{(\phi, \mathbf{II})\} \in \Delta$, and in the latter case we get $S \cup \{(\psi, \mathbf{II})\} \in \Delta$, which proves condition (iv).

(v) If $(\phi \vee \psi, \mathbf{I}) \in S \in \Delta$, then $\mathcal{M}_S \models \neg(\phi \vee \psi)$, so both $\mathcal{M}_S \models \neg\phi$ and $\mathcal{M}_S \models \neg\psi$. Thus we get $S \cup \{(\phi, \mathbf{I})\} \in \Delta$ and $S \cup \{(\psi, \mathbf{I})\} \in \Delta$, which proves condition (v).

(vi) If $(\exists x_n \phi(x_n), \mathbf{II}) \in S \in \Delta$, then $\mathcal{M}_S \models \exists x_n \phi(x_n)$, so there is some $a \in \mathcal{M}_S$ with $\mathcal{M}_S \models_{\{x_n \mapsto a\}} \phi(x_n)$. By (a) there is some $c \in C$ such that $\mathcal{M}_S \models \phi(c)$. Thus $S \cup \{(\phi(c), \mathbf{II})\} \in \Delta$, which proves condition (vi).

(vii) If $(\exists x_n \phi(x_n), \mathbf{I}) \in S \in \Delta$, then $\mathcal{M}_S \models \neg\exists x_n \phi(x_n)$, so for all $a \in \mathcal{M}_S$ we have $\mathcal{M}_S \models_{\{x_n \mapsto a\}} \neg\phi(x_n)$. By (a), for all $c \in C$ we have $\mathcal{M}_S \models \neg\phi(c)$. Thus $S \cup \{(\phi(c), \mathbf{I})\} \in \Delta$ for all $c \in C$, which proves condition (vii).

(viii) If $S \in \Delta$ and $t$ is a constant $L'$-term, then by (a) we have $t^{\mathcal{M}_S} = c^{\mathcal{M}_S}$ for some $c \in C$. Thus $\mathcal{M}_S \models t = c$, so $S \cup \{(t = c, \mathbf{II})\} \in \Delta$, which proves condition (viii).

(ix) Assume there was some $S \in \Delta$ such that $(\phi, \mathbf{II}) \in S$ and $(\phi, \mathbf{I}) \in S$. Then $\mathcal{M}_S \models \phi$ and $\mathcal{M}_S \models \neg\phi$, which is a contradiction. Thus the case of the assumption can never happen, which proves condition (ix).

This concludes the proof that $\Delta$ is a consistency property.

## Exercise 6.42

Let $T$ be an $L$-theory. Assume that $\sigma$ is a winning strategy for player **II** in MEG($T, L$). We are to prove that there is a consistency property $\Delta$ for $T$. The idea of the proof is that we create a set $\Delta$ that imitates plays of MEG($T, L$) when player **II** plays with her winning strategy $\sigma$ against all the possible strategies of player **I**. Since all sets $S \in \Delta$ must be finite in order to $\Delta$ being a consistency property, we cannot take $S$ to be all the positions in some play (might be infinite). Instead, we take only positions up to some finite number of moves.

Precisely, let $\Delta$ be the set such that $S \in \Delta$ iff there is some strategy $\tau$ of player **I** in MEG($T, L$) and natural number $n < \omega$ such that $S$ is the set of positions of the play of MEG($T, L$) when player **I** plays according to strategy $\tau$ and player **II** plays according to strategy $\sigma$. We prove that $\Delta$ is a consistency property for theory $T$ by going through conditions (i)–(ix) of the definition of a consistency property.

(i) Let $S \in \Delta$, obtained by some strategy $\tau$ for player **I** by playing up to some move $n$, and let $t$ be a constant $L' = L \cup C$-term. At this point in the game, player **I** can make an identity move on term $t$ and the game moves to position $(t = t, \mathbf{II})$. Thus $S \cup \{(t = t, \mathbf{II})\} \in \Delta$.

(ii) Let $S \in \Delta$, corresponding to some play at some phase, and let $(\phi(t), \alpha) \in S$, where $\phi(t)$ is atomic, and let $(t = t', \mathbf{II}) \in S$. Now player **I** can continue by making a substitution move, and the game continues from position $(\phi(t'), \alpha)$. Thus $S \cup \{(\phi(t'), \alpha)\} \in \Delta$.

(iii) Let $S \in \Delta$, corresponding to some play at some phase. If $(\neg\phi, \mathbf{II}) \in S$, then player **I** can make a negation move, and the game continues from position $(\phi, \mathbf{I})$. Thus $S \cup \{(\phi, \mathbf{I})\} \in \Delta$. On the other hand, if $(\neg\phi, \mathbf{I}) \in S$, then player **I** can make a negation move such that the game continues from position $(\phi, \mathbf{II})$. Thus $S \cup \{(\phi, \mathbf{II})\} \in \Delta$.

(iv) Let $S \in \Delta$, corresponding to some play at some phase, and let $(\phi \vee \psi, \mathbf{II}) \in S$. Player **I** can make a disjunction move, and the game continues from position $(\phi, \mathbf{II})$ or $(\psi, \mathbf{II})$, depending on player **II**'s choice. Thus either $S \cup \{(\phi, \mathbf{II})\} \in \Delta$ or $S \cup \{(\psi, \mathbf{II})\} \in \Delta$.

(v) Let $S \in \Delta$, corresponding to some play at some phase, and let $(\phi \vee \psi, \mathbf{I}) \in S$. Player **I** can make a disjunction move so that the game continues from position $(\phi, \mathbf{I})$. On the other hand, he can also choose the game to continue from position $(\psi, \mathbf{I})$. Thus $S \cup \{(\phi, \mathbf{I})\} \in \Delta$ and $S \cup \{(\psi, \mathbf{I})\} \in \Delta$.

(vi) Let $S \in \Delta$, corresponding to some play at some phase. Suppose $(\exists x_n \phi(x_n), \mathbf{II}) \in S$. Player **I** can make an existential move, and the game continues from position $(\phi(c), \mathbf{II})$ for some $c \in C$, depending on the choice of player **II**. Thus $S \cup \{(\phi(c), \mathbf{II})\} \in \Delta$ for some $c \in C$.

(vii) Let $S \in \Delta$, corresponding to some play at some phase. Suppose $(\exists x_n \phi(x_n), \mathbf{I}) \in S$. For each $c \in C$, player **I** is able to make an existential move so that the game continues from position $(\phi(c), \mathbf{I})$. Thus $S \cup \{(\phi(c), \mathbf{II})\} \in \Delta$ for all $c \in C$.

(viii) Let $S \in \Delta$, corresponding to some play at some phase, and let $t$ be a constant $L'$-term. Player **I** can make a constant move, and the game continues from position $(t = c, \mathbf{II})$ for some $c \in C$, depending on the choice of player **II**.

(ix) There cannot be any set $S \in \Delta$ and atomic formula $\phi$ such that $(\phi, \mathbf{II}) \in S$ and $(\phi, \mathbf{I}) \in S$, because $S$ contains exactly the game positions up to some $n$ moves in some play where player $\mathbf{II}$ has followed her winning strategy $\sigma$.

Lastly, we check the requirement that the consistency property $\Delta$ is a consistency property *for theory $T$*. Assume $\phi \in T$ and $S \in \Delta$. Then there is some strategy $\tau$ for player $\mathbf{I}$ in $\mathrm{MEG}(T, L)$ such that, after some $n$ rounds of playing, the game has produced the set $S$ of game positions. Certainly there is a strategy $\tau$ for player $\mathbf{I}$ such that on his next move he will perform a theory move and choose $(\phi, \mathbf{II})$ to become the new game position. Therefore $S \cup \{(\phi, \mathbf{II})\} \in \Delta$. This concludes the whole proof.

## Exercise 6.44

Let $H$ be a Hintikka set. Define, for $c, c' \in C$, $c \sim c'$ if $(c = c', \mathbf{II}) \in H$. We show that $\sim$ is an equivalence relation.

Firstly, for any $c \in C$, we have $(c = c, \mathbf{II}) \in H$ by condition (i) in the definition of a Hintikka set (Definition 6.30) because $c$ is a constant $L'$-term. Thus $\sim$ is reflexive.

Secondly, assume $(c = c', \mathbf{II}) \in H$. By the above, we have $(c = c, \mathbf{II}) \in H$. Now, by condition (ii) in Definition 6.30 we have $(c' = c, \mathbf{II}) \in H$. Thus $\sim$ is symmetric.

Thirdly, assume $(c = c', \mathbf{II}) \in H$ and $(c' = c'', \mathbf{II}) \in H$. Again, by condition (ii) we have $(c = c'', \mathbf{II}) \in H$. Thus $\sim$ is transitive.

## Exercise 6.45

The proof follows conditions (i)–(ix) in the definition of a consistency property (Definition 6.28). In fact we have to generalize the definition slightly. Namely, in the original definition, all sentences that appear in $\Delta$ are of some fixed vocabulary $L \cup C$. In this exercise we do not have this situation. All sentences that occur in our $\Delta$ are surely of the vocabulary $L_1 \cup L_2 \cup C$, but we avoid considering all such sentences. We have only $L_1 \cup C$- and $L_2 \cup C$-sentences. We must reflect this in the definition by slightly altering conditions (i) and (viii) which prompt for appearance of some sentence that contains an arbitrary term. In the following proof, we assume that the proper adjustments to the definition have been made.

(i) Let $S \in \Delta$ and let $t$ be a constant $L_i \cup C$-term for some $i \in \{1, 2\}$. Denote $S' = S \cup \{(t = t, \mathbf{II})\}$. If there were some $L_1 \cap L_2$-sentence $\theta$ that would separate $T(S') {\restriction} (L_1 \cup C)$ and $T(S') {\restriction} (L_2 \cup C)$, then $\theta$ would also separate $T(S) {\restriction} (L_1 \cup C)$ and $T(S) {\restriction} (L_2 \cup C)$. Namely, if some model $\mathcal{M}$ has $\mathcal{M} \models T(S) {\restriction} (L_1 \cup C)$, then $\mathcal{M} \models T(S') {\restriction} (L_1 \cup C)$, so $\mathcal{M} \models \theta$. Furthermore, if some $\mathcal{M}$ has $\mathcal{M} \not\models \theta$, then $\mathcal{M} \not\models T(S') {\restriction} (L_2 \cup C)$, so $\mathcal{M} \not\models T(S) {\restriction} (L_2 \cup C)$. Therefore we have a contradiction to the fact that $S \in \Delta$, which says that $T(S) {\restriction} (L_1 \cup C)$ and $T(S) {\restriction} (L_2 \cup C)$ are inseparable. Thus $S' \in \Delta$.

(ii) Let $(\phi(t), \alpha) \in S \in \Delta$, let $\phi(t)$ be atomic, and let $(t = t', \mathbf{II}) \in S$. Denote $S' = S \cup \{(\phi(t'), \alpha)\}$. Assume $\phi(t)$ is an $L_1 \cup C$-sentence or an $L_2 \cup C$-sentence (it does not matter in this proof of condition (ii) which one it is). If there were some $L_1 \cap L_2$-sentence $\theta$ that would separate $T(S') {\restriction} (L_1 \cup C)$ and $T(S') {\restriction} (L_2 \cup C)$, then – with a proof similar to that of condition (i) – $\theta$ would also separate $T(S) {\restriction} (L_1 \cup C)$ and $T(S) {\restriction} (L_2 \cup C)$, contrary to the fact that $S \in \Delta$. Thus $S' \in \Delta$.

(iii) Let $(\neg\phi, \mathbf{II}) \in S \in \Delta$. Assume that $\neg\phi$ is an $L_i \cup C$-sentence for some $i \in \{1, 2\}$ (does not matter which). Denote $S' = S \cup \{(\phi, \mathbf{I})\}$. If it was the case that $T(S')\restriction(L_1 \cup C)$ and $T(S')\restriction(L_2 \cup C)$ were separable by some $L_1 \cap L_2$-sentence $\theta$, then also $T(S)\restriction(L_1 \cup C)$ and $T(S)\restriction(L_2 \cup C)$ would be separable by $\theta$. The basic idea is that both $(\neg\phi, \mathbf{II})$ and $(\phi, \mathbf{I})$ contribute the same sentence to $T(S')$. More precisely, if we have a model $\mathcal{M}$ with $\mathcal{M} \models T(S)\restriction(L_1 \cup C)$, then $\mathcal{M} \models T(S')\restriction(L_1 \cup C)$, so $\mathcal{M} \models \theta$. Furthermore, if some model $\mathcal{M}$ has $\mathcal{M} \models \theta$, then $\mathcal{M} \not\models T(S')\restriction(L_2 \cup C)$, so $\mathcal{M} \not\models T(S)\restriction(L_2 \cup C)$. This is contrary to $S \in \Delta$. Thus we have $S' \in \Delta$.

(iv) Let $(\phi \vee \psi, \mathbf{II}) \in S \in \Delta$. Assume that

$$\phi \vee \psi \text{ is an } L_1 \cup C\text{-sentence.} \tag{A.26}$$

Denote $S' = S \cup \{(\phi, \mathbf{II})\}$ and $S'' = S \cup \{(\psi, \mathbf{II})\}$. If it were the case that we had $L_1 \cap L_2$-sentences $\theta'$ and $\theta''$ such that

$$\begin{aligned} &T(S')\restriction(L_1 \cup C) \text{ and } T(S')\restriction(L_2 \cup C) \text{ are separable by } \theta', \text{ and} \\ &T(S'')\restriction(L_1 \cup C) \text{ and } T(S'')\restriction(L_2 \cup C) \text{ are separable by } \theta'', \end{aligned} \tag{A.27}$$

then $\theta' \vee \theta''$ would separate $T(S)\restriction(L_1 \cup C)$ and $T(S)\restriction(L_2 \cup C)$. Namely, assume that for some model $\mathcal{M}$ we have $\mathcal{M} \models T(S)\restriction(L_1 \cup C)$. Then $\mathcal{M} \models \phi \vee \psi$ because of Eq. (A.26). Thus either $\mathcal{M} \models \phi$ or $\mathcal{M} \models \psi$. Then either $\mathcal{M} \models T(S')\restriction(L_1 \cup C)$ or $\mathcal{M} \models T(S'')\restriction(L_1 \cup C)$, so by Eq. (A.27), $\mathcal{M} \models \theta'$ or $\mathcal{M} \models \theta''$. In both cases we get $\mathcal{M} \models \theta' \vee \theta''$. Assume then that some model $\mathcal{M}$ has $\mathcal{M} \models \theta' \vee \theta''$. Then either $\mathcal{M} \models \theta'$ or $\mathcal{M} \models \theta''$, so, by Eq. (A.27), either $\mathcal{M} \not\models T(S')\restriction(L_2 \cup C)$ or $\mathcal{M} \not\models T(S'')\restriction(L_2 \cup C)$. Because of Eq. (A.26), we have $T(S')\restriction(L_2 \cup C) = T(S)\restriction(L_2 \cup C)$ and $T(S'')\restriction(L_2 \cup C) = T(S)\restriction(L_2 \cup C)$. Thus we get, in both cases, $\mathcal{M} \not\models T(S)\restriction(L_2 \cup C)$. This is a contradiction to $S \in \Delta$. Thus we have either $S' \in \Delta$ or $S'' \in \Delta$. The other case when $\phi \vee \psi$ is an $L_2 \cup C$-sentence can be proved similarly.

(v) Let $(\phi \vee \psi, \mathbf{I}) \in S \in \Delta$. Assume Eq. (A.26). (The other case is proved similarly.) Denote $S' = S \cup \{(\phi, \mathbf{I})\}$ and $S'' = S \cup \{(\psi, \mathbf{I})\}$. Assume there was some $L_1 \cap L_2$-sentence $\theta$ such that either Eq. (A.28) or Eq. (A.29) holds:

$$T(S')\restriction(L_1 \cup C) \text{ and } T(S')\restriction(L_2 \cup C) \text{ are separable by } \theta, \tag{A.28}$$

$$T(S'')\restriction(L_1 \cup C) \text{ and } T(S'')\restriction(L_2 \cup C) \text{ are separable by } \theta. \tag{A.29}$$

Then $\theta$ would separate $T(S)\restriction(L_1 \cup C)$ and $T(S)\restriction(L_2 \cup C)$. For a proof, let us assume Eq. (A.28). The case for Eq. (A.29) is proved symmetrically. Now, if $\mathcal{M} \models T(S)\restriction(L_1 \cup C)$, then $\mathcal{M} \models T(S')\restriction(L_1 \cup C)$ because of Eq. (A.26) and the fact that $\neg(\phi \vee \psi) \models \neg\phi$. Thus, by Eq. (A.28), $\mathcal{M} \models \theta$. Furthermore, if some model $\mathcal{M}$ has $\mathcal{M} \models \theta$, then by Eq. (A.28) we have $\mathcal{M} \not\models T(S')\restriction(L_2 \cup C)$. Now, because of Eq. (A.26), $T(S')\restriction(L_2 \cup C) = T(S)\restriction(L_2 \cup C)$, so $\mathcal{M} \not\models T(S)\restriction(L_2 \cup C)$. Therefore we get a contradiction with $S \in \Delta$. Thus we have $S' \in \Delta$ and $S'' \in \Delta$.

(vi) Let $(\exists x_n \phi(x_n), \mathbf{II}) \in S \in \Delta$. Assume that

$$\exists x_n \phi(x_n) \text{ is an } L_1 \cup C\text{-sentence.} \tag{A.30}$$

(The other case is proved similarly.) Let $c \in C$ be a constant that does not occur in $S$ (such constants exist because $S$ is finite). Denote $S' = S \cup \{(\phi(c), \mathbf{II})\}$. If there were some $L_1 \cap L_2$-sentence $\theta$ such that

$$\theta \text{ separates } T(S') {\upharpoonright} (L_1 \cup C) \text{ and } T(S') {\upharpoonright} (L_2 \cup C), \tag{A.31}$$

then $\theta$ would separate also $T(S) {\upharpoonright} (L_1 \cup C)$ and $T(S) {\upharpoonright} (L_2 \cup C)$, contrary to $S \in \Delta$. More precisely, if some model $\mathcal{M}$ has $\mathcal{M} \models T(S) {\upharpoonright} (L_1 \cup C)$, then $\mathcal{M} \models \exists x_n \phi(x_n)$. Therefore there is some $a \in M$ with $(\mathcal{M}, a) \models \phi(c)$, so $(\mathcal{M}, a) \models T(S') {\upharpoonright} (L_1 \cup C)$. By Eq. (A.31), $(\mathcal{M}, a) \models \theta$. Because $\theta$ does not mention $c$, we get $\mathcal{M} \models \theta$. Furthermore, if we have a model $\mathcal{M}$ such that $\mathcal{M} \models \theta$, then, by Eq. (A.31), $\mathcal{M} \not\models T(S') {\upharpoonright} (L_2 \cup C)$. Because of Eq. (A.30), $T(S') {\upharpoonright} (L_2 \cup C) = T(S) {\upharpoonright} (L_2 \cup C)$, so we have $\mathcal{M} \not\models T(S) {\upharpoonright} (L_2 \cup C)$. This gives a contradiction with $S \in \Delta$. Thus we have $S' \in \Delta$.

(vii) Let $(\exists x_n \phi(x_n), \mathbf{I}) \in S \in \Delta$. Assume Eq. (A.30). (The other case is proved similarly.) Denote, for each $c \in C$, $S_c = S \cup \{(\phi(c), \mathbf{I})\}$. If there were some $L_1 \cap L_2$-sentence $\theta$ and some $c \in C$ such that

$$\theta \text{ separates } T(S_c) {\upharpoonright} (L_1 \cup C) \text{ and } T(S_c) {\upharpoonright} (L_2 \cup C), \tag{A.32}$$

then $\theta$ would separate also $T(S) {\upharpoonright} (L_1 \cup C)$ and $T(S) {\upharpoonright} (L_2 \cup C)$. Namely, if some model $\mathcal{M}$ has $\mathcal{M} \models T(S) {\upharpoonright} (L_1 \cup C)$, then $\mathcal{M} \models \neg \exists x_n \phi(x_n)$. This gives us $\mathcal{M} \models \neg \phi(c)$, so we have $\mathcal{M} \models T(S_c) {\upharpoonright} (L_1 \cup C)$. By Eq. (A.32) we get $\mathcal{M} \models \theta$. On the other hand, if some model $\mathcal{M}$ has $\mathcal{M} \models \theta$, then, by Eq. (A.32), we get $\mathcal{M} \not\models T(S_c) {\upharpoonright} (L_2 \cup C)$. Because of Eq. (A.30), we have $T(S_c) {\upharpoonright} (L_2 \cup C) = T(S) {\upharpoonright} (L_2 \cup C)$, so $\mathcal{M} \not\models T(S) {\upharpoonright} (L_2 \cup C)$. This is contrary to $S \in \Delta$. Thus we have $S' \in \Delta$.

(viii) Let $S \in \Delta$ and assume that

$$t \text{ is a constant } L_1 \cup C\text{-term.} \tag{A.33}$$

Let $c \in C$ be a constant that does not occur in $S$ (such constants exist because $S$ is finite). Denote $S' = S \cup \{(t = c, \mathbf{II})\}$. Now assume there is some $L_1 \cap L_2$-sentence $\theta$ such that

$$\theta \text{ separates } T(S') {\upharpoonright} (L_1 \cup C) \text{ and } T(S') {\upharpoonright} (L_2 \cup C). \tag{A.34}$$

We prove that then $\theta$ separates also $T(S) {\upharpoonright} (L_1 \cup C)$ and $T(S) {\upharpoonright} (L_2 \cup C)$, contrary to assumption $S \in \Delta$. Then Eq. (A.34) is false, and that gives us $S' \in \Delta$. So, assume that some model $\mathcal{M}$ has $\mathcal{M} \models T(S) {\upharpoonright} (L_1 \cup C)$. Denote $a = t^{\mathcal{M}} \in M$. Then $(\mathcal{M}, a) \models t = c$, so $(\mathcal{M}, a) \models T(S') {\upharpoonright} (L_1 \cup C)$. By Eq. (A.34), $(\mathcal{M}, a) \models \theta$. Because $\theta$ does not contain $c$, we get $\mathcal{M} \models \theta$. Assume then that some model $\mathcal{M}$ has $\mathcal{M} \models \theta$. Then, by Eq. (A.34), $\mathcal{M} \not\models T(S') {\upharpoonright} (L_2 \cup C)$. Because of Eq. (A.33), $T(S') {\upharpoonright} (L_2 \cup C) = T(S) {\upharpoonright} (L_2 \cup C)$, so $\mathcal{M} \not\models T(S) {\upharpoonright} (L_2 \cup C)$.

(ix) Let $S \in \Delta$ and assume there is some $L_1 \cup C$-sentence $\phi$ such that $(\phi, \mathbf{II}) \in S$ and $(\phi, \mathbf{I}) \in S$. Let $\theta$ be the $L_1 \cap L_2$-sentence $\neg \forall x_0 (x_0 = x_0)$. Now, if some model $\mathcal{M}$ has $\mathcal{M} \models T(S) {\upharpoonright} (L_1 \cup C)$, then $\mathcal{M} \models \theta$. This is so because $\phi$ is an $L_1 \cup C$-sentence, so $T(S) {\upharpoonright} (L_1 \cup C)$ contains both sentences $\phi$ and $/\phi$. Thus $T(S) {\upharpoonright} (L_1 \cup C)$ has no models. Similarly, because $\theta$ has no models, we get that $\theta$ and $T(S) {\upharpoonright} (L_2 \cup C)$ have no models in common. This proves that $\theta$ separates $T(S) {\upharpoonright} (L_1 \cup C)$ and $T(S) {\upharpoonright} (L_2 \cup C)$, contrary to $S \in \Delta$. The case that $\phi$ is an $L_2 \cup C$-sentence is similar.

This concludes the whole proof.

Table A.11. *Team X for Exercise 6.54*

| $x_0$ | $x_1$ | $x_2$ |
|-------|-------|-------|
| 0 | 2 | 1 |
| 1 | 0 | 0 |
| 2 | 1 | 1 |

Table A.12. *Teams X and Y for Exercise 6.57*

| X | | | Y | | |
|---|---|---|---|---|---|
| $x_0$ | $x_1$ | $x_2$ | $x_0$ | $x_1$ | $x_2$ |
| 0 | 2 | 2 | a | c | b |
| 1 | 0 | 0 | b | b | c |
| 2 | 1 | 1 | c | a | a |
| | | | d | d | d |

# Exercise 6.54

The set $Fml_3^0$ contains infinitely many formulas, but only a finite number of them are non-equivalent. We confine ourselves to list only one formula from each equivalence class that $X$ (see Table A.11) satisfies in $\mathcal{M}$:

$$\top \qquad \neg(x_0 = x_1) \qquad \neg(x_0 = x_2)$$
$$=(x_0, x_1) \qquad =(x_0, x_2) \qquad =(x_1, x_0) \qquad =(x_1, x_2).$$

In addition to these, there are all the conjunctions of any subset of these listed formulas, such as $\neg(x_0 = x_1) \wedge =(x_0, x_1) \wedge =(x_1, x_2)$. The team $X$ also satisfies trivial consequences of these listed formulas. For example, because $=(x_0, x_2)$ is satisfied, then also $=(x_0, x_1, x_2)$ is satisfied because we have $=(x_0, x_2) \Rightarrow =(x_0, x_1, x_2)$.

Remember that formulas such as $=(x_0, x_0)$, $=(x_0, x_1, x_0)$, $=(x_0, x_0, x_0, x_0)$ and so on are satisfied by any team in any model, thus they are equivalent with $\top$. Also remember that formulas such as $\neg =(x_2, x_0)$ are not satisfied by $X$ in $\mathcal{M}$. In fact, negations of determination formulas are always false: remember the truth definition! $Fml_3^0$ also does not contain any disjunctions because their rank would be greater than 0.

Table A.13. Teams $X = X_0 \cup X_1$ and $Y$ for
Exercise 6.59

| X | | $X_0$ | | $X_1$ | | Y | |
|---|---|---|---|---|---|---|---|
| $x_0$ | $x_1$ | $x_0$ | $x_1$ | $x_0$ | $x_1$ | $x_0$ | $x_1$ |
| 0 | 2 | 1 | 3 | 0 | 2 | a | b |
| 1 | 3 | 2 | 3 | 3 | 2 | b | c |
| 2 | 3 | | | | | c | d |
| 3 | 2 | | | | | d | a |

Table A.14. Teams $X$, $X(F/x_2)$, and $Y$ for
Exercise 6.60

| X | | $X(F/x_2)$ | | | Y | |
|---|---|---|---|---|---|---|
| $x_0$ | $x_1$ | $x_0$ | $x_1$ | $x_2$ | $x_0$ | $x_1$ |
| 0 | 2 | 0 | 2 | 0 | a | b |
| 1 | 1 | 1 | 1 | 0 | b | b |
| 2 | 1 | 2 | 1 | 0 | c | a |

## Exercise 6.57

Player **I** won the game because, for example, $\mathcal{M} \models_X x_1 = x_2$ and $\mathcal{N} \not\models_Y x_1 = x_2$. Another reason why player **I** won is that $\mathcal{M} \models_X \neg(x_0 = x_1)$ and $\mathcal{N} \not\models_Y \neg(x_0 = x_1)$. There may be other similar reasons too. See Table A.12.

## Exercise 6.59

A good splitting move for player **I** is to present $X$ as $X_0 \cup X_1$, as seen in Table A.13. Then player **II** must present $Y$ as $Y_0 \cup Y_1$. After this player **I** necessarily wins, because $\mathcal{M} \models_{X_0} =(x_1)$ and $\mathcal{M} \models_{X_1} =(x_1)$, but whatever $Y_0$ and $Y_1$ are, for one of them we have $\mathcal{N} \not\models_{Y_i} =(x_1)$. You can see this by checking all the relevant 16 different splits $Y_0 \subset Y$, $Y_1 = Y \setminus Y_0$.

## Exercise 6.60

A good supplementing move for player **I** is to consider the function $F : X \to M$ that maps all $s \in X$ as $F(s) = 0$ (see Table A.14). When player **I** plays $X(F/x_2)$, he wins no matter which team $Y(G/x_2)$ player **II** plays. Namely, after this move we have $\mathcal{M} \models_{X(F/x_2)} P x_2$ but $\mathcal{N} \not\models_{Y(G/x_2)} P x_2$. The latter comes simply from the fact that $P^{\mathcal{N}} = \emptyset$.

## Exercise 6.63

Let $\mathcal{N} = (\mathbb{N}, +, \cdot, 0, 1, <)$, the standard model of arithmetic. Recall that there is a sentence $\Phi_\mathbb{N}$ that is true in a model iff the model is not isomorphic to $\mathcal{N}$. Now, if $\mathcal{M}$ is any uncountable model, it cannot be isomorphic to $\mathcal{N}$. Thus $\mathcal{M} \models \Phi_\mathbb{N}$, but $\mathcal{N}$ is isomorphic to itself, so $\mathcal{N} \not\models \Phi_\mathbb{N}$. Thus we do not have $\mathcal{N} \equiv_\mathcal{D} \mathcal{M}$.

## Exercise 6.64

We can in principle show $(\mathbb{R}, \mathbb{N}) \Rrightarrow_\mathcal{D} (\mathbb{Q}, \mathbb{N})$ by giving, for each natural number $n$, a winning strategy for player **II** in $\text{EF}_n((\mathbb{R}, \mathbb{N}), (\mathbb{Q}, \mathbb{N}))$. However, winning strategies in the Ehrenfeucht–Fraïssé game for $\mathcal{D}$ are in general very difficult to describe. Luckily, there is an easier proof. First we have to generalize the Löwenheim–Skolem Theorem of $\mathcal{D}$ (Theorem 6.5) so that it speaks about a whole theory instead of just a single sentence.

**Löwenheim–Skolem Theorem of** $\mathcal{D}$ *Suppose $T$ is a set of $\mathcal{D}$-sentences such that $T$ has an infinite model or arbitrarily large finite models. Then $T$ has models of all infinite cardinalities, in particular $T$ has a countable model and an uncountable model.*

*Proof* Enumerate the theory as $T = \{\phi_k : k < \omega\}$. For each $k < \omega$, let $\tau_{1,\phi_k}$ be the $\Sigma_1^1$-translation of $\phi_k$, $\tau_{1,\phi_k} = \exists S_1^k \cdots \exists S_n^k \psi_k$, where $\psi_k$ is first order in the vocabulary $L \cup \{S_1^k, \ldots, S_n^k\}$. We can assume that all the relation symbols $S_i^k$ are distinct. Denote $L' = L \cup \{S_i^k : k < \omega, i < n\}$. By the Löwenheim–Skolem Theorem of first order logic, there is an $L'$-model $\mathcal{M}'$ of the theory $\{\psi_k : k < \omega\}$ such that $\mathcal{M}'$ is of cardinality $\kappa$. The reduction $\mathcal{M} = \mathcal{M}'{\upharpoonright}L$ of $\mathcal{M}'$ to the original vocabulary $L$ is a model of $T$ of cardinality $\kappa$. $\quad\square$

With this formulation of the Löwenheim–Skolem Theorem, we get that, letting $T$ denote the $\mathcal{D}$-theory of the model $(\mathbb{R}, \mathbb{N})$, $T = \{\phi : (\mathbb{R}, \mathbb{N}) \models \phi\}$, there is a countable model $\mathcal{M}$ of $T$. Recall that there is a $\mathcal{D}$-sentence $\Phi_\infty$ that is true in a model $\mathcal{N}$ iff $\mathcal{N}$ is infinite. With slight modifications we can also write down a sentence $\Psi$ that is true in a model $\mathcal{N}$ of language $\{P\}$ iff both $P^\mathcal{N}$ and $N \setminus P^\mathcal{N}$ are infinite. This sentence $\Psi$ is then in the theory $T$. Thus $\mathcal{M} \models \Psi$, and hereby $\mathcal{M}$ is isomorphic to $(\mathbb{Q}, \mathbb{N})$. Thus also $(\mathbb{Q}, \mathbb{N}) \models T$. This shows $(\mathbb{R}, \mathbb{N}) \Rrightarrow_\mathcal{D} (\mathbb{Q}, \mathbb{N})$.

## Exercise 6.71

Let $P$ and $P'$ be dependence back-and-forth sets for $\mathcal{M} \succsim_\mathcal{D} \mathcal{M}'$ and $\mathcal{M}' \succsim_\mathcal{D} \mathcal{M}''$, respectively. To show $\mathcal{M} \succsim_\mathcal{D} \mathcal{M}''$, define $P''$ as follows:

$$P'' = \{(X, Z) : \text{there is a team } Y \text{ such that } (X, Y) \in P \text{ and } (Y, Z) \in P'\}.$$

We will now show that $P''$ is a dependence back-and-forth set for $\mathcal{M} \succsim_\mathcal{D} \mathcal{M}''$.

Firstly, $P'' \subseteq \text{Part}(\mathcal{M}, \mathcal{M}'')$ because if $(X, Z) \in P''$ then we have some $(X, Y) \in P$ and $(Y, Z) \in P'$, so there is $n$ such that $X \subseteq M^n$ and $Y \subseteq M'^n$, and $m$ such that $Y \subseteq M'^m$ and $Z \subseteq M''^m$. This means $m = n$, so $X \subseteq M^n$ and $Z \subseteq M''^n$. Also, if $\mathcal{M} \models_X \phi$ for some atomic formula $\phi$, then by assumptions, $\mathcal{M}' \models_Y \phi$, and further, $\mathcal{M}'' \models_Z \phi$.

To see that $P''$ is non-empty, first show that $(\emptyset, \emptyset) \in P$ and $(\emptyset, \emptyset) \in P'$. We know that $P$ is non-empty, so there is some $(X, Y) \in P$. Because $X = X \cup \emptyset$, there is some split $Y = Y_1 \cup Y_2$ such that $(\emptyset, Y_2) \in P$. Because $\mathcal{M} \models_\emptyset \bot$, we have $\mathcal{M}' \models_{Y_2} \bot$. This means that $Y_2 = \emptyset$. Thus we have $(\emptyset, \emptyset) \in P'$. Similarly for $P'$. Thus we get $(\emptyset, \emptyset) \in P''$, which shows that $P''$ is non-empty.

Condition (i)   Let $(X, Z) \in P''$ and let $X = X_1 \cup X_2$. By definition of $P''$, we then have some $(X, Y) \in P$ and $(Y, Z) \in P'$. By properties of $P$ there is some split $Y = Y_1 \cup Y_2$ such that $(X_1, Y_1) \in P$ and $(X_2, Y_2) \in P$. Again, by properties of $P'$ there is some split $Z = Z_1 \cup Z_2$ such that $(Y_1, Z_1) \in P'$ and $(Y_2, Z_2) \in P'$. By definition of $P''$ we get $(X_1, Z_1) \in P''$ and $(X_2, Z_2) \in P''$.

Condition (ii)   Let $(X, Z) \in P''$ and let $n$ be a natural number. By definition of $P''$ we then have some $(X, Y) \in P$ and $(Y, Z) \in P'$. By properties of $P$ and $P'$ we have $(X(M/x_n), Y(M'/x_n)) \in P$, and further, $(Y(M'/x_n), Z(M''/x_n)) \in P'$. By definition of $P''$ we get $(X(M/x_n), Z(M''/x_n)) \in P$.

Condition (iii)   Let $(X, Z) \in P''$, let $n$ be a natural number, and let $F : M \to X$ be some function. By definition of $P''$ we then have some $(X, Y) \in P$ and $(Y, Z) \in P'$. By properties of $P$ there is some function $G : M' \to Y$ such that $(X(F/x_n), Y(G/x_n)) \in P$. By properties of $P'$ there is some function $H : M'' \to Z$ such that $(Y(G/x_n), Z(H/x_n)) \in P'$. Thus, by definition of $P''$ we get $(X(F/x_n), Z(H/x_n)) \in P''$.

This completes the proof.

# Chapter 7

## Exercise 7.1

Here are the definitions as $\Sigma_0$-formulas. For brevity, we write $\forall x \in y \phi$ for $\forall x(x \in y \to \phi)$, and we write $\exists x \in y \phi$ for $\exists x(x \in y \land \phi)$:

(i)   $x = y \cup z$   $\overbrace{\forall v \in x(v \in y \lor v \in z)}^{x \subseteq y \cup z} \land \overbrace{\forall v \in y(v \in x)}^{y \subseteq x} \land \overbrace{\forall v \in z(v \in x)}^{z \subseteq x}$,

(ii)   $x = \{y, z\}$   $\forall v \in x(v = y \lor v = z) \land y \in x \land z \in x$,

(iii)   $x$ is transitive   $\forall y \in x \forall z \in y(z \in x)$,

(iv)   $x$ is an ordinal   "$x$ is transitive" $\land \forall y \in x$("$y$ is transitive"),

(v)   $x : y \to z$   $\overbrace{\forall v \in x \exists u \in y \exists w \in z(v = (u, w))}^{x \text{ is a relation between } y \text{ and } z} \land$

$\overbrace{\forall u \in y \exists w \in z \exists v \in x(v = (u, w))}^{\text{every element in } y \text{ has an image}} \land$

$\underbrace{\forall u \in y \forall w \in z \exists v \in x(v = (u, w))}_{\text{images are unique}} \land$

$\forall u \in y \forall r \in z \forall s \in z\big((\exists v \in x(v = (u, r)) \land \exists v \in x(v = (u, s))) \to r = s\big).$

Definition (iv) uses definition (iii). For definition (v) we also need to define "$v = (u, w)$" as a $\Sigma_0$-formula. A suitable formula is $z = \{u, \{u, w\}\}$, which in turn uses definition (ii) above.

## Exercise 7.2

We prove the claim by induction on $\phi$.

If $\phi$ is an atomic formula, it is of the form $x_i = x_j$ or $x_i \in x_j$. In both cases the claim holds. For example, $(M, \in) \models a_i \in a_j$ iff $a_i \in a_j$, simply because the model $(M, \in)$ interprets the binary relation symbol as the usual "belongs to" relation.

The cases for negation, disjunction and existential quantifier go as follows:

$$(M, \in) \models \neg\phi(a_1, \ldots, a_n)$$
$$\iff (M, \in) \not\models \phi(a_1, \ldots, a_n)$$
$$\iff \text{not } \phi(a_1, \ldots, a_n)$$
$$\iff \neg\phi(a_1, \ldots, a_n);$$

$$(M, \in) \models (\phi \vee \psi)(a_1, \ldots, a_n)$$
$$\iff (M, \in) \models \phi(a_1, \ldots, a_n) \text{ or } (M, \in) \models \psi(a_1, \ldots, a_n)$$
$$\iff \phi(a_1, \ldots, a_n) \text{ or } \psi(a_1, \ldots, a_n)$$
$$\iff (\phi \vee \psi)(a_1, \ldots, a_n);$$

$$(M, \in) \models \exists x (x \in a_i \wedge \phi)(x, a_1, \ldots, a_n)$$
$$\iff \text{there is } a \in M \text{ such that } (M, \in) \models a \in a_i \wedge \phi(a, a_1, \ldots, a_n)$$
$$\iff \text{there is } a \in M : (M, \in) \models a \in a_i \text{ and } (M, \in) \models \phi(a, a_1, \ldots, a_n)$$
$$\iff \text{there is } a \in M \text{ such that } a \in a_i \text{ and } \phi(a, a_1, \ldots, a_n) \quad \text{(A.35)}$$
$$\Rightarrow \text{there is } a \text{ such that } a \in a_i \text{ and } \phi(a, a_1, \ldots, a_n) \quad \text{(A.36)}$$
$$\iff \text{there is } a \text{ such that } a \in a_i \wedge \phi(a_1, \ldots, a_n)$$
$$\iff \exists x (x \in a_i \wedge \phi)(x, a_1, \ldots, a_n).$$

To achieve the implication Eq. (A.36) $\Rightarrow$ Eq. (A.35), assume Eq. (A.36). Because $a_i \in M$ and $M$ is transitive, also $a \in M$. Thus we get Eq. (A.35).

## Exercise 7.3

Let $(M, E)$ be a well-founded model that satisfies the Axiom of Extensionality, $\forall x_0 \forall x_1 (\forall x_2 (x_2 \in x_0 \leftrightarrow x_2 \in x_1) \rightarrow x_0 = x_1)$. First of all, just as we have defined the rank of an element, $\text{rk}(x) = \sup\{\text{rk}(y) + 1 : y \in x\}$, we will now define the $E$-rank of an element $x \in M$ by $\text{rk}^E(x) = \sup\{\text{rk}^E(y) + 1 : yEx\}$.

Because $(M, E)$ is well-founded, we can define a mapping $\pi$ by the equation $\pi(x) = \{\pi(y) : yEx\}$ by induction on the $E$-rank of elements $x \in M$. Namely, if $\text{rk}^E(x) = 0$, then there are no $y \in M$ with $yEx$, so we can set $\pi(x) = \emptyset$. Let $x \in M$ and assume that $\pi$ has been defined for elements with $E$-rank lower than $\text{rk}^E(x)$, in particular for $y \in M$ with $yEx$. Then the equation $\pi(x) = \{\pi(y) : yEx\}$ is well defined, so we can use it for defining $\pi(x)$.

Denote by $N$ the image of $M$ under $\pi$, $N = \pi``M$. We claim that $N$ is a transitive set. To prove it, let $x \in N$ and $y \in x$. Then $y = \pi(z)$ for some $z \in M$, so $y \in N$, by definition of $\pi$.

Next we claim that $\pi$ is one-to-one. As an assumption towards contradiction, assume $\pi(x) = \pi(y)$ for some $x, y \in M$ with $x \neq y$. In fact, we can and will pick $\pi(x)$ and $\pi(y)$ so that they are also minimal in respect of their rank, $\text{rk}(\pi(x))$. Because $(M, E)$ is extensional, we get from $x \neq y$ that there is $u \in M$ with $u \in x$ and $u \notin y$ (or $u \notin x$ and

$u \in y$, but that is a symmetric case). Of course $\pi(u) \in \pi(x)$, and because $\pi(x) = \pi(y)$ we also get that $\pi(u) \in \pi(y)$. But by definition of $\pi$ there is some $v \in y$ such that $\pi(v) = \pi(u)$. Now we have $u \neq v$ and $\pi(u) = \pi(v)$, and because $\pi(u) \in \pi(x)$ we also have $\mathrm{rk}(\pi(u)) < \mathrm{rk}(\pi(x))$. This is contrary to our choice of $\pi(x)$ to be minimal in respect of rank. Thus $\pi$ is one-to-one. As $N$ is merely the image of $M$ under $\pi$, $\pi : M \to N$ is even a bijection.

To see that $\pi$ is an isomorphism, we need to show $x E y \iff \pi(x) \in \pi(y)$. The implication to the right is clear from the definition of $\pi$. For the other direction, assume $\pi(x) \in \pi(y)$. By definition of $\pi$, there is some $z \in M$ such that $z E y$ and $\pi(z) = \pi(x)$. Because $\pi$ is one-to-one, we get that $z = x$. Thus $x E y$.

Our final claim is that if $E$ is actually $\in$, the usual "belongs to" relation, then for any transitive set $A \subseteq M$ we have $\pi(x) = x$ for all $x \in A$. We prove the claim by induction on the rank of elements. Note: because $E$ is $\in$, we can write things like $y \in x$ instead of $y E x$ for elements $x, y \in M$. In general, for all $x \in M$ we have $y E x$ iff $y \in x$ and $y \in M$ iff $y \in x \cap M$. In particular, the definition of $\pi$ becomes $\pi(x) = \{\pi(y) : y \in x \cap M\}$. Now, let $x \in A$. Because $A$ is transitive, we get $x \subseteq A$, and thus $x = x \cap A = x \cap M$. If $\mathrm{rk}(x) = 0$, then $x = \emptyset$, and by definition of $\pi$, $\pi(x) = \emptyset$. Otherwise, we assume that the claim holds for elements $y \in A$ with $\mathrm{rk}(y) < \mathrm{rk}(x)$, in particular all $y \in A$ with $y \in x$, that is, all $y \in x \cap A = x \cap M$. Then we have the following equation:

$$\pi(x) = \{\pi(y) : y \in x \cap M\} = \{y : y \in x \cap M\} = x \cap M = x.$$

This completes the proof.

## Exercise 7.4

Our goal is to find a $\Sigma_2$-definition $\phi(x)$ for the property of $x$ being equal to $V_\kappa$, where $\kappa = \beth_\kappa$. Let us first express $\phi(x)$ as a formula and then explain it in words:

$$\phi := \phi_1 \wedge \phi_2 \wedge \forall z \phi_3 \wedge \exists k \exists V \exists f (\phi_4 \wedge \phi_5);$$

$$\phi_1 := \mathrm{Trans}(x) \wedge \exists y \in x (y = \omega);$$

$$\phi_2 := \forall u \in x \exists y \in x (y = \mathcal{P}_x(u));$$

$$\phi_3 := \forall y \in x (z \subseteq y \to z \in x);$$

$$\phi_4 := \forall y \in x \big(\mathrm{Ord}(y) \to y \in k\big) \wedge \forall y \in k \big(y \in x \wedge \mathrm{Ord}(y)\big) \wedge \mathrm{Bij}(f, x, k);$$

$$\phi_5 := \mathrm{Fun}(V, k, x) \wedge \forall z \in x \exists y \in k \big(z \in V(y)\big) \wedge$$
$$\forall y \in k \big(y = \emptyset \to V(y) = \emptyset\big) \wedge$$
$$\forall y \in k \forall z \in k \big(z = y^+ \to V(z) = \mathcal{P}_x(V(y))\big) \wedge$$
$$\forall y \in k \big(\mathrm{Limit}(y) \to \big(\forall z \in y(V(z) \subseteq V(y)) \wedge$$
$$\forall u \in V(y) \exists z \in y(u \in V(z))\big)\big).$$

The formulas $\mathrm{Trans}(x)$, $\mathrm{Ord}(x)$, and $\mathrm{Fun}(x, y, z)$ are $\Sigma_0$-definitions for the properties "$x$ is transitive," "$x$ is an ordinal," and "$x$ is a function $y \to z$," respectively, as in Exercise 7.1. The other relevant $\Sigma_0$-formulas and informal expressions are as follows:

$$x = \emptyset \qquad \forall y \in x (y \neq y);$$

$$x \subseteq y \qquad \forall u \in x (u \in y);$$

$$y = \mathcal{P}_x(z) \qquad \forall u \in x (u \in y \leftrightarrow u \subseteq z);$$

$$x = y^+ \qquad \forall u \in x (u \in y \vee u = y) \wedge y \subseteq x \wedge y \in x;$$

$$\text{Limit}(x) \qquad \neg(x = \emptyset) \wedge \text{Ord}(x) \wedge \forall y \in x \big(\text{Ord}(y) \rightarrow \exists z \in x (z = y^+)\big);$$

$$x = \omega \qquad \text{Limit}(x) \wedge \forall y \in x \big(\text{Limit}(y) \rightarrow y = \emptyset\big);$$

$$\text{Bij}(x, y, z) \qquad \text{Fun}(x, y, z) \wedge \forall u \in y \forall v \in y \big(x(u) = x(v) \rightarrow u = v\big)$$
$$\qquad \wedge \forall u \in z \exists v \in y (x(v) = u);$$

$$\psi(f(y)) \qquad \exists p \in f \exists v \in p \big(p = (y, v) \wedge \psi(v)\big).$$

On the last line, $\psi(x)$ is a $\Sigma_0$-formula, and $f$ is a variable symbol for which $\text{Fun}(f, r, s) \wedge y \in r$ holds for some variable symbols $r$ and $s$.

We see that $\phi_1, \ldots, \phi_5$ are $\Sigma_0$-formulas. Therefore $\phi$ is equivalent to a $\Sigma_2$-formula. Note that $\phi$ itself is not, strictly speaking, a $\Sigma_2$-formula because the unrestricted quantification does not appear only at the very beginning of the formula. However, this can be cured simply by pulling out the unrestricted quantifiers from the conjuncts.

The subformula $\phi_1(x)$ says that $x$ is transitive and contains $\omega$; $\phi_2(x)$ says that $x$ satisfies the power set axiom; $\forall z \phi_3(x)$ says that $x$ contains all subsets of its elements. We will need this to ensure that the power sets that $x$ contains are really power sets in the usual sense. Further, $\phi_4(x)$ says that $k$ is the set of ordinals in $x$, and that $f$ is a bijection between the ordinals of $x$ and $x$ itself; $\phi_5(x)$ says that $V : k \rightarrow x$ is a restriction of the function $\alpha \mapsto V_\alpha$ for ordinals $\alpha$, and that $x$ is covered by the values of the function $V$.

We see that, when $\kappa = \beth_\kappa$, $V_\kappa$ satisfies $\phi(x)$. For example, $V_\kappa$ satisfies the formula $\exists k \exists V \exists f \big(\phi_4(x)\big)$ because the ordinals in $V_\kappa$ are exactly all $\alpha \in \kappa$, and, on the other hand, it holds in general (because $\kappa$ is an ordinal) that $|V_{\omega + \kappa}| = \beth_\kappa$. Because $\omega \neq \beth_\omega$, we have $\omega \in \kappa$, so $\omega + \kappa = \kappa$. Now we have $|V_\kappa| = |V_{\omega + \kappa}| = \beth_\kappa = \kappa$, so there is a bijection between $V_\kappa$ and $\kappa$, the ordinals in $V_\kappa$.

To see that no other sets satisfy $\phi(x)$, assume that $\phi(M)$ holds for some set $M$. From $\phi_4(M)$ we get that $k$ is the set of ordinals in $M$. From $\phi_1(M)$ we get that $M$ is transitive. Because also the class ON of ordinals is transitive, the set $k = M \cap \text{ON}$ is transitive, and thus $k$ is an ordinal itself. From $\phi_1(M)$ we get $\omega < k$. We also get from $\phi_4(M)$ that $|M| = |k|$.

We claim that for all ordinals $\alpha \in k$ we have $V(\alpha) = V_\alpha$, and prove it by induction on $\alpha$. From $\phi_5(M)$ we get that $V(\emptyset) = \emptyset = V_\emptyset$. For a successor ordinal $\alpha^+ \in k$, assuming that $V(\alpha) = V_\alpha$, we get $V(\alpha^+) = \{z \in M : z \subseteq V_\alpha\}$. From $\forall z \phi_3(M)$ we get $\{z : z \subseteq V_\alpha\} \subseteq \{z \in M : z \subseteq V_\alpha\}$. Thus we get $V(\alpha^+) = \{z \in M : z \subseteq V_\alpha\} = \{z : z \subseteq V_\alpha\} = \mathcal{P}(V_\alpha) = V_{\alpha^+}$. For a limit ordinal $\gamma \subset k$, assuming that $V(\alpha) = V_\alpha$ for all $\alpha < \gamma$, we get from $\phi_5(M)$ that $V(\gamma) = \bigcup \{V(\alpha) : \alpha < \gamma\} = \bigcup \{V_\alpha : \alpha < \gamma\} = V_\gamma$. This completes the induction.

We show that $k$ is not a successor ordinal. Let $\alpha \in k$. Then $V_\alpha \in M$. By $\phi_2(M)$ and $\forall z \phi_3(M)$, $V_{\alpha^+} = \mathcal{P}(V_\alpha) = \mathcal{P}_M(V_\alpha) \in M$. Because $\alpha^+ \subseteq V_{\alpha^+}$, we get by $\forall z \phi_3(M)$ that $\alpha^+ \in M$, whence $\alpha^+ \in k$. Hereby $k$ must be a limit ordinal.

Now we have $V_k = \bigcup \{V_\alpha : \alpha < k\}$, and for each $\alpha < k$ we have $V_\alpha \subseteq M$ by the facts that $V_\alpha \in M$ and $M$ is transitive. Therefore $V_k \subseteq M$. On the other hand, by $\phi_5(M)$, $M \subseteq \bigcup \{V(\alpha) : \alpha \in k\} = \bigcup \{V_\alpha : \alpha < k\} = V_k$. Thus $M = V_k$.

This, combined with the previous results, yields $|V_k| = |k|$. This in addition to the fact that $\omega < k$ shows that $k$ is an uncountable ordinal. Therefore $\omega + k = k$, whence $\beth_k = |V_{\omega+k}| = |V_k| = |k|$. Thus $k$ is a cardinal and $k = \beth_k$. We have shown that $M = V_\kappa$ for some $\kappa = \beth_\kappa$.

## Exercise 7.5

To solve the exercise it suffices to prove the more general claim that, for all $\mathcal{D}$-formulas $\phi$, all limit ordinals $\alpha$, all $M \in V_\alpha$, and all $X \in V_\alpha$, we have

$$\text{Sat}_\phi(M, X) \quad \text{iff} \quad V_\alpha \models \text{Sat}_\phi(M, X). \tag{A.37}$$

Here, $\text{Sat}_\phi(M, X)$ is the first order conjunction "$M$ is a model" $\wedge$ "$X$ is a team for $M$" $\wedge$ "$\text{Fr}(\phi) \subseteq \text{dom}(X)$" $\wedge$ "$M \models_X \phi$."

To be very very precise, we would have to write out $\text{Sat}_\phi(M, X)$ explicitly as a first order formula and prove Eq. (A.37), like we did in Exercise 7.2 for all $\Sigma_0$-formulas. The difference is that this time we will face some formulas that are not $\Sigma_0$, but luckily we also have a stronger assumption, namely that $M, X \in V_\alpha$ and $\alpha$ is a limit ordinal. However, we shall avoid this tedious task for the most part, and show just an outline and a few excerpts from the proof.

We can express "$M$ is a model (in the language of $\phi$)" as a $\Sigma_0$-formula by writing something like "$M$ is a tuple $(a_1, \ldots, a_n)$, where $a_2, \ldots, a_n$ are constants, relations and functions of the correct arities in respect of the language of $\phi$." From Exercise 7.2, we get that

$$\text{"}M \text{ is a model"} \quad \text{iff} \quad V_\alpha \models \text{"}M \text{ is a model."}$$

Borrowing the formula $\text{Fun}(x, y, z)$ from Exercise 7.4, we can express "$X$ is a team for $M$" by writing $\exists d \forall s \in X \big(\text{Fun}(s, d, M)\big)$. This is not a $\Sigma_0$-formula, but $d$ corresponds to $\text{dom}(X)$ and we know that $\text{rk}(\text{dom}(X)) \leq \text{rk}(X)$. Thus from $X \in V_\alpha$ we get that if we find a suitable value for $d$ from the set theoretical universe, we find this value in $V_\alpha$. This yields

$$\text{"}X \text{ is a team for } M\text{"} \quad \text{iff} \quad V_\alpha \models \text{"}X \text{ is a team for } M.\text{"}$$

We can write "$\text{Fr}(\phi) \subseteq \text{dom}(X)$" in a similar way. Note that $\phi$ is fixed relative to the formula $\text{Sat}_\phi(M, X)$, and so is the finite set $\text{Fr}(\phi)$. Therefore it is very simple to write a formula that defines $\text{Fr}(\phi)$.

The most demanding part is to define, for each $\mathcal{D}$-formula $\phi$, a first order formula $\text{Mod}_\phi(M, X)$ that expresses "$M \models_X \phi$." We can define $\text{Mod}_\phi(M, X)$ inductively on $\phi$. The definitions mimic the truth definition of $\mathcal{D}$. A few examples are as follows:

$$\text{Mod}_{=(t_1, \ldots, t_n)} := \forall s \in X \forall s' \in X \big((t_1^{\mathcal{M}}\langle s \rangle = t_1^{\mathcal{M}}\langle s' \rangle \wedge \cdots \wedge t_{n-1}^{\mathcal{M}}\langle s \rangle = t_{n-1}^{\mathcal{M}}\langle s' \rangle)$$
$$\rightarrow t_n^{\mathcal{M}}\langle s \rangle = t_n^{\mathcal{M}}\langle s' \rangle);$$

$$\text{Mod}_{\phi \vee \psi} := \exists Y \exists Z \big(X = Y \cup Z \wedge \text{Mod}_\phi(M, Y) \wedge \text{Mod}_\psi(M, Z)\big);$$

$$\text{Mod}_{\exists x \phi} := \exists F \exists Y \big(\text{Fun}(F, X, M) \wedge Y = X(F/x_x) \wedge \text{Mod}_\phi(M, Y)\big).$$

We now proceed to a proof by induction on $\phi$ that

$$\mathrm{Mod}_\phi(M, X) \iff V_\alpha \models \mathrm{Mod}_\phi(M, X). \tag{A.38}$$

We shall only present the proof for the three cases mentioned above.

If $\phi$ is $=(t_1, \ldots, t_n)$, then $\mathrm{Mod}_\phi(M, X)$ is a $\Sigma_0$-formula as soon as we write out a $\Sigma_0$-formula expressing "$x = t^M \langle s \rangle$" for a term $t$. It is possible but takes some space and is therefore omitted here. Now Exercise 7.2 applies, and we get Eq. (A.38).

If $\phi$ is $\psi \vee \theta$, then any suitable values for the variables $Y$ and $Z$ in the formula $\mathrm{Mod}_\phi(M, X)$ are bound to be in $V_\alpha$ because $X \in V_\alpha$, $\mathrm{rk}(Y) \leq \mathrm{rk}(X)$ and $\mathrm{rk}(Z) \leq \mathrm{rk}(X)$. Thus we get Eq. (A.38).

Let $\phi$ be $\exists x \psi$. We will use the fact that if $\mathrm{rk}(\mathrm{dom}(f)) \leq \beta$ and $\mathrm{rk}(\mathrm{rng}(f)) \leq \beta$ then $\mathrm{rk}(f) < \beta + \omega$. Because $M, X \in V_\alpha$ and $\alpha$ is a limit, there is some $\beta < \alpha$ such that $\mathrm{rk}(M) \leq \beta$ and $\mathrm{rk}(X) \leq \beta$. Therefore any suitable value for the variable $F$ in $\mathrm{Mod}_\phi(M, X)$ has $\mathrm{rk}(F) < \beta + \omega \leq \alpha$. Thus $F \in V_\alpha$. We need also $Y = X(F/x) \in V_\alpha$. Because $\mathrm{dom}(s)$ is a finite collection of variable symbols, we can assume that all $s \in Y$ have $\mathrm{rk}(\mathrm{dom}(s)) = m < \omega$, where $m$ does not depend on $s$. We also know that $\mathrm{rk}(\mathrm{rng}(s)) \leq \beta$, as $\mathrm{rng}(s) \subseteq M$. Thus $\mathrm{rk}(s) \leq \beta + n$, where $n < \omega$ does not depend on $s$. Combining these facts, we get that $\mathrm{rk}(Y) = \sup\{\mathrm{rk}(s) + 1 : s \in Y\} \leq \beta + n + 1 < \alpha$. Thus any suitable value for the variable $Y$ in $\mathrm{Mod}_\phi(M, X)$ has $Y \in V_\alpha$. Therefore we get Eq. (A.38).

## Exercise 7.6

Our goal is to write down a $\mathcal{D}$-sentence $\psi$ such that, for all models $\mathcal{M}$, $\mathcal{M} \models \psi$ iff $\mathcal{M}$ is not isomorphic to a model $(V_\kappa, \in, K, V, f)$, where $\kappa = \beth_\kappa$, $K = \{\alpha : \alpha < \kappa\}$, $V : K \to V_\kappa$ maps $\alpha \mapsto V_\alpha$, and $f : M \to K$ is a bijection. Note that this property is in close correspondence with the one in Exercise 7.4, where we wrote down a $\Sigma_2$-formula $\phi(x)$ that is satisfied only by $V_\kappa$, where $\kappa = \beth_\kappa$.

Let us examine deeper the analogy between Exercises 7.4 and 7.6. The variable symbol $x$ in Exercise 7.4 corresponds to the arbitrary model $\mathcal{M}$ in Exercise 7.6. Bounded quantification in the formula $\phi(x)$ corresponds to picking an element in the model $\mathcal{M}$. This can be done by first order quantification in the $\mathcal{D}$-sentence $\psi$. Unbounded quantification in $\phi(x)$ corresponds to picking some object outside the universe $M$. This can, in certain cases, be seen as second order quantification, i.e. the use of dependency statements $=(t_1, \ldots, t_n)$ in $\psi$. However, we can only express second order existential quantification in $\mathcal{D}$, and there are both existential and universal unbounded quantifiers in $\phi(x)$. More precisely, we know how to express the existence of functions $f : M \to M$ and subsets $A \subseteq M$ in $\mathcal{D}$. Unbounded universal quantifiers in $\phi(x)$ can be expressed by an extension of the language (i.e. vocabulary) of $\mathcal{M}$. Because $\mathcal{M}$ is arbitrary, the interpretations of any extra non-logical symbols are also arbitrary, so this will correspond to universal quantification of certain kind of elements outside the universe $M$, and those elements are functions $f : M \to M$ and subsets $A \subseteq M$. Luckily, this will be enough.

This said, we take as a starting point the formula $\phi(x)$ from the solution of Exercise 7.4. Remember that $\psi$ is to express practically $\neg\phi(x)$. Let the language of $\psi$ be $\{E, K, V, f\}$. The binary relation symbol $E$ shall denote the "belongs to" relation of $\mathcal{M}$, and the other symbols are as in Exercise 7.4; $K$ is a unary relation symbol, and

$V$ and $f$ are unary function symbols. Let $\psi$ be as follows (further explanations can be found later):

$$\psi := \neg\psi_1 \vee \neg\psi_2 \vee \psi_3 \vee \neg\psi_4 \vee \neg\psi_5 \vee \neg\psi_6 \vee \Phi_{\mathrm{wf}};$$

$$\psi_1 := \exists x(x = \omega);$$

$$\psi_2 := \forall x \exists y(y = \mathcal{P}_M(x));$$

$$\psi_3 := \exists x \exists w \forall u \exists v \big(\forall z((u = z \to v = w) \to zEx)$$
$$\wedge\, \forall y \exists z(=(y, z) \wedge \neg((u = z \to v = w) \leftrightarrow zEy))\big);$$

$$\psi_4 := \forall x(Ky \leftrightarrow \mathrm{Ord}(y)) \wedge \mathrm{Bij}(f, K);$$

$$\psi_5 := \forall x \exists y\big(Ky \wedge xE(Vy)\big) \wedge \forall x(x = \emptyset \to Vx = \emptyset)$$
$$\wedge\, \forall x \forall y\big((Kx \wedge y = x^+) \to Vy = \mathcal{P}_M(Vx)\big)$$
$$\wedge\, \forall x\big(\mathrm{Limit}(x) \to \big(\forall y(yEx \to Vy \subseteq Vx) \wedge \forall z \exists y(z \in Vx \to z \in Vy)\big)\big);$$

$$\psi_6 := \forall x \forall y\big(\forall z(zEx \leftrightarrow zEy) \to x = y\big).$$

Note that $\Phi_{\mathrm{wf}}$ is the formula that holds in a model $\mathcal{M} = (M, E)$ iff the binary relation $E^{\mathcal{M}}$ is not well-founded. The abbreviations used above are written in $\mathcal{D}$ as follows:

| | |
|---|---|
| $x = \emptyset$ | $\forall y \neg(yEx);$ |
| $x \subseteq y$ | $\forall u(uEx \to uEy);$ |
| $y = \mathcal{P}_M(x)$ | $\forall u(uEy \leftrightarrow u \subseteq x);$ |
| $x = y^+$ | $\forall u(uEx \leftrightarrow (uEy \vee u = y));$ |
| $\mathrm{Trans}(x)$ | $\forall y \forall z(zEy \to zEx);$ |
| $\mathrm{Ord}(x)$ | $\mathrm{Trans}(x) \wedge \forall y(yEx \to \mathrm{Trans}(y));$ |
| $\mathrm{Limit}(x)$ | $\neg(x = \emptyset) \wedge \mathrm{Ord}(x) \wedge \forall y\big((yEx \wedge \mathrm{Limit}(y)) \to \exists z(zEx \wedge z = y^+)\big);$ |
| $x = \omega$ | $\mathrm{Limit}(x) \wedge \forall y\big((yEx \wedge \mathrm{Limit}(y)) \to y = \emptyset\big);$ |
| $\mathrm{Bij}(f, P)$ | $\forall u(Pfu) \wedge \forall u \forall v(fu = fv \to u = v) \wedge \forall u \exists v(Pu \to fv = u).$ |

Note that $\mathrm{Bij}(f, P)$ expresses that the unary function $f^{\mathcal{M}}$ is a bijection $M \to P^{\mathcal{M}}$.

The subformula $\psi_2$ is the power set axiom; $\psi_3$ expresses that "$\mathcal{M}$ misses a subset," i.e. there is an element $x \in M$ and a set $A \subseteq M$ of elements $z$ that belong to $x$ in the sense of the model $\mathcal{M}$, but there is no element $y \in M$ such that the elements of $A$ would be exactly the elements of $y$ in the sense of the model $\mathcal{M}$; $\psi_4$ expresses that $K^{\mathcal{M}}$ is the set of ordinals in $M$ in the sense of $\mathcal{M}$ and $f^{\mathcal{M}}$ is a bijection $M \to K^{\mathcal{M}}$; $\psi_5$ expresses that $M$ is covered by the values of the function $V^{\mathcal{M}}$ and that $V^{\mathcal{M}}$ is a restriction of the function $\alpha \mapsto V_\alpha$ for ordinals $\alpha$ in the sense of $\mathcal{M}$; $\psi_6$ is the axiom of extensionality.

We see that $(V_\kappa, \in, K, V, f) \not\models \psi$ when $\kappa = \beth_\kappa$, $K = \{\alpha : \alpha < \kappa\}$, $V : K \to V_\kappa$ maps $\alpha \mapsto V_\alpha$, and $f : M \to K$ is a bijection. Let us check, for example, how $(V_\kappa, \in, K, V, f) \not\models \psi_3$. Let $x, w \in V_\kappa$, and let $F : V_\kappa \to V_\kappa$ choose for each value $u \in V_\kappa$ some value $v = F(u)$ such that $u \in x$ whenever $F(u) = w$. Let $y = \{z \in V_\kappa : F(z) = w\}$. We have $\mathrm{rk}(z) < \mathrm{rk}(x) < \kappa$ for all $z \in y$, whence $\mathrm{rk}(y) < \kappa$, so $y \in V_\kappa$. Now, if $z \in V_\kappa$, we have $F(z) = w$ iff $z \in y$. This shows that $\psi_3$ fails.

Assume then that $\mathcal{M} = (M, E, K, V, f)$ is a model for which $\mathcal{M} \not\models \psi$. We want to show that $\mathcal{M}$ is isomorphic to some model $(V_\kappa, \in, K', V', f')$ with $\kappa = \beth_\kappa$. Because formulas $\psi_i$ for $i = 1, 2, 4, 5, 6$ are first order, we have $\mathcal{N} \not\models \neg\psi_i$ iff $\mathcal{N} \models \psi_i$ for these *i*. From $\mathcal{M} \models \psi_6$ we get that $\mathcal{M}$ is extensional, and from $\mathcal{M} \not\models \Phi_{wf}$ we get that $\mathcal{M}$ is well-founded. By Mostowski's Collapsing Lemma, $\mathcal{M}$ is isomorphic to a transitive model $\mathcal{N} = (N, \in, K', V', f')$. From $\mathcal{N} \models \psi_4$ we get that $K'$ is the set of ordinals in $N$, and thus $K'$ is an ordinal itself. From $\mathcal{N} \models \psi_1$ we get that $\omega < K'$. We also get from $\mathcal{N} \models \psi_4$ that $|N| = |K'|$.

We claim that for all ordinals $\alpha \in K'$ we have $V'(\alpha) = V_\alpha$, and prove it by induction on $\alpha$. From $\mathcal{N} \models \psi_5$ we get that $V'(\emptyset) = \emptyset = V_\emptyset$. For a successor ordinal $\alpha^+ \in K'$, assuming that $V'(\alpha) = V_\alpha$, we get $V'(\alpha^+) = \{z \in N : z \subset V_\alpha\}$. From $\mathcal{N} \not\models \psi_3$ we get that whenever $z \subset V_\alpha$, then $z \in N$. Thus we get $V'(\alpha^+) = \{z \in N : z \subset V_\alpha\} = \{z : z \subset V_\alpha\} = \mathcal{P}(V_\alpha) = V_{\alpha^+}$. For a limit ordinal $\gamma \in K'$, assuming that $V'(\alpha) = V_\alpha$ for all $\alpha < \gamma$, we get from $\mathcal{N} \models \psi_5$ that $V'(\gamma) = \bigcup\{V'(\alpha) : \alpha < \gamma\} = \bigcup\{V_\alpha : \alpha < \gamma\} = V_\gamma$. This completes the induction.

We show that $K'$ is not a successor ordinal. Let $\alpha \in K'$. Then $V_\alpha \in N$. By $\mathcal{N} \models \psi_2$ and $\mathcal{N} \not\models \psi_3$, $V_{\alpha^+} = \mathcal{P}(V_\alpha) \in N$. Because $\alpha^+ \subseteq V_{\alpha^+}$, we get by $\mathcal{N} \not\models \psi_3$ that $\alpha^+ \in N$, whence $\alpha^+ \in K'$. Hereby $K'$ must be a limit ordinal.

Now we have $V_{K'} = \bigcup\{V_\alpha : \alpha < K'\}$, and for each $\alpha < K'$ we have $V_\alpha \subseteq N$ by the facts that $V_\alpha \in N$ and $N$ is transitive. Therefore $V_{K'} \subseteq M$. On the other hand, by $\mathcal{N} \models \psi_5$, $N \subseteq V_{K'}$. Thus $N = V_{K'}$.

This, combined with the previous results, yields $|V_{K'}| = |K'|$, which in turn with the fact that $\omega < K'$ shows that $K'$ is uncountable. Therefore $\omega + K' = K'$, whence $\beth_{K'} = |V_{\omega+K'}| = |V_{K'}| = |K'|$. Thus $K'$ is a cardinal and $K' = \beth_{K'}$. We have shown that $\mathcal{M}$ is isomorphic to $\mathcal{N} = (V_\kappa, \in, K', V', f')$ for some $\kappa = \beth_\kappa$.

# Chapter 8

## Exercise 8.3

Because (total) independence and (total) dependence are defined in terms of determination, we must first find the minimal (in respect of $\subseteq$) sets $W \subseteq \{x_2, x_3\}$ that determine $\{x_1\}$.

First consider $\{x_2\}$. If $\{x_2\} \geq \{x_1\}$ was to hold, it would mean the existence of a function $f$ that maps $1 \mapsto 0$ (as required by agent $s_1$ of the team), $0 \mapsto 0$ (as required by $s_2$), and $1 \mapsto 1$ (as required by $s_3$). But such a function cannot exist as the requirements by agents $s_1$ and $s_3$ are contradictory. (By the way, this definition of determination in terms of functions is equivalent to the one given in the course material, as one can prove.) Thus $\{x_2\} \not\geq \{x_1\}$.

Now consider $\{x_3\}$. Because there is no function that maps $0 \mapsto 0, 1 \mapsto 0$, and $1 \mapsto 1$, we can conclude that $\{x_3\} \not\geq \{x_1\}$.

Now consider $\{x_2, x_3\}$. To establish $\{x_2, x_3\} \geq \{x_1\}$, we need to find a function that maps $(1, 0) \mapsto 0, (0, 1) \mapsto 0$, and $(1, 1) \mapsto 1$. But this is possible, so we have $\{x_2, x_3\} \geq \{x_1\}$.

By these three observations we can now say that $\{x_2, x_3\}$ is the only minimal subset of $\{x_2, x_3\}$ that determines $\{x_1\}$. In fact, it is the only such subset, not only the only minimal one.

Now we can say that $\{x_1\}$ is dependent on $\{x_2, x_3\}, \{x_2\}, \{x_3\}$, and $\emptyset$, i.e. all subsets of $\{x_2, x_3\}$. This is so because each of these is included in some minimal set that determines $\{x_1\}$, namely $\{x_2, x_3\}$.

We can see that $\{x_1\}$ is also totally dependent on $\{x_2, x_3\}$, $\{x_2\}$, $\{x_3\}$, and $\emptyset$, because each of them is included in all of the minimal determining sets. Well, there was only one minimal determining set in the first place, namely $\{x_2, x_3\}$.

Note that $\{x_1\}$ is independent of $\emptyset$ and nothing else because only for $\emptyset$ can we find a minimal determining set that it avoids. All other $W \subseteq \{x_2, x_3\}$ intersect all the minimal sets that determine $\{x_1\}$.

Note that $\{x_1\}$ is also totally independent of $\emptyset$ and nothing else because of the same reason why it is independent of only that set, and the fact that $\{x_2, x_3\}$ is the only minimal set that determines $\{x_1\}$.

## Exercise 8.4

By concentrated checking of all possible choices – as in Exercise 8.3 – we can find out that the only minimal sets $W \subseteq \{x_2, x_3, x_4, x_5\}$ that determine $\{x_1\}$ are $W_1 = \{x_2, x_3\}$ and $W_2 = \{x_3, x_4, x_5\}$. All the other sets either do not determine $\{x_1\}$ or they include $W_1$ or $W_2$. Now to the actual questions.

Note that $\{x_1\}$ is dependent on every set that is included in some minimal set that determines $\{x_1\}$; i.e. $\{x_1\}$ is dependent on every $W \subseteq W_1$ and every $W \subseteq W_2$, i.e. $\{x_2, x_3\}$, $\{x_2\}$, $\{x_3\}$, $\{x_3, x_4, x_5\}$, $\{x_3, x_4\}$, $\{x_3, x_5\}$, $\{x_4, x_5\}$, $\{x_4\}$, $\{x_5\}$, and $\emptyset$.

Note that $\{x_1\}$ is totally dependent on every set $W$ that is included in all the sets that determine $\{x_1\}$. In fact it suffices that this $W$ is included in all minimal determining sets, because every determining set contains a minimal determining set, which then contains $W$. Thus, $\{x_1\}$ is totally dependent on every $W \subseteq W_1 \cap W_2$, that is, $\{x_3\}$, and $\emptyset$.

Note that $\{x_1\}$ is independent of all the sets that do not intersect some set that determines $\{x_1\}$. If a set does not intersect some determining set, then it does not intersect some minimal determining set either. Thus $\{x_1\}$ is independent of every $W \subseteq \{x_2, x_3, x_4, x_5\} \setminus W_1$ and every $W \subseteq \{x_2, x_3, x_4, x_5\} \setminus W_2$, i.e. $\{x_4, x_5\}$, $\{x_4\}$, $\{x_5\}$, $\{x_2\}$, and $\emptyset$.

Note that $\{x_1\}$ is totally independent of every set that does not intersect any minimal set that determines $\{x_1\}$; i.e. $\{x_1\}$ is totally independent of every $W \subseteq (\{x_2, x_3, x_4, x_5\} \setminus W_1) \cap (\{x_2, x_3, x_4, x_5\} \setminus W_2)$, i.e. $\emptyset$ alone.

## Exercise 8.5

If, for example, $a = 0$ then $\{x_3, x_4\}$ determines $\{x_1\}$. This is verified by the mapping $f$ : $(3, 6) \mapsto 15, (1, 1) \mapsto 1, (2, 0) \mapsto 6, (2, 2) \mapsto 4, (0, 5) \mapsto 0$.

If $a = 2$ then $\{x_3, x_4\}$ does not determine $\{x_1\}$ because $s_3$ and $s_4$ are identical in $\{x_3, x_4\}$ but different in $\{x_1\}$. Or, in other words, there is no function that maps $(3, 6) \mapsto 15$, $(1, 1) \mapsto 1, (2, 2) \mapsto 6, (2, 2) \mapsto 4$, and $(0, 5) \mapsto 0$.

If, for example, $a = 0$ then $\{x_2, x_4\}$ determines $\{x_1\}$. This is verified by the mapping $g : (15, 6) \mapsto 15, (1, 1) \mapsto 1, (0, 0) \mapsto 6, (3, 2) \mapsto 4, (0, 5) \mapsto 0$.

If $a = 5$ then $\{x_2, x_4\}$ does not determine $\{x_1\}$ because $s_3$ and $s_5$ are identical in $\{x_2, x_4\}$ but different in $\{x_1\}$. Or, in other words, there is no function that maps $(15, 6) \mapsto 15$, $(1, 1) \mapsto 1, (0, 5) \mapsto 6, (3, 2) \mapsto 4$, and $(0, 5) \mapsto 0$.

## Exercise 8.6

In team (a) in Table A.15, $\{x_1\}$ determines $\{x_4\}$, and so does $\{x_3\}$. Therefore $\{x_4\}$ is dependent but not totally dependent on $\{x_1\}$.

Table A.15. *Teams (a), (b), (c), and (d) for Exercise 8.6*

| | (a) | | | | | (b) | | | |
|---|---|---|---|---|---|---|---|---|---|
| | $x_1$ | $x_2$ | $x_3$ | $x_4$ | | $x_1$ | $x_2$ | $x_3$ | $x_4$ |
| $s_1$ | 0 | 0 | 0 | 0 | $s_1$ | 0 | 0 | 0 | 0 |
| $s_2$ | 0 | 1 | 0 | 0 | $s_2$ | 0 | 1 | 0 | 0 |
| $s_3$ | 1 | 1 | 1 | 1 | $s_3$ | 1 | 1 | 0 | 1 |

| | (c) | | | | | (d) | | | |
|---|---|---|---|---|---|---|---|---|---|
| | $x_1$ | $x_2$ | $x_3$ | $x_4$ | | $x_1$ | $x_2$ | $x_3$ | $x_4$ |
| $s_1$ | 0 | 0 | 0 | 0 | $s_1$ | 0 | 0 | 0 | 0 |
| $s_2$ | 0 | 1 | 1 | 1 | $s_2$ | 0 | 1 | 0 | 1 |
| $s_3$ | 1 | 1 | 0 | 1 | $s_3$ | 1 | 1 | 0 | 1 |

In team (b), $\{x_1\}$ is the only minimal $W \subseteq \{x_1, x_2, x_3\}$ that determines $\{x_4\}$. Therefore $\{x_4\}$ is totally dependent on $\{x_1\}$.

In team (c), $\{x_2\}$ is a minimal set that determines $\{x_4\}$, and so is $\{x_1, x_3\}$. Because $\{x_1\}$ intersects only the latter one of these two, $\{x_4\}$ is independent but not totally independent of $\{x_1\}$.

In team (d), $\{x_2\}$ is the only minimal $W \subseteq \{x_1, x_2, x_3\}$ that determines $\{x_4\}$, and $\{x_1\}$ does not intersect in it. Therefore $\{x_4\}$ is totally independent of $\{x_1\}$.

Note that there exist also other correct answers, not only the ones shown here.

## Exercise 8.7

We want to prove for all TL-formulas $\phi$ where $\sim$ does not occur, all models $\mathcal{M}$ and all teams $X$ with $\mathrm{Fr}(\phi) \subseteq \mathrm{dom}(X)$ that there is a $\mathcal{D}$-formula $\phi^*$ such that

$$\mathcal{M} \models_X \phi \iff \mathcal{M} \models_X \phi^*. \tag{A.39}$$

Note that on the left side of Eq. (A.39) we use TL-semantics and on the right side we use $\mathcal{D}$-semantics.

We prove the claim by induction on $\phi$ by using the following translation from $\sim$-free TL-formulas to $\mathcal{D}$-formulas:

$$(t = t')^* \mapsto t = t', \qquad\qquad (\phi \otimes \psi)^* \mapsto \phi^* \vee \psi^*,$$
$$(\neg t = t')^* \mapsto \neg t = t', \qquad\qquad (\phi \wedge \psi)^* \mapsto \phi^* \wedge \psi^*,$$
$$(Rt_1 \ldots t_n)^* \mapsto Rt_1 \ldots t_n, \qquad\qquad (\exists x_n \phi)^* \mapsto \exists x_n \phi^*,$$
$$(\neg Rt_1 \ldots t_n)^* \mapsto \neg Rt_1 \ldots t_n, \qquad\qquad (!x_n \phi)^* \mapsto \forall x_n \phi^*.$$
$$(=(t_1, \ldots, t_n))^* \mapsto =(t_1, \ldots, t_n),$$

The claim in Eq. (A.39) can now be proved by following definitions because the TL-semantics of $\phi$ is exactly the same as the $\mathcal{D}$-semantics of $\phi^*$.

## Exercise 8.8

Recall the definitions of the four dependence values:

$$\top \text{ is } =(); \quad \bot \text{ is } {\sim}=(); \quad \mathbf{0} \text{ is } \neg=(); \quad \mathbf{1} \text{ is } {\sim}\neg=().$$

To prove items (i) and (ii) of Example 8.7, first note that $\mathcal{M} \models_X =()$ holds for all teams $X$. This gives $\mathbf{0} \Rightarrow \top$ and $\mathbf{1} \Rightarrow \top$. We also get that $\mathcal{M} \models_X {\sim}=()$ holds for no teams $X$. This gives $\bot \Rightarrow \mathbf{0}$ and $\bot \Rightarrow \mathbf{1}$.

Then we prove items (iii) and (iv). We see immediately that $\mathbf{1} = {\sim}\mathbf{0}$ and $\bot = {\sim}\top$. It is also the case that $\mathcal{M} \models_X {\sim}{\sim}\phi$ iff $\mathcal{M} \models_X \phi$ for any TL-formula $\phi$. Therefore we get $\mathbf{0} \equiv {\sim}{\sim}\mathbf{0} = {\sim}\mathbf{1}$, and $\top \equiv {\sim}{\sim}\top = {\sim}\bot$.

Finally, for item (v), we observe that $\mathbf{0} = \neg\top$ by definition.

## Exercise 8.9

Let $\mathcal{M}$ be a model and let $X$ be a team for $\mathcal{M}$. Because $\mathcal{M} \models_X \top$ holds (for any team), we get

$$\mathcal{M} \models_X \phi \wedge \top \iff \mathcal{M} \models_X \phi \text{ and } \mathcal{M} \models_X \top \iff \mathcal{M} \models_X \phi.$$

Recall that $\phi \vee \psi$ that is shorthand for ${\sim}({\sim}\phi \wedge {\sim}\psi)$ and that $\phi \oplus \psi$ is shorthand for ${\sim}({\sim}\phi \otimes {\sim}\psi)$. We have the following chains of equivalences:

$$\mathcal{M} \models_X {\sim}({\sim}\phi \wedge {\sim}\top)$$
$$\iff \mathcal{M} \not\models_X {\sim}\phi \wedge {\sim}\top$$
$$\iff \mathcal{M} \not\models_X {\sim}\phi \text{ or } \mathcal{M} \not\models_X {\sim}\top$$
$$\iff \mathcal{M} \models_X \phi \text{ or } \mathcal{M} \models_X \top$$
$$\iff \mathcal{M} \models_X \top.$$

Note that when a team $X$ is given, and $Y \cup Z = X$ holds for some sets $Y$ and $Z$, then $\text{dom}(Y) = \text{dom}(Z)$ necessarily holds. Thus in these cases we can omit this condition. Now,

$$\mathcal{M} \models_X \top \otimes \top$$
$$\iff \mathcal{M} \models_Y \top \text{ and } \mathcal{M} \models_Z \top \text{ for some } Y \cup Z = X$$
$$\iff \mathcal{M} \models_X \top.$$

The last equivalence above holds because both sides are always true. Further,

$$\mathcal{M} \models_X {\sim}({\sim}\phi \otimes {\sim}\top)$$
$$\iff \mathcal{M} \not\models_X {\sim}\phi \otimes {\sim}\top$$
$$\iff \text{for all } Y \cup Z = X : \mathcal{M} \not\models_Y {\sim}\phi \text{ or } \mathcal{M} \not\models_Z {\sim}\top$$
$$\iff \text{for all } Y \cup Z = X : \mathcal{M} \models_Y \phi \text{ or } \mathcal{M} \models_Z \top$$
$$\iff \mathcal{M} \models_X \top.$$

The last equivalence above holds because both sides are always true.

## Exercise 8.10

Let $\mathcal{M}$ be a model and $X$ a team for $\mathcal{M}$. Because $\mathcal{M} \models_X \bot$ does not hold (for any team), we get

$$\mathcal{M} \models_X \phi \wedge \bot \iff \mathcal{M} \models_X \phi \text{ and } \mathcal{M} \models_X \bot \iff \mathcal{M} \models_X \bot.$$

Recall that $\phi \vee \psi$ is shorthand for $\sim(\sim\phi \wedge \sim\psi)$ and that $\phi \oplus \psi$ is shorthand for $\sim(\sim\phi \otimes \sim\psi)$. We have the following chains of equivalences:

$$\mathcal{M} \models_X \sim(\sim\phi \wedge \sim\bot)$$
$$\iff \mathcal{M} \not\models_X \sim\phi \wedge \sim\bot$$
$$\iff \mathcal{M} \not\models_X \sim\phi \text{ or } \mathcal{M} \not\models_X \sim\bot$$
$$\iff \mathcal{M} \models_X \phi \text{ or } \mathcal{M} \models_X \bot$$
$$\iff \mathcal{M} \models_X \phi;$$

and

$$\mathcal{M} \models_X \phi \otimes \bot$$
$$\iff \mathcal{M} \models_Y \phi \text{ and } \mathcal{M} \models_Z \bot \text{ for some } Y \cup Z = X$$
$$\iff \mathcal{M} \models_X \bot.$$

The last equivalence above holds because both sides are always false. Finally,

$$\mathcal{M} \models_X \sim(\sim\bot \otimes \sim\bot)$$
$$\iff \mathcal{M} \not\models_X \sim\bot \otimes \sim\bot$$
$$\iff \text{for all } Y \cup Z = X : \mathcal{M} \not\models_Y \sim\bot \text{ or } \mathcal{M} \not\models_Z \sim\bot$$
$$\iff \text{for all } Y \cup Z = X : \mathcal{M} \models_Y \bot \text{ or } \mathcal{M} \models_Z \bot$$
$$\iff \mathcal{M} \models_X \bot.$$

The last equivalence above holds because both sides are always false.

## Exercise 8.11

Let $\mathcal{M}$ be a model and let $X$ be a team for $\mathcal{M}$. Assume first that $\phi$ is a $\mathcal{D}$-formula (that is, $\phi$ is a TL-formula, and $\sim$ does not occur in $\phi$). Then we have the old result that $\mathcal{M} \models_\emptyset \phi$. Also recall that $\mathcal{M} \models_X \mathbf{0}$ iff $X = \emptyset$. With these observations we get the following chain of equivalences:

$$\mathcal{M} \models_X \phi \wedge \mathbf{0}$$
$$\iff \mathcal{M} \models_X \phi \text{ and } \mathcal{M} \models_X \mathbf{0} \tag{A.40}$$
$$\iff X = \emptyset \tag{A.41}$$
$$\iff \mathcal{M} \models_X \mathbf{0}.$$

The implication Eq. (A.41) $\Rightarrow$ Eq. (A.40) is based on $\phi$ being a $\mathcal{D}$-formula. Also,

$$\mathcal{M} \models_X \sim(\sim \phi \wedge \sim \mathbf{0})$$
$$\Longleftrightarrow \mathcal{M} \not\models_X \sim \phi \wedge \sim \mathbf{0}$$
$$\Longleftrightarrow \mathcal{M} \not\models_X \sim \phi \text{ or } \mathcal{M} \not\models_X \sim \mathbf{0}$$
$$\Longleftrightarrow \mathcal{M} \models_X \phi \text{ or } \mathcal{M} \models_X \mathbf{0}$$
$$\Longleftrightarrow \mathcal{M} \models_X \phi \text{ or } X = \emptyset \tag{A.42}$$
$$\Longleftrightarrow \mathcal{M} \models_X \phi. \tag{A.43}$$

The implication Eq. (A.42) $\Rightarrow$ Eq. (A.43) is based on $\phi$ being a $\mathcal{D}$-formula:

Now we let $\phi$ be any TL-formula:

$$\mathcal{M} \models_X \phi \otimes \mathbf{0}$$
$$\Longleftrightarrow \mathcal{M} \models_Y \phi \text{ and } \mathcal{M} \models_Z \mathbf{0} \text{ for some } Y \cup Z = X$$
$$\Longleftrightarrow \mathcal{M} \models_Y \phi \text{ and } Z = \emptyset \text{ for some } Y \cup Z = X$$
$$\Longleftrightarrow \mathcal{M} \models_X \phi;$$

$$\mathcal{M} \models_X \sim(\sim \mathbf{0} \otimes \sim \mathbf{0})$$
$$\Longleftrightarrow \text{ for all } Y \cup Z = X : \mathcal{M} \not\models_Y \sim \mathbf{0} \text{ or } \mathcal{M} \not\models_Z \sim \mathbf{0}$$
$$\Longleftrightarrow \text{ for all } Y \cup Z = X : \mathcal{M} \models_Y \mathbf{0} \text{ or } \mathcal{M} \models_Z \mathbf{0}$$
$$\Longleftrightarrow \text{ for all } Y \cup Z = X : Y = \emptyset \text{ or } Z = \emptyset$$
$$\Longleftrightarrow X = \emptyset$$
$$\Longleftrightarrow \mathcal{M} \models_X \mathbf{0}.$$

## Exercise 8.13

Let $X = \{\emptyset\}$ and let $\mathcal{M}$ be a model with at least two elements. Choose $\phi$ to be the TL-formula $\mathbf{1} \otimes \mathbf{1}$. Then $\mathcal{M} \models_X !x_0\phi$. To see why, observe that $\mathcal{M} \models_X !x_0\phi$ iff $\mathcal{M} \models_Y \phi$, where $Y = X(M/x_0)$ has at least two elements (as many as $M$ has). We can split $Y = Y_1 \cup Y_2$ so that both $Y_1$ and $Y_2$ are non-empty, whence $\mathcal{M} \models_{Y_i} \mathbf{1}$ for both $i = 1, 2$. Thus $\mathcal{M} \models_Y \phi$.

On the other hand, $\mathcal{M} \models_X \forall x_0\phi$ does not hold. To see why, observe that $\mathcal{M} \models_X \forall x_0\phi$ iff $\mathcal{M} \models_Z \phi$, where $Z = X(F/x_0)$ has only one element (as many as $X$ has). Thus any split $Z = Z_1 \cup Z_2$ is bound to have $Z_i = \emptyset$ for one $i = 1, 2$. For this $i$ we have $\mathcal{M} \not\models_{Z_i} \mathbf{1}$, so we get $\mathcal{M} \not\models_Z \phi$.

## Exercise 8.15

Let $\phi$ be a TL-formula, let $\mathcal{M}$ be a model, and let $X$ be a team for $\mathcal{M}$. We want to express that for all subsets $Y \subseteq X$ there is a subset $Z \subseteq Y$ for which $\phi$ holds. Let $\psi$ be the TL-formula $(\phi \otimes \top) \oplus \bot$. Then we have the following:

$$\mathcal{M} \models_X \psi$$
$$\iff \text{for all } Y \cup Z = X : \mathcal{M} \models_Y \phi \otimes \top \text{ or } \mathcal{M} \models_Z \bot$$
$$\iff \text{for all } Y \subseteq X : \mathcal{M} \models_Y \phi \otimes \top$$
$$\iff \text{for all } Y \subseteq X \text{ there is } Z \cup V = Y : \mathcal{M} \models_Z \phi \text{ and } \mathcal{M} \models_V \top$$
$$\iff \text{for all } Y \subseteq X \text{ there is } Z \subseteq Y : \mathcal{M} \models_Z \phi.$$

# References

[1]   W. W. Armstrong. Dependency structures of data base relationships. In *Information Processing 74. Proc. IFIP Congress*, Stockholm (Amsterdam: North-Holland, 1974), pp. 580–583.

[2]   A. Blass and Y. Gurevich. Henkin quantifiers and complete problems. *Ann. Pure Appl. Logic*, **32**:3 (1986), 1–16.

[3]   J. P. Burgess. A remark on Henkin sentences and their contraries. *Notre Dame J. Formal Logic*, **44**:3 (2003), 185–188.

[4]   X. Caicedo and M. Krynicki. Quantifiers for reasoning with imperfect information and $\Sigma_1^1$-logic. In W. A. Carnielli and I. M. L. D'Ottaviano, eds., *Advances in Contemporary Logic and Computer Science* (Providence, RI: American Mathematical Society, 1999), pp. 17–31. 1999.

[5]   P. Cameron and W. Hodges. Some combinatorics of imperfect information. *J. Symbolic Logic*, **66**:2 (2001), 673–684.

[6]   W. Craig. Linear reasoning. A new form of the Herbrand-Gentzen theorem. *J. Symbolic Logic*, **22** (1957), 250–268.

[7]   H. B. Enderton. Finite partially-ordered quantifiers. *Z. Math. Logik Grundlagen Math.*, **16**, (1970), 393–397.

[8]   H. B. Enderton. *A Mathematical Introduction to Logic*, 2nd edn. (Burlington, MA: Harcourt/Academic Press, 2001).

[9]   T. Frayne, A. C. Morel, and D. S. Scott. Reduced direct products. *Fund. Math.*, **51**, (1962/3), 195–228.

[10]  D. Gale and F. M. Stewart. Infinite games with perfect information. In H. W. Kuhn and A. W. Tucker, eds., *Contributions to the Theory of Games*, vol. 2, (Princeton: Princeton University Press, 1953), pp. 245–266.

[11]  J.-Y. Girard. Linear logic. *Theor. Comp. Sci.*, **50**:1 (1987), 101 pp.

[12]  K. Gödel. *Collected Works. Vol. I.* (New York: Oxford University Press, 1986), Publications 1929–1936. Edited and with a preface by Solomon Feferman.

[13]  D. Harel. Characterizing second-order logic with first-order quantifiers, *Z. Math. Logik Grundlag. Math.*, **25**:5 (1979), 419–422.

[14]  L. Henkin. Some remarks on infinitely long formulas. In *Infinitistic Methods. Proc. Symposium Foundations of Math.*, Warsaw, (London: Pergamon, 1961), pp. 167–183.

[15] L. Henkin. The completeness of the first-order functional calculus. *J. Symbolic Logic*, **14** (1949), 159–166.

[16] J. Hintikka and G. Sandu. Informational independence as a semantical phenomenon. In J. E. Fenstad, I. T. Frolov, and R. Hilpinen, eds., *Logic, Methodology and Philosophy of Science, VIII. Stud. Logic Found. Math.*, vol. *126* (Amsterdam: North-Holland, 1989), pp. 571–589.

[17] J. Hintikka and G. Sandu. Game-theoretic semantics. In J. F. A. K. van Benthem and G. B. A. ter Meulen, eds., *Handbook of Logic and Language*, vol. *1* (Boston, MA: MIT Press, 1997), pp. 361–410.

[18] J. Hintikka. Form and content in quantification theory. *Acta Philos. Fenn.*, **8**, (1955), 7–55.

[19] J. Hintikka. *The Principles of Mathematics Revisited* (Cambridge: Cambridge University Press, 1996).

[20] J. Hintikka. Hyperclassical logic (a.k.a. IF logic) and its implications for logical theory. *Bull. Symbolic Logic*, **8**:3 (2002), 404–423.

[21] W. Hodges. Compositional semantics for a language of imperfect information. *Logic J. IGPL*, **5**:4 (1997), 539–563.

[22] W. Hodges. Some strange quantifiers. In J. Mycielski, G. Rozenberg, and A. Salomaa, eds., *Structures in Logic and Computer Science. Lecture Notes in Computer Sci.*, vol. *1261* (Berlin: Springer, 1997), pp. 51–65.

[23] T. Janssen and F. Dechesne. Signalling: a tricky business. In J. van Benthem, G. Heinzmann, M. Rebuschi, and H. Visser, eds., *The Age of Alternative Logics: Assessing Philosophy of Logic and Mathematics Today* (Berlin: Springer, 2006).

[24] T. M. V. Janssen. Independent choices and the interpretation of IF logic. *J. Logic Lang. Inform.*, **11**:3 (2002), 367–387.

[25] T. Jech. *Set Theory*, 2nd edn., (Berlin: Springer-Verlag, 1997).

[26] S. Krajewski. Mutually inconsistent satisfaction classes. *Bull. Acad. Polon. Sci. Sér. Sci. Math. Astron. Phys.*, **22** (1974), 983–987.

[27] G. Kreisel and J.-L. Krivine. *Elements of Mathematical Logic. Model Theory.* (Amsterdam: North-Holland Publishing Co., 1967).

[28] S. Kripke. An outline of a theory of truth. *J. Phil.*, **72** (1975), 690–715.

[29] M. Krynicki and A. H. Lachlan. On the semantics of the Henkin quantifier, *J. Symbolic Logic*, **44**:2 1999), 184–200.

[30] A. Lévy. A hierarchy of formulas in set theory. *Mem. Am. Math. Soc.*, **57** (1965), 76 pp.

[31] E. G. K. Lopez-Escobar. A non-interpolation theorem. *Bull. Acad. Polon. Sci. Sér. Sci. Math. Astronom. Phys.*, 17:109–112, 1969.

[32] M. Magidor. On the role of supercompact and extendible cardinals in logic. *Israel J. Math.*, **10** (1971), 147–157.

[33] Y. N. Moschovakis. *Elementary Induction on Abstract Structures.* Studies in Logic and the Foundations of Mathematics, vol. 77 (Amsterdam: North-Holland, 1974).

[34] A. Robinson. A result on consistency and its application to the theory of definition. *Nederl. Akad. Wetensch. Proc. Ser. A. Math.*, **18** (1956), 47–58.

[35] G. Sandu and T. Hyttinen. IF logic and the foundations of mathematics. *Synthese*, **126**: 1–2 (2001), 37–47.

[36] T. Skolem. Logisch-kombinatorische Untersuchungen über die Erfüllbarkeit oder Beweisbarkeit mathematischer Sätze nebst einem Theoreme über dichte

Mengen. *Videnskapsselskapets skrifter I, Matematisknaturvidenskabelig Klasse, Videnskappsselskapet i Kristiania*, **4** (1920), 1–36.

[37]  T. Skolem. *Selected Works in Logic* (Oslo: Scandinavian University Press, 1970).

[38]  A. Tarski. The semantic conception of truth and the foundations of semantics. *Phil. Phenomenol. Res.*, **4** (1944), 341–376.

[39]  A. Tarski and R. L. Vaught. Arithmetical extensions of relational systems. *Compositio Math*, **13** (1958), 81–102.

[40]  J. Väänänen. Second-order logic and foundations of mathematics. *Bull. Symbolic Logic*, **7**:4 (2001), 504–520.

[41]  J. Väänänen. On the semantics of informational independence. *Logic J. IGPL*, **10**:3 (2002), 339–352.

[42]  J. Väänänen. A remark on nondeterminacy in IF logic. *Acta Philosophica Fennica*, **78** (2006), 71–77.

[43]  J. von Neumann and O. Morgenstern. *Theory of Games and Economic Behavior* (Princeton: Princeton University Press, 1944).

[44]  W. J. Walkoe, Jr. Finite partially-ordered quantification. *J. Symbolic Logic*, **35** (1970), 535–555.

# Index

Printed in the United States
by Baker & Taylor Publisher Services

Printed in the United States
by Baker & Taylor Publisher Services